CONTENTS

D1239965

Introduction

The *Modern Chemistry* Test Generator and Assessment Item Listing

The *Modern Chemistry* Test Generator consists of a comprehensive bank of more than 2000 test items and the ExamView Pro 3.0 Test Builder software, which enables you to produce your own tests based on these items and those items you create yourself. Both Macintosh® and Windows® versions of the Test Generator are included on the *Modern Chemistry One-Stop Planner CD-ROM with Test Generator*. Directions on pp. v-vii of this book explain how to install the program on your computer. This Assessment Item Listing is a printout of all the test items in the Test Generator.

ExamView Pro 3.0 Software

The ExamView Pro 3.0 Test Builder program enables you to quickly create printed tests. You can enter your own questions and customize the appearance of the tests you create. The ExamView Pro 3.0 Test Builder program offers many unique features and provides numerous options that allow you to customize the content and appearance of the tests you create.

Test Items

The *Modern Chemistry* Test Generator contains a file of test items for each chapter of the textbook. The test items are in a variety of formats, including true/false, multiple choice, completion, problem, and essay. Each item is correlated to the chapter objectives in the textbook and by difficulty level.

Item Codes

As you browse through this Assessment Item Listing, you will see that all test items of the same type are gathered together under an identifying head. Each item is coded to assist you with item selection. Following is an explanation of the codes.

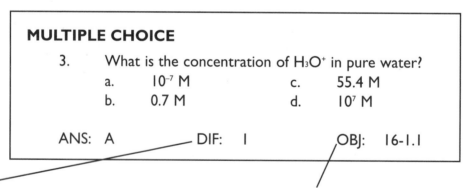

MULTIPLE CHOICE

3. What is the concentration of H_3O^+ in pure water?
    a. 10^{-7} M   c.   55.4 M
    b.   0.7 M   d. 10^7 M

    ANS: A   DIF: I   OBJ: 16-1.1

DIF defines the difficulty of the item.
I requires recall of information
II requires analysis and interpretation of known information
III requires application of knowledge to new situations

OBJ lists the chapter number, section number, and objective.
(16-1.1 = Chapter 16, Section 1, Objective 1)

INSTALLATION AND STARTUP

The ExamView Pro 3.0 test generator software is provided on the One-Stop Planner CD-ROM. The ExamView test generator includes the program and all of the questions for the corresponding textbook. Your ExamView software includes three components: Test Builder, Question Bank Editor, and Test Player. The **Test Builder** includes options to create, edit, print, and save tests. The **Question Bank Editor** lets you create or edit question banks. The **Test Player** is a separate program that your students can use to take on-line* (computerized or LAN-based) tests. Please refer to the ExamView User's Guide on the One-Stop Planner CD-ROM for complete instructions.

Before you can use the test generator, you must install the program and the test banks on your hard drive. The system requirements, installation instructions, and startup procedures are provided below.

SYSTEM REQUIREMENTS

To use the ExamView Pro 3.0 test generator, your computer must meet or exceed the following minimum hardware requirements:

Windows®
- Pentium computer
- Windows 95®, Windows 98®, Windows 2000® (or a more recent version)
- color monitor (VGA-compatible)
- CD-ROM and/or high-density floppy disk drive
- hard drive with at least 7 MB space available
- 8 MB available memory (*16 MB memory recommended*)
- *If you wish to use the on-line test player, you must have an Internet connection to access the Internet testing features.**

Macintosh®
- PowerPC processor, 100 MHz computer
- System 7.5 (or a more recent version)
- color monitor (VGA-compatible)
- CD-ROM and/or high-density floppy disk drive
- hard drive with at least 7 MB space available
- 8 MB available memory (*16 MB memory recommended*)
- *If you wish to use the on-line test player, you must have an Internet connection with System 8.6 (or more recent version) to access the Internet testing features.**

You can use the online test player to host tests on your personal or school Web site or local area network (LAN) at no additional charge. The ExamView Web site's Internet test-hosting service must be purchased separately. Visit www.examview.com to learn more.

INSTALLATION

Instructions for installing ExamView from the CD-ROM:

Windows®
Step 1
Turn on your computer.
Step 2
Insert the One-Stop Planner disc with the ExamView test generator into the CD-ROM drive.
Step 3
Click the **Start** button on the *Taskbar,* and choose the *Run* option.
Step 4
The ExamView software is provided on the One-Stop Planner CD-ROM under the drive letter that corresponds to the CD-ROM drive on your computer (e.g., **d:\setup.exe**). The setup program will automatically install everything you need to use ExamView.
Step 5
Follow the prompts on the screen to complete the installation process.

Macintosh®
Step 1
Turn on your computer.
Step 2
Insert the One-Stop Planner disc with the ExamView test generator into the CD-ROM drive.
Step 3
Double-click the *ExamView* installation icon to start the program.
Step 4
Follow the prompts on the screen to complete the installation process.

*Instructions for installing **ExamView** from the One-Stop Planner Menu (Macintosh® or Windows®):*

Follow steps 1 and 2 from above.
Step 3
Click on **One-Stop.pdf.** (If you do not have Adobe Acrobat® Reader installed on your computer, install it before proceeding by clicking on **Reader Installer.**)
Step 4
Click on the **Test Generator** button.
Step 5
Click on **Install ExamView.**
Step 6
Select the operating system you are using (Macintosh® or Windows®).
Step 7
ExamView will launch the installer. Follow the prompts on the screen to complete the installation process.

GETTING STARTED

After you complete the installation process, follow these instructions to start the ExamView test generator software. See the ExamView User's Guide on the One-Stop Planner for further instructions on the options for creating a test and editing a question bank.

Startup Instructions

Step 1
Turn on the computer.

Step 2
Windows®: Click the **Start** button on the *Taskbar*. Highlight the **Programs** menu, and locate the *ExamView Test Generator* folder. Select the *ExamView Pro* option to start the software.
Macintosh®: Locate and open the *ExamView* folder. Double-click the *ExamView Pro* program icon.

Step 3
The first time you run the software, you will be prompted to enter your name, school/institution name, and city/state. You are now ready to begin using the ExamView software.

Step 4
Each time you start ExamView, the **Startup** menu appears. Choose one of the options shown.

Step 5
Use ExamView to create a test or edit questions in a question bank.

Technical Support

If you have any questions about the Test Generator or need assistance, call the Holt, Rinehart and Winston technical support line at 1-800-323-9239, Monday through Friday, 7:00 A.M. to 6:00 P.M., Central Standard Time. You can contact the Technical Support Center on the Internet at http://www.hrwtechsupport.com or by e-mail at tsc@hrwtechsupport.com.

MULTIPLE CHOICE

1. The study of matter and changes in matter best describes the science of
   a. biology.
   b. physics.
   c. microbiology.
   d. chemistry.

   ANS: D   DIF: I   OBJ: 1-1.1

2. Chemistry may be least useful in studying
   a. matter.
   b. synthetic fibers.
   c. falling bodies.
   d. medicine.

   ANS: C   DIF: II   OBJ: 1-1.1

3. Chemistry is
   a. a biological science.
   b. a physical science.
   c. concerned mostly with living things.
   d. the study of electricity.

   ANS: B   DIF: I   OBJ: 1-1.1

4. Chemistry is defined as the study of the composition and structure of materials and
   a. the categories of matter.
   b. the changes in matter.
   c. the electrical currents in matter.
   d. molecules in living things.

   ANS: B   DIF: I   OBJ: 1-1.1

5. Chemistry is the study of all of the following EXCEPT
   a. matter.
   b. changes in matter.
   c. energy associated with changes in matter.
   d. projectile motion.

   ANS: D   DIF: I   OBJ: 1-1.1

6. Study of the composition and structure of materials and the changes that materials undergo best describes the science of
   a. chemistry.
   b. biology.
   c. physics.
   d. engineering.

   ANS: A   DIF: I   OBJ: 1-1.1

7. Chemistry may be most useful in studying
   a. the movement of asteroids.
   b. why materials corrode.
   c. eating habits of ducks.
   d. streamlining of race cars.

   ANS: B   DIF: II   OBJ: 1-1.1

8. The branch of chemistry that includes the study of materials and processes that occur in living things is
   a. organic chemistry.
   b. physical chemistry.
   c. analytical chemistry.
   d. biochemistry.

   ANS: D   DIF: I   OBJ: 1-1.2

9. The branch of chemistry that is concerned with the identification and composition of materials is
   a. analytical chemistry.
   b. inorganic chemistry.
   c. physical chemistry.
   d. organic chemistry.

   ANS: A   DIF: I   OBJ: 1-1.2

10. The study of substances containing carbon is
    a. organic chemistry.
    b. inorganic chemistry.
    c. nuclear chemistry.
    d. analytical chemistry.

    ANS: A   DIF: I   OBJ: 1-1.2

11. Organic chemistry, inorganic chemistry, and physical chemistry are NOT
    a. biological sciences.
    b. physical sciences.
    c. quantitative branches of chemistry.
    d. concerned primarily with nonliving things.

    ANS: A   DIF: I   OBJ: 1-1.2

12. The branch of chemistry concerned with the properties, changes, and relationships between energy and matter is
    a. inorganic chemistry.
    b. analytical chemistry.
    c. physical chemistry.
    d. theoretical chemistry.

    ANS: C   DIF: I   OBJ: 1-1.2

13. Technology is the
    a. application of chemical principles to predict events.
    b. application of scientific knowledge to solve problems.
    c. study of scientific processes.
    d. analysis of chemical behavior.

    ANS: B   DIF: I   OBJ: 1-1.3

14. An example of technology is the
    a. addition of a side group to an organic molecule during synthesis.
    b. use of a new antibiotic to fight an infection.
    c. measurement of iron concentration in a water sample.
    d. study of atomic fusion reactions.

    ANS: B   DIF: II   OBJ: 1-1.3

15. Basic research is
    a. the production and use of products that improve our quality of life.
    b. carried out to solve a problem.
    c. the identification of the components and composition of materials.
    d. carried out for the sake of increasing knowledge.

    ANS: D   DIF: I   OBJ: 1-1.3

16. Applied research is
    a. the production and use of products that improve our quality of life.
    b. carried out to solve a problem.
    c. the use of mathematics and computers to design and predict the properties of new compounds.
    d. carried out for the sake of increasing knowledge.

    ANS: B   DIF: I   OBJ: 1-1.3

17. Which statement is NOT true about applied research?
    a. It is conducted to meet goals defined by specific needs.
    b. It is usually carried out to solve a practical problem.
    c. It is the study of how and why a specific reaction occurs.
    d. It may not be driven primarily by scientific curiosity or a desire to know.

    ANS: C   DIF: I   OBJ: 1-1.3

18. Which statement is NOT true about basic research?
    a. It is carried out for the sake of increasing knowledge.
    b. It is carried out to solve a specific problem.
    c. It is the study of how and why a specific reaction occurs.
    d. It may be driven by scientific curiosity alone.

    ANS: B   DIF: I   OBJ: 1-1.3

19. Matter includes all of the following EXCEPT
    a. air.   c. smoke.
    b. light.   d. water vapor.

    ANS: B   DIF: I   OBJ: 1-2.1

20. A physical property may be investigated by
    a. melting ice.   c. allowing silver to tarnish.
    b. letting milk turn sour.   d. burning wood.

    ANS: A   DIF: II   OBJ: 1-2.1

21. Chemical properties
    a. include changes of state of a substance.
    b. include mass and color.
    c. include changes that alter the identity of a substance.
    d. can be observed without altering the identity of a substance.

    ANS: C   DIF: I   OBJ: 1-2.1

22. Two features that distinguish matter are
    a. mass and velocity.   c. mass and volume.
    b. weight and velocity.   d. weight and volume.

    ANS: C   DIF: I   OBJ: 1-2.1

23. One chemical property of matter is
    a. boiling point.
    b. texture.
    c. reactivity.
    d. density.

    ANS: C   DIF: II   OBJ: 1-2.1

24. An example of an extensive physical property is
    a. mass.
    b. density.
    c. color.
    d. boiling point.

    ANS: A   DIF: II   OBJ: 1-2.1

25. Which of the following is an intensive physical property?
    a. volume
    b. length
    c. color
    d. mass

    ANS: C   DIF: II   OBJ: 1-2.1

26. A chemical change occurs when
    a. dissolved minerals solidify to form a crystal.
    b. ethanol is purified through distillation.
    c. salt deposits form from evaporated sea water.
    d. a leaf changes color.

    ANS: D   DIF: II   OBJ: 1-2.2

27. The melting of candle wax is classified as a physical change because it
    a. produces no new substances.
    b. transfers energy.
    c. absorbs heat.
    d. changes the chemical properties of wax.

    ANS: A   DIF: II   OBJ: 1-2.2

28. An example of a chemical change is
    a. sanding wood.
    b. melting ice.
    c. milk going sour.
    d. vaporizing gasoline.

    ANS: C   DIF: II   OBJ: 1-2.2

29. A physical change occurs when a
    a. peach spoils.
    b. copper bowl tarnishes.
    c. bracelet turns your wrist green.
    d. glue gun melts a glue stick.

    ANS: D   DIF: II   OBJ: 1-2.2

30. The particles in a solid are
    a. packed closely together.
    b. very far apart.
    c. constantly in motion.
    d. able to slide past each other.

    ANS: A   DIF: I   OBJ: 1-2.3

31. The state of matter in which a material is most likely to resist compression is the
    a. solid state.   c. gaseous state.
    b. liquid state.   d. vaporous state.

    ANS: A   DIF: I   OBJ: 1-2.3

32. The state of matter in which a material has definite shape and definite volume is the
    a. liquid state.   c. gaseous state.
    b. solid state.   d. vaporous state.

    ANS: B   DIF: I   OBJ: 1-2.3

33. The state of matter in which a material has neither a definite shape nor a definite volume is the
    a. gaseous state.   c. elemental state.
    b. liquid state.   d. solid state.

    ANS: A   DIF: I   OBJ: 1-2.3

34. The state of matter in which particles are rigidly held in fixed positions is the
    a. gaseous state.   c. vaporous state.
    b. liquid state.   d. solid state.

    ANS: D   DIF: I   OBJ: 1-2.3

35. A substance classified as a fluid contains particles that
    a. quickly expand into any available space.   c. may slide past each other.
    b. are held in fixed positions.   d. are very far from each other.

    ANS: C   DIF: I   OBJ: 1-2.3

36. The state of matter in which a material has a definite volume but no definite shape is the
    a. gaseous state.   c. frozen state.
    b. solid state.   d. liquid state.

    ANS: D   DIF: I   OBJ: 1-2.3

37. Under ordinary conditions of temperature and pressure, the particles in a gas are
    a. closely packed.   c. held in fixed positions.
    b. very far from each other.   d. able to slide past each other.

    ANS: B   DIF: I   OBJ: 1-2.3

38. A list of pure substances could include
    a. bread dough.   c. vitamin C (ascorbic acid).
    b. vinegar (5% acetic acid).   d. sea water.

    ANS: C   DIF: II   OBJ: 1-2.4

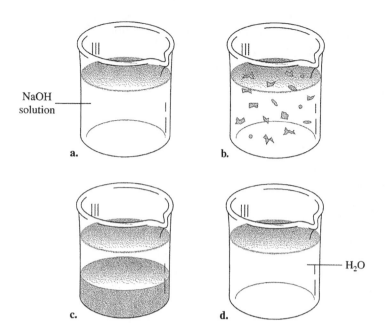

NaOH solution

a.

b.

c.

d.   H$_2$O

39. The homogeneous mixture in the illustration above is in container
    a. a.   c. c.
    b. b.   d. d.

    ANS: A   DIF: II   OBJ: 1-2.4

40. The substances that are chemically bound together are
    a. the gases in the air.   c. dust particles in air.
    b. the elements that compose water.   d. substances in blood.

    ANS: B   DIF: II   OBJ: 1-2.4

41. Physical means can be used to separate
    a. elements.   c. mixtures.
    b. pure substances.   d. compounds.

    ANS: C   DIF: I   OBJ: 1-2.4

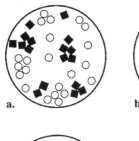
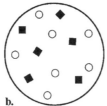
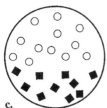
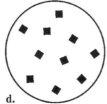

42. Which part of the illustration above shows the particles in a heterogeneous mixture?

a. a

b. b

c. c

d. d

ANS: A   DIF: II   OBJ: 1-2.4

	Group 1	Group 2		Group 13	Group 14	Group 15	Group 16	Group 17	Group 18
1	1 **H** Hydrogen 1.01								2 **He** Helium 4.00
2	3 **Li** Lithium 6.94	4 **Be** Beryllium 9.01		5 **B** Boron 10.81	6 **C** Carbon 12.01	7 **N** Nitrogen 14.01	8 **O** Oxygen 16.00	9 **F** Fluorine 19.00	10 **Ne** Neon 20.18
3	11 **Na** Sodium 22.99	12 **Mg** Magnesium 24.30		13 **Al** Aluminum 26.98	14 **Si** Silicon 28.08	15 **P** Phosphorus 30.97	16 **S** Sulfur 32.07	17 **Cl** Chlorine 35.45	18 **Ar** Argon 39.95
4	19 **K** Potassium 39.10	20 **Ca** Calcium 40.08		31 **Ga** Gallium 69.72	32 **Ge** Germanium 72.61	33 **As** Arsenic 74.92	34 **Se** Selenium 78.96	35 **Br** Bromine 79.90	36 **Kr** Krypton 83.80
5	37 **Rb** Rubidium 85.47	38 **Sr** Strontium 87.62		49 **In** Indium 114.82	50 **Sn** Tin 118.71	51 **Sb** Antimony 121.76	52 **Te** Tellurium 127.60	53 **I** Iodine 126.90	54 **Xe** Xenon 131.29
6	55 **Cs** Cesium 132.90	56 **Ba** Barium 137.33		81 **Tl** Thallium 204.38	82 **Pb** Lead 207.2	83 **Bi** Bismuth 208.98	84 **Po** Polonium (208.98)	85 **At** Astatine (209.99)	86 **Rn** Radon (222.02)
7	87 **Fr** Francium (223.02)	88 **Ra** Radium (226.02)							

43. Group _____ in the figure above contains only metals.

a. 2

b. 13

c. 17

d. 18

ANS: A   DIF: II   OBJ: 1-3.2

Modern Chemistry Assessment Item Listing

7

44. Based on their location in the figure above, oxygen and selenium have
    a. the same number of neutrons.   c. similar properties.
    b. the same conductivity.   d. the same number of electron orbitals.

    ANS: C   DIF: II   OBJ: 1-3.3

45. Use the figure above. Which element has properties most similar to those of sodium?
    a. boron   c. sulfur
    b. calcium   d. nitrogen

    ANS: B   DIF: II   OBJ: 1-3.3

46. Based on its location in the figure above, you could infer that _____ is very unreactive.
    a. Ca   c. Si
    b. P   d. Ar

    ANS: D   DIF: II   OBJ: 1-3.3

47. Based on their location in the figure above, boron and antimony might be good elements to use as
    a. semiconductors.   c. construction materials.
    b. fuels.   d. catalysts.

    ANS: A   DIF: II   OBJ: 1-3.4

48. The most useful source of chemical information about the elements is a
    a. calculator.   c. periodic table.
    b. table of metric equivalents.   d. table of isotopes.

    ANS: C   DIF: I   OBJ: 1-3.3

49. A horizontal row of blocks in the periodic table is called a(n)
    a. group.   c. family.
    b. period.   d. octet.

    ANS: B   DIF: I   OBJ: 1-3.3

50. Elements in a group in the periodic table can be expected to have similar
    a. atomic masses.   c. numbers of neutrons.
    b. atomic numbers.   d. properties.

    ANS: D   DIF: I   OBJ: 1-3.3

51. A vertical column of blocks in the periodic table is called a(n)
    a. group.   c. property.
    b. period.   d. octet.

    ANS: A   DIF: I   OBJ: 1-3.3

52. The elements that border the zigzag line in the periodic table are
    a. inactive.
    b. metals.
    c. metalloids.
    d. nonmetals.

    ANS: C   DIF: I   OBJ: 1-3.3

53. Which is NOT a property of metals?
    a. malleability
    b. ability to conduct heat and electricity
    c. unreactivity
    d. tensile strength

    ANS: C   DIF: I   OBJ: 1-3.4

54. Which statement is NOT true of nonmetals?
    a. They have characteristics of both metals and nonmetals.
    b. Many are gases at room temperature.
    c. They have low conductivity.
    d. There are fewer nonmetals than metals.

    ANS: A   DIF: I   OBJ: 1-3.4

55. Which statement is NOT true of most metalloids?
    a. They are used in computers and calculators.
    b. They are semiconductors of electricity.
    c. They are generally unreactive.
    d. They have characteristics of both metals and nonmetals.

    ANS: C   DIF: I   OBJ: 1-3.4

COMPLETION

	Group 1	Group 2		Group 13	Group 14	Group 15	Group 16	Group 17	Group 18
1	1 **H** Hydrogen 1.01								2 **He** Helium 4.00
2	3 **Li** Lithium 6.94	4 **Be** Beryllium 9.01		5 **B** Boron 10.81	6 **C** Carbon 12.01	7 **N** Nitrogen 14.01	8 **O** Oxygen 16.00	9 **F** Fluorine 19.00	10 **Ne** Neon 20.18
3	11 **Na** Sodium 22.99	12 **Mg** Magnesium 24.30		13 **Al** Aluminum 26.98	14 **Si** Silicon 28.08	15 **P** Phosphorus 30.97	16 **S** Sulfur 32.07	17 **Cl** Chlorine 35.45	18 **Ar** Argon 39.95
4	19 **K** Potassium 39.10	20 **Ca** Calcium 40.08		31 **Ga** Gallium 69.72	32 **Ge** Germanium 72.61	33 **As** Arsenic 74.92	34 **Se** Selenium 78.96	35 **Br** Bromine 79.90	36 **Kr** Krypton 83.80
5	37 **Rb** Rubidium 85.47	38 **Sr** Strontium 87.62		49 **In** Indium 114.82	50 **Sn** Tin 118.71	51 **Sb** Antimony 121.76	52 **Te** Tellurium 127.60	53 **I** Iodine 126.90	54 **Xe** Xenon 131.29
6	55 **Cs** Cesium 132.90	56 **Ba** Barium 137.33		81 **Tl** Thallium 204.38	82 **Pb** Lead 207.2	83 **Bi** Bismuth 208.98	84 **Po** Polonium (208.98)	85 **At** Astatine (209.99)	86 **Rn** Radon (222.02)
7	87 **Fr** Francium (223.02)	88 **Ra** Radium (226.02)							

1. Use the periodic table to write the name for the element that has the symbol Pb.

   ANS: lead   DIF: I   OBJ: 1-3.1

2. Use the periodic table to write the name for the element that has the symbol Mg.

   ANS: magnesium   DIF: I   OBJ: 1-3.1

3. Use the periodic table to write the name for the element that has the symbol H.

   ANS: hydrogen   DIF: I   OBJ: 1-3.1

4. Use the periodic table to write the name for the element that has the symbol P.

   ANS: phosphorus   DIF: I   OBJ: 1-3.1

5. Use the periodic table to write the name for the element that has the symbol Al.

   ANS: aluminum   DIF: I   OBJ: 1-3.1

6. Use the periodic table to write the name for the element that has the symbol Bi.

   ANS: bismuth   DIF: I   OBJ: 1-3.1

7. Use the periodic table to write the symbol for the element helium.

ANS: He   DIF: I   OBJ: 1-3.2

8. Use the periodic table to write the symbol for the element tin.

ANS: Sn   DIF: I   OBJ: 1-3.2

9. Use the periodic table to write the symbol for the element neon.

ANS: Ne   DIF: I   OBJ: 1-3.2

10. Use the periodic table to write the symbol for the element iodine.

ANS: I   DIF: I   OBJ: 1-3.2

11. Use the periodic table to write the symbol for the element chlorine.

ANS: Cl   DIF: I   OBJ: 1-3.2

12. Use the periodic table to write the symbol for the element argon.

ANS: Ar   DIF: I   OBJ: 1-3.2

SHORT ANSWER

1. In one experiment, magnesium metal is melted. In a second experiment, magnesium metal is burned. Classify the change in each experiment as chemical or physical. Explain your reasoning.

ANS:
In the first experiment, a physical change occurred. The chemical properties of magnesium were unchanged. In the second experiment, a chemical change occurred. A new substance with its own chemical properties was formed.

DIF: II   OBJ: 1-2.2

2. Explain the difference between a pure substance and a homogeneous mixture. Use an example.

ANS:
A homogeneous mixture can be separated by physical means, whereas a pure substance cannot. For example, salt can be removed from a salt-water mixture by evaporating the water, but to separate water into hydrogen and oxygen requires chemical means.

DIF: II   OBJ: 1-2.4

3. Xenon is generally unreactive. How is its low reactivity related to its position in the periodic table?

ANS:
Xenon is a noble gas. It is located in the same group in the periodic table as other noble gases. All noble gases have low reactivities.

DIF: II   OBJ: 1-3.3

4. How can you use the periodic table to make a prediction about the properties of xenon and helium, both in Group 18?

ANS:
In the periodic table, elements in the same column have similar properties. Because helium and xenon are located in the same group, their properties are similar.

DIF: II   OBJ: 1-3.3

5. Give an example of a region of the periodic table that acts as a bridge between two other regions. What does this tell you about the properties of the bridging elements?

ANS:
Metalloids form a region that bridges metals and nonmetals. Metalloids have properties of both metals and nonmetals.

DIF: II   OBJ: 1-3.4

ESSAY

1. Compare and contrast solids, liquids, and gases by explaining the behavior of their particles. Draw models to illustrate your answer.

ANS:
The arrangement of the particles in the three states account for their different properties. Particles in a solid move very little; particles in a liquid move more; and gas particles move the most. In the drawn models, particles in solids should appear closely packed and structured; particles in liquids should appear able to flow randomly past one another; and particles in gases should appear sparsely and randomly spaced.

DIF: III   OBJ: 1-2.3

Element name	Symbol	Element name	Symbol	Element name	Symbol
Carbon	C	Boron	B	Sulfur	S
Calcium	Ca	Beryllium	Be	Silicon	Si
Cadmium	Cd	Barium	Ba	Silver	Ag
Cobalt	Co	Bismuth	Bi	Sodium	Na
Copper	Cu	Bromine	Br	Strontium	Sr

2. Refer to the figure above to write a general rule for naming elements. There will be exceptions, such as sodium, to your rule. What might account for these exceptions?

ANS:
The symbols of most elements correspond to one or two letters of their names, beginning with the first letter. Exceptions are elements that take their symbols from foreign words. For example, the symbol for sodium, Na, comes from the Latin word for sodium, natrium.

DIF: III   OBJ: 1-3.3

3. Use examples to show how the properties and classifications of elements change as you move across a period.

ANS:
The closer two elements are within a period, the more similar their properties are. Moving across a period, elements progress from metals to metalloids to nonmetals to noble gases.

DIF: II   OBJ: 1-3.3

4. What properties can be predicted from the position of a metallic element in the periodic table?

ANS:
The element displays the properties characteristic of metals, such as solid physical state, grayish color, shiny surface, and good conductivity.

DIF: II   OBJ: 1-3.4

MULTIPLE CHOICE

1. All of the following are steps in the scientific method EXCEPT
   a. observing and recording data.
   b. forming a hypothesis.
   c. discarding data inconsistent with the hypothesis.
   d. developing a model.

   ANS: C   DIF: I   OBJ: 2-1.1

2. The reason for organizing, analyzing, and classifying data is
   a. so that computers can be used.
   b. to prove a law.
   c. to find relationships among the data.
   d. to separate qualitative and quantitative data.

   ANS: C   DIF: I   OBJ: 2-1.1

3. Which of the following observations is quantitative?
   a. The liquid turns blue litmus paper red.   c. The liquid tastes bitter.
   b. The liquid boils at 100°C.   d. The liquid is cloudy.

   ANS: B   DIF: II   OBJ: 2-1.2

4. Which of the following observations is qualitative?
   a. A chemical reaction is complete in 2.3 seconds.
   b. The solid has a mass of 23.4 grams.
   c. The pH of a liquid is 5.
   d. Salt deposits form from an evaporated liquid.

   ANS: D   DIF: II   OBJ: 2-1.2

5. Quantitative observations are recorded using
   a. numerical information.   c. non-numerical information.
   b. a control.   d. a system.

   ANS: A   DIF: I   OBJ: 2-1.2

6. Qualitative observations are recorded using
   a. numerical information.   c. non-numerical information.
   b. a control.   d. a system.

   ANS: C   DIF: I   OBJ: 2-1.2

7. A testable statement used for making predictions and carrying out further experiments is a
   a. law.   c. generalization.
   b. theory.   d. hypothesis.

   ANS: D   DIF: I   OBJ: 2-1.3

8. A theory is best described as a
   a. series of experimental observations.
   b. generalization that explains a body of known facts or phenomena.
   c. scientifically proven fact.
   d. testable statement.

   ANS: B   DIF: I   OBJ: 2-1.3

9. A plausible explanation of a body of observed natural phenomena is a scientific
   a. principle.   c. law.
   b. experiment.   d. theory.

   ANS: D   DIF: I   OBJ: 2-1.3

10. The validity of scientific concepts is evaluated by
    a. collecting facts.   c. voting by scientists.
    b. providing explanations.   d. testing hypotheses.

    ANS: D   DIF: I   OBJ: 2-1.3

11. A theory is an accepted explanation of an observed phenomenon until
    a. one study conflicts with the theory.
    b. repeated data and observation conflict with the theory.
    c. scientists disagree about the methods used to gather the data.
    d. an eminent scientist declares that it is inadequate.

    ANS: B   DIF: I   OBJ: 2-1.3

12. Standards are chosen because they
    a. have units that can be converted to other units.
    b. are reproducible in another laboratory.
    c. cannot be destroyed by any common physical or chemical means.
    d. are easily changed.

    ANS: B   DIF: I   OBJ: 2-2.1

13. A quantity does not have
    a. magnitude.   c. size.
    b. measurement.   d. amount.

    ANS: B   DIF: I   OBJ: 2-2.1

14. All of the following describe measurement standards EXCEPT
    a. measurement standards avoid ambiguity.
    b. measurement standards must be unchanging.
    c. a standard need not agree with a previously defined size.
    d. confusion is eliminated when the correct measurement is applied.

    ANS: C   DIF: I   OBJ: 2-2.1

15. All of the following describe a unit EXCEPT
    a. a unit compares what is being measured with a previously defined size.
    b. a unit is usually preceded by a number.
    c. a unit is usually not important in finding a solution to a problem.
    d. the choice of unit depends on the quantity being measured.

    ANS: C   DIF: I   OBJ: 2-2.1

16. All of the following are examples of units EXCEPT
    a. weight.   c. gram.
    b. kilometer.   d. teaspoon.

    ANS: A   DIF: I   OBJ: 2-2.1

17. All but one of these units are SI base units. The exception is the
    a. kilogram.   c. liter.
    b. second.   d. Kelvin.

    ANS: C   DIF: I   OBJ: 2-2.2

18. The SI standard units for length and mass are
    a. centimeter and gram.   c. centimeter and kilogram.
    b. meter and gram.   d. meter and kilogram.

    ANS: D   DIF: I   OBJ: 2-2.2

19. The metric unit for length that is closest to the thickness of a dime is the
    a. micrometer.   c. centimeter.
    b. millimeter.   d. decimeter.

    ANS: B   DIF: II   OBJ: 2-2.2

20. The symbol mm represents
    a. micrometer.   c. milliliter.
    b. millimeter.   d. meter.

    ANS: B   DIF: I   OBJ: 2-2.2

21. The symbols for units of length in order from smallest to largest are
    a. m, cm, mm, km.   c. km, mm, cm, m.
    b. mm, m, cm, km.   d. mm, cm, m, km.

    ANS: D   DIF: I   OBJ: 2-2.2

22. The symbol for the metric unit used to measure mass is
    a. m.   c. g.
    b. mm.   d. L.

    ANS: C   DIF: I   OBJ: 2-2.2

23. The quantity of matter per unit volume is
    a. mass.   c. inertia.
    b. weight.   d. density.

    ANS: D   DIF: I   OBJ: 2-2.2

24. A quantity that describes the concentration of matter is
    a. weight.   c. volume.
    b. density.   d. mass.

    ANS: B   DIF: I   OBJ: 2-2.2

25. The unit m^3 measures
    a. length.   c. volume.
    b. mass.   d. density.

    ANS: C   DIF: I   OBJ: 2-2.2

26. The liter is defined as
    a. $1000\ m^3$.   c. $1000\ g^3$.
    b. $1000\ cm^3$.   d. $1000\ c^3$.

    ANS: B   DIF: I   OBJ: 2-2.2

27. The standard unit for mass is the
    a. gram.   c. meter.
    b. cubic centimeter.   d. kilogram.

    ANS: D   DIF: I   OBJ: 2-2.2

28. A volume of 1 cubic centimeter is equivalent to
    a. 1 milliliter.   c. 1 liter.
    b. 1 gram.   d. 10^{-1} cubic decimeters.

    ANS: A   DIF: I   OBJ: 2-2.2

29. The symbol that represents the measured unit for volume is
    a. mL.   c. mm.
    b. mg.   d. cm.

    ANS: A   DIF: I   OBJ: 2-2.2

30. The SI base unit for time is the
    a. day.   c. minute.
    b. hour.   d. second.

    ANS: D   DIF: I   OBJ: 2-2.2

31. The unit abbreviation for time is
    a. hr.   c. sec.
    b. h.   d. s.

    ANS: D   DIF: I   OBJ: 2-2.2

32. The most appropriate SI unit for measuring the length of an automobile is the
    a. centimeter.   c. meter.
    b. kilometer.   d. liter.

    ANS: C   DIF: II   OBJ: 2-2.2

33. The SI base unit for length is the
    a. meter.   c. centimeter.
    b. millimeter.   d. kilometer.

    ANS: A   DIF: I   OBJ: 2-2.2

34. All of the following are SI units for density EXCEPT
    a. kg/m^3.   c. g/cm^3.
    b. g/mL.   d. g/m^2.

    ANS: D   DIF: I   OBJ: 2-2.2

35. A change in the force of Earth's gravity on an object will affect its
    a. mass.   c. weight.
    b. density.   d. kinetic energy.

    ANS: C   DIF: I   OBJ: 2-2.3

36. A measure of Earth's gravitational pull on matter is
    a. density.   c. volume.
    b. weight.   d. mass.

    ANS: B   DIF: I   OBJ: 2-2.3

37. A measure of the quantity of matter is
    a. density.   c. volume.
    b. weight.   d. mass.

    ANS: D   DIF: I   OBJ: 2-2.3

38. A true statement about mass is that
    a. mass is often measured with a spring scale.
    b. mass is expressed in pounds.
    c. as the force of Earth's gravity on an object increases, the object's mass increases.
    d. mass is determined by comparing the mass of an object with a set of standard masses that
        are part of a balance.

    ANS: D   DIF: I   OBJ: 2-2.3

39. To determine density, the quantities that must be measured are
    a. mass and weight.   c. volume and concentration.
    b. volume and weight.   d. volume and mass.

    ANS: D   DIF: I   OBJ: 2-2.4

40. The relationship between the mass m of a material, its volume V, and its density D is
    a. $V = mD$.
    b. $Vm = D$.
    c. $DV = m$.
    d. $D + V = m$.

    ANS: C   DIF: I   OBJ: 2-2.4

41. To calculate the density of an object,
    a. multiply its mass and its volume.
    b. divide its mass by its volume.
    c. divide its volume by its mass.
    d. divide its mass by its area.

    ANS: B   DIF: I   OBJ: 2-2.4

42. When density is measured,
    a. a balance is always used.
    b. The units are always kg/m^3.
    c. the temperature should be specified.
    d. the mass and volume do not need to be measured.

    ANS: C   DIF: I   OBJ: 2-2.4

43. Which statement about density is true?
    a. Two samples of a pure substance may have different densities.
    b. Density is a chemical property.
    c. Density is a physical property.
    d. The density of a sample depends on its location on Earth.

    ANS: C   DIF: I   OBJ: 2-2.4

44. The density of aluminum is 2.70 g/cm^3. The volume of a solid piece of aluminum is 1.50 cm^3. Find its mass.
    a. 1.50 g
    b. 1.80 g
    c. 2.70 g
    d. 4.05 g

    ANS: D   DIF: III   OBJ: 2-2.4

45. The mass of a 5.00 cm^3 sample of gold is 96.5 g. The density of gold is
    a. 0.0518 g/cm^3.
    b. 19.3 g/cm^3.
    c. 101.5 g/cm^3.
    d. 483 g/cm^3.

    ANS: B   DIF: III   OBJ: 2-2.4

46. The density of pure diamond is 3.5 g/cm^3. The mass of a diamond is 0.25 g. Find its volume.
    a. 0.071 cm^3
    b. 0.875 cm^3
    c. 3.5 cm^3
    d. 14 cm^3

    ANS: A   DIF: III   OBJ: 2-2.4

47. What is the density of 37.72 g of matter whose volume is 6.80 cm^3?
    a. 0.18 g/cm^3
    b. 5.55 g/cm^3
    c. 30.92 g/cm^3
    d. 256.4 g/cm^3

    ANS: B   DIF: III   OBJ: 2-2.4

48. The density of sugar is 1.59 g/cm^3. The mass of a sample is 4.0 g. Find the volume of the sample.
   a. 2.5 cm^3
   b. 6.36 cm^3
   c. 0.39 cm^3
   d. 2.5 g/cm^3

   ANS: A   DIF: III   OBJ: 2-2.4

49. The mass of a 5.00 cm^3 sample of clay is 11 g. What is the density of the clay?
   a. 0.45 g/cm^3
   b. 2.2 g/cm^3
   c. 6 g/cm^3
   d. 55 g/cm^3

   ANS: B   DIF: III   OBJ: 2-2.4

50. The mass of a 6.0 mL sample of kerosene is 4.92 g. The density of kerosene is
   a. 0.82 g/mL.
   b. 0.92 g/cm^3.
   c. 1.2 g/mL.
   d. 1.5 g/cm^3.

   ANS: A   DIF: III   OBJ: 2-2.4

51. 100 milliliters is equivalent to
   a. 1 hectoliter.
   b. 1 microliter.
   c. 1 centiliter.
   d. 1 deciliter.

   ANS: D   DIF: II   OBJ: 2-2.5

52. 10^{-2} meter is the same as
   a. 1 hectometer.
   b. 10 millimeters.
   c. 0.1 centimeter.
   d. 1000 micrometers.

   ANS: B   DIF: II   OBJ: 2-2.5

53. 0.25 g is equivalent to
   a. 250 kg.
   b. 250 mg.
   c. 0.025 mg.
   d. 0.025 kg.

   ANS: B   DIF: II   OBJ: 2-2.5

54. 0.05 cm is the same as
   a. 0.000 05 m.
   b. 0.005 mm.
   c. 0.05 m.
   d. 0.5 mm.

   ANS: D   DIF: II   OBJ: 2-2.5

55. 1.06 L of water is equivalent to
   a. 0.001 06 mL.
   b. 10.6 mL.
   c. 106 mL.
   d. 1060 mL.

   ANS: D   DIF: II   OBJ: 2-2.5

56. The number of grams equal to 0.5 kg is
   a. 0.0005.
   b. 0.005.
   c. 500.
   d. 5000.

   ANS: C   DIF: II   OBJ: 2-2.5

57. 30ºC equals
    a. −243.15 K.   c. 243.15 K.
    b. 9.1 K.   d. 303.15 K.

    ANS: D   DIF: II   OBJ: 2-2.5

58. Convert −25ºC to the kelvin scale.
    a. −323.15 K   c. 248.15 K
    b. −248.15 K   d. 323.15 K

    ANS: C   DIF: II   OBJ: 2-2.5

59. How many minutes are in 1 week?
    a. 168 min   c. 10 080 min
    b. 1440 min   d. 100 800 min

    ANS: C   DIF: II   OBJ: 2-2.5

60. If 1 inch equals 2.54 cm, how many centimeters equal 1 yard?
    a. 0.07 cm   c. 36 cm
    b. 14.17 cm   d. 91.4 cm

    ANS: D   DIF: II   OBJ: 2-2.5

61. A measurement that closely agrees with accepted values is said to be
    a. precise.   c. significant.
    b. reliable.   d. accurate.

    ANS: D   DIF: I   OBJ: 2-3.1

62. A measurement is said to have good precision if it
    a. agrees closely with an accepted standard.
    b. agrees closely with other measurements of the same quantity.
    c. has a small number of significant figures.
    d. has a large number of significant figures.

    ANS: B   DIF: I   OBJ: 2-3.1

63. If some measurements agree closely but differ widely from the actual value, these measurements
    are
    a. neither precise nor accurate.
    b. accurate, but not precise.
    c. acceptable as a new standard of accuracy.
    d. precise, but not accurate.

    ANS: D   DIF: II   OBJ: 2-3.1

64. Poor precision in scientific measurement may arise from
    a. the standard being too strict.
    b. human error.
    c. limitations of the measuring instrument.
    d. both human error and the limitations of the measuring instrument.

    ANS: D   DIF: II   OBJ: 2-3.1

65. Precision pertains to all of the following EXCEPT
    a. reproducibility of measurements.
    b. agreement among numerical values.
    c. sameness of measurements.
    d. closeness of a measurement to an accepted value.

    ANS: D   DIF: I   OBJ: 2-3.1

66. These values were obtained as the mass of products from the same reaction: 8.83 g; 8.84 g; 8.82 g. The known mass of products from that reaction is 8.60 g. The values are
    a. accurate.   c. both accurate and precise.
    b. precise.   d. neither accurate nor precise.

    ANS: B   DIF: II   OBJ: 2-3.1

67. Five darts strike near the center of the target. Whoever threw the darts is
    a. accurate.   c. both accurate and precise.
    b. precise.   d. neither accurate nor precise.

    ANS: C   DIF: II   OBJ: 2-3.1

68. A chemist who frequently carries out a complex experiment is likely to have high
    a. accuracy, but low precision.   c. precision.
    b. accuracy.   d. precision, but low accuracy.

    ANS: C   DIF: II   OBJ: 2-3.1

69. When applied to scientific measurements, the words *accuracy and precision*
    a. are used interchangeably.   c. can cause uncertainty in experiments.
    b. have limitations.   d. have distinctly different meanings.

    ANS: D   DIF: I   OBJ: 2-3.1

70. Using the same balance, a chemist obtained the values 5.224 g, 5.235 g, and 5.25 g for the mass of a sample. These measurements have
    a. good precision.   c. poor precision.
    b. good accuracy.   d. poor accuracy.

    ANS: C   DIF: II   OBJ: 2-3.1

Mass Data of Sample

	Trial 1	Trial 2	Trial 3	Trial 4
Student A	1.43 g	1.52 g	1.47 g	1.42 g
Student B	1.43 g	1.40 g	1.46 g	1.44 g
Student C	1.54 g	1.56 g	1.58 g	1.50 g
Student D	0.86 g	1.24 g	1.52 g	1.42 g

71. Four students each measured the mass of one 1.43 g sample four times. The results in the table above indicate that the data collected by _____ reflect the greatest accuracy and precision.
    a. Student A
    b. Student B
    c. Student C
    d. Student D

    ANS: B   DIF: II   OBJ: 2-3.1

72. When determining the number of significant digits in a measurement,
    a. all zeros are significant.
    b. all nonzero digits are significant.
    c. all zeros between two nonzero digits are not significant.
    d. all nonzero digits are not significant.

    ANS: B   DIF: I   OBJ: 2-3.2

73. For numbers less than 0.1, such as 0.06, the zeros to the right of the decimal point but before the first nonzero digit
    a. are significant.
    b. show the decimal place of the first digit.
    c. show that the zero on the left side of the decimal is not significant.
    d. show uncertainty.

    ANS: B   DIF: I   OBJ: 2-3.2

74. To two significant figures, the measurement 0.0255 g should be reported as
    a. 0.02 g.
    b. 0.025 g.
    c. 0.026 g.
    d. 2.5×10^2 g.

    ANS: C   DIF: II   OBJ: 2-3.2

75. In division and multiplication, the answer must not have more significant figures than the
    a. number in the calculation with fewest significant figures.
    b. number in the calculation with most significant figures.
    c. average number of significant figures in the calculation.
    d. total number of significant figures in the calculation.

    ANS: A   DIF: I   OBJ: 2-3.2

76. A sum or difference of whole numbers should be rounded so that the final digit is in the same place as the

    a.   rightmost uncertain digit.   c.   leftmost uncertain digit.
    b.   last digit in the longest number.   d.   leftmost certain digit.

    ANS: C   DIF: I   OBJ: 2-3.2

77. The number of significant figures in the measurement 0.000 305 kg is

    a.   3.   c.   5.
    b.   4.   d.   6.

    ANS: A   DIF: II   OBJ: 2-3.2

78. The number of significant figures in the measured value 0.003 20 g is

    a.   2.   c.   5.
    b.   3.   d.   6.

    ANS: B   DIF: II   OBJ: 2-3.2

79. The measurement that has been expressed to three significant figures is

    a.   0.052 g.   c.   3.065 g.
    b.   0.202 g.   d.   5000 g.

    ANS: B   DIF: II   OBJ: 2-3.2

80. The number of significant figures in the measurement 170.040 km is

    a.   3.   c.   5.
    b.   4.   d.   6.

    ANS: D   DIF: II   OBJ: 2-3.2

81. The measurement that has been expressed to four significant figures is

    a.   0.0020 mm.   c.   30.00 mm.
    b.   0.004 02 mm.   d.   402.10 mm.

    ANS: C   DIF: II   OBJ: 2-3.2

82. The number of significant figures in the measurement 210 cm is

    a.   1.   c.   3.
    b.   2.   d.   4.

    ANS: B   DIF: II   OBJ: 2-3.2

83. The measurement that has only nonsignificant zeros is

    a.   0.0037 mL.   c.   400. mL.
    b.   60.0 mL.   d.   506 mL.

    ANS: A   DIF: II   OBJ: 2-3.2

84. The number that has five significant figures is
    a. 23 410.
    b. 0.006 52.
    c. 0.017 83.
    d. 10.292.

    ANS: D   DIF: II   OBJ: 2-3.2

85. Using a metric ruler with 1 mm divisions, you find the sides of a rectangular piece of plywood are 3.54 cm and 4.85 cm. You calculate that the area is 17.1690 cm². To the correct number of significant figures, the result should be expressed as
    a. 17.1 cm².
    b. 17.169 cm².
    c. 17.17 cm².
    d. 17.2 cm².

    ANS: D   DIF: II   OBJ: 2-3.2

86. When 64.4 is divided by 2.00, the correct number of significant figures in the result is
    a. 1.
    b. 2.
    c. 3.
    d. 4.

    ANS: C   DIF: III   OBJ: 2-3.3

87. The dimensions of a rectangular solid are measured to be 1.27 cm, 1.3 cm, and 2.5 cm. The volume should be recorded as
    a. 4.128 cm³.
    b. 4.12 cm³.
    c. 4.13 cm³.
    d. 4.1 cm³.

    ANS: D   DIF: III   OBJ: 2-3.3

88. Three samples of 0.12 g, 1.8 g, and 0.562 g are mixed together. The combined mass of all three samples, expressed to the correct number of significant figures, should be recorded as
    a. 2.4 g.
    b. 2.48 g.
    c. 2.482 g.
    d. 2.5 g.

    ANS: D   DIF: III   OBJ: 2-3.3

89. Divide 5.7 m by 2 m. The quotient is correctly reported as
    a. 2.8 m.
    b. 2.85 m.
    c. 2.9 m.
    d. 3 m.

    ANS: D   DIF: III   OBJ: 2-3.3

90. The sum of 314.53 km and 32 km is correctly expressed as
    a. 346 km.
    b. 346.5 km.
    c. 346.53 km.
    d. 347 km.

    ANS: D   DIF: III   OBJ: 2-3.3

91. The product of 13 cm and 5.7 cm is correctly reported as
    a. 74 cm².
    b. 74.0 cm².
    c. 74.1 cm².
    d. 75 cm².

    ANS: A   DIF: III   OBJ: 2-3.3

92. Round 1.245 633 501 × 10^8 to four significant figures.
   a. 1246
   b. 1.2456 × 10^8
   c. 1.246 × 10^8
   d. 1.246 × 10^4

   ANS: C   DIF: III   OBJ: 2-3.3

93. The correct number of significant figures that should appear in the answer to the calculation 3.475 × 1.97 + 2.4712 is
   a. 2.
   b. 3.
   c. 4.
   d. 5.

   ANS: B   DIF: III   OBJ: 2-3.3

94. How many significant digits should be shown in the product of 1.6 cm and 2.4 cm?
   a. 1
   b. 2
   c. 3
   d. 4

   ANS: B   DIF: III   OBJ: 2-3.3

95. Written in scientific notation, the measurement 0.000 065 cm is
   a. 65 × 10^{-4} cm.
   b. 6.5 × 10^{-5} cm.
   c. 6.5 × 10^{-6} cm.
   d. 6.5 × 10^{-4} cm.

   ANS: B   DIF: III   OBJ: 2-3.4

96. The measurement 0.020 L is the same as
   a. 2.0 × 10^{-3} L.
   b. 2.0 × 10^2 L.
   c. 2.0 × 10^{-2} L.
   d. 2.0 × 10^{-1} L.

   ANS: C   DIF: III   OBJ: 2-3.4

97. Expressed in scientific notation, 0.0930 m is
   a. 93 × 10^{-3} m.
   b. 9.3 × 10^{-3} m.
   c. 9.30 × 10^{-2} m.
   d. 9.30 × 10^{-4} m.

   ANS: C   DIF: III   OBJ: 2-3.4

98. The speed of light is 300 000 km/s. In scientific notation, this speed is
   a. 3 × 10^5 km/s.
   b. 3.00 × 10^5 km/s.
   c. 3.0 × 10^6 km/s.
   d. 3.00 × 10^6 km/s.

   ANS: A   DIF: III   OBJ: 2-3.4

99. The average distance between the Earth and the moon is 386 000 km. Expressed in scientific notation, this distance is
   a. 386 × 10^3 km.
   b. 38 × 10^4 km.
   c. 3.8 × 10^5 km.
   d. 3.86 × 10^5 km.

   ANS: D   DIF: III   OBJ: 2-3.4

100. An analytical balance can measure mass to the nearest 1/10 000 of a gram, 0.0001 g. In scientific notation, the accuracy of the balance would be expressed as
   a. 1.0×10^{-3} g.
   b. 1×10^{3} g.
   c. 1×10^{4} g.
   d. 1×10^{-4} g.

   ANS: D   DIF: III   OBJ: 2-3.4

101. When 1.92×10^{-6} kg is divided by 6.8×10^{2} mL, the quotient in kg/mL equals
   a. 2.8×10^{-4}.
   b. 2.8×10^{-5}.
   c. 2.8×10^{-8}.
   d. 2.8×10^{-9}.

   ANS: D   DIF: III   OBJ: 2-3.4

102. When 6.02×10^{23} is multiplied by 9.1×10^{-31}, the product is
   a. 5.5×10^{-8}.
   b. 5.5×10^{54}.
   c. 5.5×10^{-7}.
   d. 5.5×10^{-53}.

   ANS: C   DIF: III   OBJ: 2-3.4

103. The result of dividing 10^{7} by 10^{-3} is
   a. 10^{-4}.
   b. $10^{2.5}$.
   c. 10^{4}.
   d. 10^{10}.

   ANS: D   DIF: III   OBJ: 2-3.4

104. The capacity of a Florence flask is 250 mL. Its capacity in liters expressed in scientific notation is
   a. 2.5×10^{-2} L.
   b. 2.5×10^{-1} L.
   c. 2.5×10^{1} L.
   d. 2.5×10^{2} L.

   ANS: B   DIF: III   OBJ: 2-3.4

105. If values for x and y vary as an inverse proportion,
   a. their quotient is a constant.
   b. their graph is a parabola.
   c. their product is a constant.
   d. their graph is a straight line.

   ANS: C   DIF: I   OBJ: 2-3.5

106. Two variables are directly proportional if their _____ has a constant value.
   a. sum
   b. difference
   c. quotient
   d. product

   ANS: C   DIF: I   OBJ: 2-3.5

107. The graph of a direct proportion is a(n)
   a. straight line.
   b. ellipse.
   c. parabola.
   d. hyperbola.

   ANS: A   DIF: I   OBJ: 2-3.5

108. Two variables are inversely proportional if their _____ has a constant value.
    a. sum
    b. difference
    c. product
    d. quotient

ANS: C   DIF: I   OBJ: 2-3.5

109. The graph of an inverse proportion is a(n)
    a. straight line.
    b. ellipse.
    c. parabola.
    d. hyperbola.

ANS: D   DIF: I   OBJ: 2-3.5

110. Which of the following statements about $y = kx$ is NOT true?
    a. y is directly proportional to x.
    b. x is a variable.
    c. The product of y and x is a constant.
    d. The graph of y versus x should be a straight line.

ANS: C   DIF: II   OBJ: 2-3.5

111. In the expression $m = DV$, where m is mass, D is density, and V is volume, density is the
    a. variable.
    b. difference of m and V.
    c. constant.
    d. product of m and V.

ANS: C   DIF: II   OBJ: 2-3.5

112. Which of the following does NOT describe an inverse proportion between x and y?
    a. $xy = k$
    b. $x = k/y$
    c. $y = k/x$
    d. $k = x/y$

ANS: D   DIF: II   OBJ: 2-3.5

113. Which of the following does NOT describe a direct proportion between x and y?
    a. $xy = k$
    b. $x/y = k$
    c. $y/x = k$
    d. $x = ky$

ANS: A   DIF: II   OBJ: 2-3.5

114. In the equation density = mass/volume, mass divided by volume has a constant ratio. This means that the
    a. equation graphs as a straight line.
    b. variables mass and volume are inversely proportional.
    c. equation graphs as a hyperbola.
    d. product of mass and volume is a constant.

ANS: A   DIF: II   OBJ: 2-3.5

SHORT ANSWER

1. What are the first steps scientists take to analyze the cause of a disease?

   ANS:
   Scientists must first observe the symptoms of the disease. Then they form a hypothesis and design an experiment with a control and variables to test their hypothesis about the probable cause of the disease.

   DIF: II   OBJ: 2-1.1

2. Note the differences between qualitative and quantitative observations.

   ANS:
   In qualitative observations, the data are descriptive and non-numerical. In quantitative observations, the data are numerical.

   DIF: II   OBJ: 2-1.2

3. Compare and contrast a model with a written theory.

   ANS:
   A written theory is a broad generalization used to explain observations. A model is a visual, verbal, or mathematical theory that illustrates or explains abstract concepts.

   DIF: II   OBJ: 2-1.3

4. How is a theory different from a hypothesis?

   ANS:
   A theory is a broad generalization based on observations, reasoning, and data that is used to explain a phenomenon and predict future events. A hypothesis is an educated assumption based on observation and data. When tested, a hypothesis may become a theory.

   DIF: II   OBJ: 2-1.3

5. Explain why pictures of atoms and molecules are models.

   ANS:
   The pictures represent the atoms and molecules and are consistent with what we know about their behavior. They are not true representations because atoms and molecules are not hard, colored spheres.

   DIF: II   OBJ: 2-1.3

6. What would be the most appropriate SI unit for expressing the mass of a single brick?

ANS:
The mass of a brick should be expressed in kilograms.

DIF: II   OBJ: 2-2.2

7. Why are only seven basic SI units needed to express almost any measured quantity?

ANS:
The base units are combined to form derived units.

DIF: II   OBJ: 2-2.2

8. Give an example to show how SI units can be adjusted to measure the mass, length, or volume of very small or very large objects.

ANS:
The prefix milli- is used with the base unit gram to express the mass of a tiny object.

DIF: III   OBJ: 2-2.2

9. Distinguish between mass and weight.

ANS:
Mass is the amount of matter in something. Weight depends on Earth's gravitational attraction on the object.

DIF: II   OBJ: 2-2.3

10. Distinguish between precision and accuracy.

ANS:
Precision is how close a set of measurements of the same quantity are. Accuracy is how close a measurement is to the true value.

DIF: II   OBJ: 2-3.1

11. Explain the importance of significant figures when working with measurements and calculated quantities based on actual measurements.

ANS:
Significant figures identify the digits in a measurement that are certain, along with one estimated digit. When using measurements in calculations, your result must reflect the precision of the least precise measurement.

DIF: II   OBJ: 2-3.2

12. In a calculation based on measured quantities, why must the number of significant figures in the result be limited?

ANS:
The number of significant figures must reflect the precision of the measurements that were used in the calculation. If significant figures are not considered, the answer might appear to be far more precise than it actually is.

DIF: II   OBJ: 2-3.2

ESSAY

1. Why should a scientist record all observations, even those that appear insignificant?

ANS:
Many important discoveries are made through mistakes or by accident. If seemingly insignificant observations are not recorded, such discoveries can not be reproduced.

DIF: II   OBJ: 2-1.1

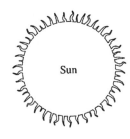
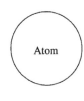

2. Evaluate the models in the figure above. Describe any ways that the models differ from the real objects.

ANS:
The model of the sun is accurate in showing that the sun is round and has a fiery surface. The model of an atom shows that the atom is a particle. The sun model is inaccurate because it is only two-dimensional, and it is smaller than the real sun. In addition, it does not show the sun's composition. The atomic model is inaccurate because it is larger than a real atom, because it is only two-dimensional, and because it does not depict the atom's composition.

DIF: III   OBJ: 2-1.3

Mass and Volume Data

Sample	Mass	Volume
1	6.02 g	2.23 mL
2	18.42 g	2.34 mL
3	35.15 g	3.10 mL

3. Explain how the samples in the table above could be identified.

ANS:
The density of each sample could be calculated. Then each value could be compared to known values to identify the substances.

DIF: III   OBJ: 2-2.4

MULTIPLE CHOICE

1. The law of conservation of mass follows from the concept that
   a. atoms are indivisible.
   b. atoms of different elements have different properties.
   c. matter is composed of atoms.
   d. atoms can be destroyed in chemical reactions.

   ANS: A   DIF: I   OBJ: 3-1.1

2. If each atom of element D has 3 mass units and each atom of element E has 5 mass units, a chemical molecule composed of one atom each of D and E has
   a. 15 mass units.   c. 35 mass units.
   b. 2 mass units.   d. 8 mass units.

   ANS: D   DIF: II   OBJ: 3-1.1

3. A certain compound is composed of elements G and H. It always has the same mass ratio of G to H because
   a. all atoms have the same mass.   c. G and H have characteristic masses.
   b. any excess of G or H will be destroyed.   d. G and H have identical masses.

   ANS: C   DIF: II   OBJ: 3-1.1

4. If 4 g of element A combine with 10 g of element B, then 12 g of element A combine with _____ g of element B.
   a. 10   c. 24
   b. 12   d. 30

   ANS: D   DIF: III   OBJ: 3-1.1

5. If 6 g of element K combine with 17 g of element L, how many grams of element K combine with 85 g of element L?
   a. 17 g   c. 30 g
   b. 23 g   d. 91 g

   ANS: C   DIF: III   OBJ: 3-1.1

6. In oxides of nitrogen, such as N_2O, NO, NO_2, and N_2O_3, atoms combine in small whole-number ratios. This evidence supports the law of
   a. conservation of mass.   c. definite composition.
   b. multiple proportions.   d. mass action.

   ANS: B   DIF: II   OBJ: 3-1.1

7. The two oxides of lead, PbO and PbO_2, are explained by the
   a. periodic law.   c. atomic law.
   b. law of multiple proportions.   d. law of conservation of mass.

   ANS: B   DIF: II   OBJ: 3-1.1

8. If two or more compounds are composed of the same two elements, the ratio of the masses of one element that combine with a fixed mass of the other element is a simple whole number. This is a statement of the law of
   a. conservation of mass.
   b. mass action.
   c. multiple proportions.
   d. definite composition.

   ANS: C   DIF: II   OBJ: 3-1.1

9. Which two compounds are examples of the law of multiple proportions?
   a. $FeCl_3$ and $Fe_2(SO_4)_3$
   b. O_2 and O_3
   c. CO and CO_2
   d. $FeCl_2$ and $Fe(NO_3)_2$

   ANS: C   DIF: II   OBJ: 3-1.1

10. The law of multiple proportions can be partly explained by the idea that
    a. elements can combine in only one way to form compounds.
    b. whole atoms of the same elements combine to form compounds.
    c. elements in a compound always occur in a 1:1 ratio.
    d. only atoms of the same element can combine.

    ANS: B   DIF: I   OBJ: 3-1.1

11. In water, H_2O, the ratio of the masses of oxygen to hydrogen is 8:1. What is the ratio of the masses of oxygen to hydrogen in hydrogen peroxide, H_2O_2?
    a. 1:1
    b. 8:1
    c. 16:1
    d. 32:1

    ANS: C   DIF: III   OBJ: 3-1.1

12. If 3 g of element C combine with 8 g of element D to form compound CD, how many grams of D are needed to form compound CD2?
    a. 8 g
    b. 16 g
    c. 11 g
    d. 19 g

    ANS: B   DIF: III   OBJ: 3-1.1

13. Oxygen can combine with carbon to form two compounds, carbon monoxide and carbon dioxide. The ratio of the masses of oxygen that combine with a given mass of carbon is 1:2. This is an example of
    a. the law of conservation of mass.
    b. Dalton's atomic theory.
    c. the law of conservation of energy.
    d. the law of multiple proportions.

    ANS: D   DIF: II   OBJ: 3-1.1

14. If 63.5 g of copper (Cu) combine with 16 g of oxygen (O) to form the compound CuO, how many grams of oxygen will be needed to combine with the same amount of copper to form the compound CuO_2?
    a. 16 g
    b. 32 g
    c. 64 g
    d. 127 g

    ANS: B   DIF: III   OBJ: 3-1.1

15. According to the law of definite proportions, any two samples of KCl have
    a. the same mass.
    b. slightly different molecular structures.
    c. the same melting point.
    d. the same ratio of elements.

    ANS: D   DIF: I   OBJ: 3-1.1

16. According to the law of conservation of mass, when sodium, hydrogen, and oxygen react to form a compound, the mass of the compound is _____ the sum of the masses of the individual elements.
    a. equal to
    b. greater than
    c. less than
    d. either greater than or less than

    ANS: A   DIF: II   OBJ: 3-1.1

17. The atomic mass of an atom of carbon is 12, and the atomic mass of an atom of oxygen is 16. To produce CO, 16 g of oxygen can be combined with 12 g of carbon. What is the ratio of oxygen to carbon when 32 g of oxygen combine with 12 g of carbon?
    a. 1:1
    b. 2:1
    c. 1:2
    d. 8:3

    ANS: B   DIF: II   OBJ: 3-1.1

18. Who was the schoolmaster who studied chemistry and proposed an atomic theory?
    a. John Dalton
    b. Jons Berzelius
    c. Robert Brown
    d. Dmitri Mendeleev

    ANS: A   DIF: I   OBJ: 3-1.2

19. Who first recognized that the ratio of the number of atoms that combine is the same as the ratio of the masses that combine?
    a. Jons Berzelius
    b. Edward Morley
    c. John Dalton
    d. Jon Newlands

    ANS: C   DIF: I   OBJ: 3-1.2

20. The principles of atomic theory recognized today were conceived by
    a. Avogadro.
    b. Bohr.
    c. Dalton.
    d. Rutherford.

    ANS: C   DIF: I   OBJ: 3-1.2

21. According to Dalton's atomic theory, atoms
    a. are destroyed in chemical reactions.
    b. can be divided.
    c. of each element are identical in size, mass, and other properties.
    d. of different elements cannot combine.

    ANS: C   DIF: I   OBJ: 3-1.2

22. Which of the following is NOT part of Dalton's atomic theory?
    a. Atoms cannot be divided, created, or destroyed.
    b. The number of protons in an atom is its atomic number.
    c. In chemical reactions, atoms are combined, separated, or rearranged.
    d. All matter is composed of extremely small particles called atoms.

    ANS: B   DIF: I   OBJ: 3-1.2

23. According to Dalton's atomic theory, atoms
    a. of different elements combine in simple whole-number ratios to form compounds.
    b. can be divided into protons, neutrons, and electrons.
    c. of all elements are identical in size and mass.
    d. can be destroyed in chemical reactions.

    ANS: A   DIF: I   OBJ: 3-1.2

24. Dalton's atomic theory did NOT explain the law of
    a. whole-number ratios.   c. conservation of mass.
    b. definite proportions.   d. conservation of energy.

    ANS: D   DIF: I   OBJ: 3-1.2

25. The law of definite proportions
    a. contradicted Dalton's atomic theory.
    b. was explained by Dalton's atomic theory.
    c. replaced the law of conservation of mass.
    d. assumes that atoms of all elements are identical.

    ANS: B   DIF: I   OBJ: 3-1.2

26. Dalton's atomic theory helped to explain the law of conservation of mass because it stated that atoms
    a. could not combine.   c. all had the same mass.
    b. could not be created or destroyed. d. were invisible.

    ANS: B   DIF: I   OBJ: 3-1.2

27. Who proposed the law of multiple proportions?
    a. Avogadro   c. Dalton
    b. Rutherford   d. Thomson

    ANS: C   DIF: I   OBJ: 3-1.2

28. Dalton's theory essentially agreed with the present atomic theory EXCEPT for the statement that
    a. all matter is made up of small particles called atoms.
    b. atoms are not divided in chemical reactions.
    c. atoms of the same element are chemically alike.
    d. all atoms of the same element have the same mass.

    ANS: D   DIF: II   OBJ: 3-1.3

29. Which of the following statements is true?
    a. Atoms of the same element may have different masses.
    b. Atoms may be divided in ordinary chemical reactions.
    c. Atoms can never combine with any other atoms.
    d. Matter is composed of large particles called atoms.

    ANS: A   DIF: II   OBJ: 3-1.3

30. Which concept in Dalton's atomic theory has been modified?
    a. All matter is composed of atoms.
    b. Atoms of different elements have different properties and masses.
    c. Atoms can combine in chemical reactions.
    d. Atoms cannot be divided.

    ANS: D   DIF: I   OBJ: 3-1.3

31. The atomic theory proposed by Dalton
    a. has been totally discarded.
    b. has been expanded and modified.
    c. has been accepted unchanged to the present day.
    d. has been found to be false.

    ANS: B   DIF: I   OBJ: 3-1.3

32. In early experiments on electricity and matter, an electrical current was passed through a glass tube containing
    a. water.   c. liquid oxygen.
    b. gas under high pressure.   d. gas under low pressure.

    ANS: D   DIF: I   OBJ: 3-2.1

33. In a glass tube, electrical current passes from the negative electrode, called the _____, to the other electrode.
    a. cathode   c. electron
    b. anode   d. millikan

    ANS: A   DIF: I   OBJ: 3-2.1

34. When an electrical current passed through a glass tube, a paddle wheel placed between the electrodes moved. Scientists concluded that
    a. a magnetic field was produced.
    b. particles were passing from the cathode to the anode.
    c. there was gas in the tube.
    d. atoms were indivisible.

    ANS: B   DIF: I   OBJ: 3-2.1

35. The rays produced in a cathode tube in early experiments were
    a. unaffected by a magnetic field.   c. found to carry a positive charge.
    b. deflected away from a negative plate.   d. striking the cathode.

    ANS: B   DIF: I   OBJ: 3-2.1

36. The behavior of cathode rays produced in a glass tube containing gas at low pressure led scientists to conclude that the rays
    a. were not composed of matter.
    b. were composed of positively charged particles.
    c. were composed of negatively charged particles.
    d. were composed of uncharged particles.

    ANS: C   DIF: I   OBJ: 3-2.1

37. Experiments with cathode rays led to the discovery of the
    a. proton.   c. neutron.
    b. nucleus.   d. electron.

    ANS: D   DIF: I   OBJ: 3-2.1

38. After measuring the ratio of the charge of a cathode-ray particle to its mass, Thomson concluded that the particles
    a. had no mass.   c. had a very large mass.
    b. had a very small mass.   d. carried a positive charge.

    ANS: B   DIF: I   OBJ: 3-2.1

39. Millikan's experiments
    a. demonstrated that the electron carried no charge.
    b. demonstrated that the electron carried the smallest possible positive charge.
    c. measured the charge on the electron.
    d. demonstrated that the electron was massless.

    ANS: C   DIF: I   OBJ: 3-2.1

40. Because any element used in the cathode produced electrons, scientists concluded that
    a. all atoms contained electrons.   c. atoms were indivisible.
    b. only metals contained electrons.   d. atoms carried a negative charge.

    ANS: A   DIF: I   OBJ: 3-2.1

41. The discovery of the electron resulted from experiments using
    a. gold foil.   c. neutrons.
    b. cathode rays.   d. alpha particles.

    ANS: B   DIF: I   OBJ: 3-2.1

42. The deflection of cathode rays in Thomson's experiments was evidence of the _____ nature of electrons.
    a. wave   c. particle
    b. charged   d. spinning

    ANS: B   DIF: I   OBJ: 3-2.1

43. Who discovered the nucleus by bombarding gold foil with positively charged particles and noting that some particles were widely deflected?
    a. Rutherford
    b. Dalton
    c. Chadwick
    d. Bohr

    ANS: A   DIF: I   OBJ: 3-2.2

44. In Rutherford's experiments, very few positively charged particles
    a. were slightly deflected as they passed through the metal.
    b. were greatly deflected back from the metal.
    c. passed straight through the metal.
    d. combined with the metal.

    ANS: B   DIF: I   OBJ: 3-2.2

45. In Rutherford's experiments, positively charged particles
    a. passed through a tube containing gas.
    b. were used to bombard a cathode plate.
    c. collided with electrons.
    d. were used to bombard thin metal foil.

    ANS: D   DIF: I   OBJ: 3-2.2

46. In Rutherford's experiments, most of the particles
    a. bounced back.
    b. passed through the foil.
    c. were absorbed by the foil.
    d. combined with the foil.

    ANS: B   DIF: I   OBJ: 3-2.2

47. Because most particles fired at metal foil passed straight through, Rutherford concluded that
    a. atoms were mostly empty space.
    b. atoms contained no charged particles.
    c. electrons formed the nucleus.
    d. atoms were indivisible.

    ANS: A   DIF: II   OBJ: 3-2.2

48. Because a few positively charged particles bounced back from the foil, Rutherford concluded that such particles were
    a. striking electrons.
    b. indivisible.
    c. repelled by densely packed regions of positive charge.
    d. magnetic.

    ANS: C   DIF: II   OBJ: 3-2.2

49. Rutherford's experiments led to the discovery of the
    a. electron.
    b. cathode ray.
    c. nucleus.
    d. neutron.

    ANS: C   DIF: I   OBJ: 3-2.2

50. Rutherford's experimental results led him to conclude that atoms contain massive central regions that have
   a. a positive charge.
   b. a negative charge.
   c. no charge.
   d. both protons and electrons.

   ANS: A   DIF: I   OBJ: 3-2.2

51. Rutherford fired positively charged particles at metal foil and concluded that most of the mass of an atom was
   a. in the electrons.
   b. concentrated in the nucleus.
   c. evenly spread throughout the atom.
   d. in rings around the atom.

   ANS: B   DIF: I   OBJ: 3-2.2

52. What did Rutherford conclude about the structure of the atom?
   a. An atom is indivisible.
   b. Electrons make up the center of an atom.
   c. An atom carries a positive charge.
   d. An atom contains a small, dense, positively charged central region.

   ANS: D   DIF: I   OBJ: 3-2.2

53. In Rutherford's experiments, the backward deflection of alpha particles gave evidence of an atom's
   a. size.
   b. electron orbitals.
   c. charge.
   d. nucleus.

   ANS: D   DIF: I   OBJ: 3-2.2

54. A positively charged particle with mass 1.673×10^{-24} g is a(n)
   a. proton.
   b. neutron.
   c. electron.
   d. positron.

   ANS: A   DIF: I   OBJ: 3-2.3

55. A nuclear particle that has about the same mass as a proton, but with no electrical charge, is called a(n)
   a. nuclide.
   b. neutron.
   c. electron.
   d. isotope.

   ANS: B   DIF: I   OBJ: 3-2.3

56. The nucleus of an atom has all of the following characteristics EXCEPT that it
   a. is positively charged.
   b. is very dense.
   c. contains nearly all of the atom's mass.
   d. contains nearly all of the atom's volume.

   ANS: D   DIF: I   OBJ: 3-2.3

57. Which part of an atom has a mass approximately equal to 1/2000 of the mass of a common hydrogen atom?
    a. nucleus
    b. electron
    c. proton
    d. electron cloud

    ANS: B   DIF: I   OBJ: 3-2.3

58. The mass of a neutron is
    a. about the same as that of a proton.
    b. about the same as that of an electron.
    c. double that of a proton.
    d. double that of an electron.

    ANS: A   DIF: I   OBJ: 3-2.3

59. The nucleus of most atoms is composed of
    a. tightly packed protons.
    b. tightly packed neutrons.
    c. tightly packed protons and neutrons.
    d. loosely connected protons and electrons.

    ANS: C   DIF: I   OBJ: 3-2.3

60. Protons and neutrons strongly attract when they
    a. are moving fast.
    b. are very close together.
    c. are at high energies.
    d. have opposite charges.

    ANS: B   DIF: I   OBJ: 3-2.3

61. Protons within a nucleus are attracted to each other by
    a. nuclear forces.
    b. opposite charges.
    c. their energy levels.
    d. electron repulsion.

    ANS: A   DIF: I   OBJ: 3-2.3

62. Protons have
    a. negative charges.
    b. an attraction for neutrons.
    c. no charges.
    d. no mass.

    ANS: B   DIF: I   OBJ: 3-2.3

63. An atom is electrically neutral because
    a. neutrons balance the protons and electrons.
    b. nuclear forces stabilize the charges.
    c. the numbers of protons and electrons are equal.
    d. the numbers of protons and neutrons are equal.

    ANS: C   DIF: I   OBJ: 3-2.4

64. Most of the volume of an atom is occupied by the
    a. nucleus.
    b. nuclides.
    c. electron cloud.
    d. protons.

    ANS: C   DIF: I   OBJ: 3-2.4

65. The charge on the electron cloud
    a. prevents compounds from forming.
    b. balances the charge on the nucleus.
    c. attracts electron clouds in other atoms to form compounds.
    d. does not exist.

    ANS: B   DIF: I   OBJ: 3-2.4

66. The smallest unit of an element that can exist either alone or in combination with other such
    particles of the same or different elements is the
    a. electron.   c. neutron.
    b. proton.   d. atom.

    ANS: D   DIF: I   OBJ: 3-2.4

67. The forces that hold the particles in the nucleus together are called
    a. nuclear forces.   c. magnetic forces.
    b. gravitational forces.   d. electron clouds.

    ANS: A   DIF: I   OBJ: 3-2.4

68. The radius of an atom extends to the outer edge of the
    a. nucleus.   c. region occupied by the neutrons.
    b. region occupied by the electrons. d. positive charges.

    ANS: B   DIF: I   OBJ: 3-2.4

69. Isotopes are atoms of the same element that have different
    a. principal chemical properties.   c. numbers of protons.
    b. masses.   d. numbers of electrons.

    ANS: B   DIF: I   OBJ: 3-3.1

70. Atoms of the same element that have different masses are called
    a. moles.   c. nuclides.
    b. isotopes.   d. neutrons.

    ANS: B   DIF: I   OBJ: 3-3.1

71. Isotopes of an element contain different numbers of
    a. electrons.   c. neutrons.
    b. protons.   d. nuclides.

    ANS: C   DIF: I   OBJ: 3-3.1

72. The most common form of hydrogen has
    a. no neutrons.   c. two neutrons.
    b. one neutron.   d. three neutrons.

    ANS: A   DIF: I   OBJ: 3-3.1

73. The only radioactive form of hydrogen is
    a. protium.   c. tritium.
    b. deuterium.   d. quadrium.

    ANS: C   DIF: I   OBJ: 3-3.1

74. The tritium atom consists of
    a. one proton, two neutrons, and two electrons.
    b. one proton, one neutron, and one electron.
    c. one proton, two neutrons, and one electron.
    d. two protons, one neutron, and one electron.

    ANS: C   DIF: I   OBJ: 3-3.1

75. What is the mass number of deuterium?
    a. 1   c. 3
    b. 2   d. 4

    ANS: B   DIF: I   OBJ: 3-3.1

76. How many isotopes of hydrogen are known?
    a. 2   c. 4
    b. 3   d. 5

    ANS: B   DIF: I   OBJ: 3-3.1

77. The hydrogen isotope with the least mass is named
    a. tritium.   c. deuterium.
    b. helium.   d. protium.

    ANS: D   DIF: I   OBJ: 3-3.1

78. Deuterium contains one proton and
    a. two neutrons.   c. no neutrons.
    b. one neutron.   d. two electrons.

    ANS: B   DIF: I   OBJ: 3-3.1

79. Deuterium differs from tritium in having one
    a. less neutron.   c. more electron.
    b. more proton.   d. more neutron.

    ANS: A   DIF: I   OBJ: 3-3.1

80. All isotopes of hydrogen contain
    a. one neutron.   c. one proton.
    b. two electrons.   d. two nuclei.

    ANS: C   DIF: I   OBJ: 3-3.1

81. Protium contains one proton and
    a. one neutron.
    b. two neutrons.
    c. no neutrons.
    d. three electrons.

    ANS: C   DIF: I   OBJ: 3-3.1

82. Helium-4 and helium-3 are
    a. isotopes.
    b. different elements.
    c. compounds.
    d. nuclei.

    ANS: A   DIF: I   OBJ: 3-3.1

83. Isotopes of each element differ in
    a. the number of neutrons in the nucleus.
    b. atomic number.
    c. the number of electrons in the highest energy level.
    d. the total number of electrons.

    ANS: A   DIF: I   OBJ: 3-3.1

84. The atomic number of oxygen, 8, indicates that there are eight
    a. protons in the nucleus of an oxygen atom.
    b. oxygen nuclides.
    c. neutrons outside the oxygen atom's nucleus.
    d. energy levels in the oxygen atom's nucleus.

    ANS: A   DIF: II   OBJ: 3-3.2

85. The total number of protons and neutrons in the nucleus of an atom is its
    a. atomic number.
    b. Avogadro constant.
    c. mass number.
    d. number of neutrons.

    ANS: C   DIF: I   OBJ: 3-3.2

86. As the mass number of the isotopes of an element increases, the number of protons
    a. decreases.
    b. increases.
    c. remains the same.
    d. doubles each time the mass number increases.

    ANS: C   DIF: I   OBJ: 3-3.2

87. As the atomic number increases, the number of electrons in an atom
    a. decreases.
    b. increases.
    c. remains the same.
    d. is undetermined.

    ANS: B   DIF: I   OBJ: 3-3.2

88. All atoms of the same element have the same
    a. atomic mass.
    b. number of neutrons.
    c. mass number.
    d. atomic number.

    ANS: D   DIF: I   OBJ: 3-3.2

89. Atoms of the same element can differ in
    a. chemical properties.
    b. mass number.
    c. atomic number.
    d. number of protons and electrons.

    ANS: B   DIF: I   OBJ: 3-3.2

90. In determining atomic mass units, the standard is the
    a. C-12 atom.
    b. C-14 atom.
    c. H-1 atom.
    d. O-16 atom.

    ANS: A   DIF: I   OBJ: 3-3.2

91. The abbreviation for atomic mass unit is
    a. amu.
    b. mu.
    c. a.
    d. μ.

    ANS: A   DIF: I   OBJ: 3-3.2

92. The relative atomic mass of an atom can be found by comparing the mass of the atom to the mass of
    a. one atom of carbon-12.
    b. one atom of hydrogen-1.
    c. a proton.
    d. uranium-235.

    ANS: A   DIF: I   OBJ: 3-3.2

93. The average atomic mass of an element is the average of the atomic masses of its
    a. naturally occurring isotopes.
    b. two most abundant isotopes.
    c. nonradioactive isotopes.
    d. artificial isotopes.

    ANS: A   DIF: I   OBJ: 3-3.2

94. The atomic mass of an isotope is its
    a. average atomic mass.
    b. isotopic mass number.
    c. atomic number.
    d. relative atomic mass.

    ANS: D   DIF: I   OBJ: 3-3.2

95. The carbon-12 atom is assigned a relative mass of exactly
    a. 1 amu.
    b. 6 amu.
    c. 12 amu.
    d. 100 amu.

    ANS: C   DIF: I   OBJ: 3-3.2

96. The average atomic mass of an element depends on both the masses of its isotopes and each isotope's
    a. atomic number.
    b. radioactivity.
    c. relative abundance.
    d. mass number.

    ANS: C   DIF: I   OBJ: 3-3.2

97. The average atomic mass of an element
    a. is the mass of the most abundant isotope.
    b. may not equal the mass of any of its isotopes.
    c. cannot be calculated.
    d. always adds up to 100.

    ANS: B   DIF: I   OBJ: 3-3.2

98. A single atom of an isotope does not have a(n)
    a. relative atomic mass.   c. mass number.
    b. atomic number.   d. average atomic mass.

    ANS: D   DIF: I   OBJ: 3-3.2

99. The atomic mass listed in the periodic table is the
    a. average atomic mass.
    b. relative atomic mass of the most abundant isotope.
    c. relative atomic mass of the most stable radioactive isotope.
    d. mass number of the most abundant isotope.

    ANS: A   DIF: I   OBJ: 3-3.2

100. An aluminum isotope consists of 13 protons, 13 electrons, and 14 neutrons. Its mass number is
    a. 13.   c. 27.
    b. 14.   d. 40.

    ANS: C   DIF: III   OBJ: 3-3.2

1																	Group 18

1
H
Hydrogen
1.01

<table>
<tr><td colspan="2">Group 18</td></tr>
<tr><td>2</td></tr>
<tr><td>He</td></tr>
<tr><td>Helium</td></tr>
<tr><td>4.00</td></tr>
</table>

Periodic Table

Group 1	Group 2		Group 13	Group 14	Group 15	Group 16	Group 17	
3 **Li** Lithium 6.94	4 **Be** Beryllium 9.01		5 **B** Boron 10.81	6 **C** Carbon 12.01	7 **N** Nitrogen 14.01	8 **O** Oxygen 16.00	9 **F** Fluorine 19.00	10 **Ne** Neon 20.18
11 **Na** Sodium 22.99	12 **Mg** Magnesium 24.30		13 **Al** Aluminum 26.98	14 **Si** Silicon 28.08	15 **P** Phosphorus 30.97	16 **S** Sulfur 32.07	17 **Cl** Chlorine 35.45	18 **Ar** Argon 39.95
19 **K** Potassium 39.10	20 **Ca** Calcium 40.08		31 **Ga** Gallium 69.72	32 **Ge** Germanium 72.61	33 **As** Arsenic 74.92	34 **Se** Selenium 78.96	35 **Br** Bromine 79.90	36 **Kr** Krypton 83.80
37 **Rb** Rubidium 85.47	38 **Sr** Strontium 87.62		49 **In** Indium 114.82	50 **Sn** Tin 118.71	51 **Sb** Antimony 121.76	52 **Te** Tellurium 127.60	53 **I** Iodine 126.90	54 **Xe** Xenon 131.29
55 **Cs** Cesium 132.90	56 **Ba** Barium 137.33		81 **Tl** Thallium 204.38	82 **Pb** Lead 207.2	83 **Bi** Bismuth 208.98	84 **Po** Polonium (208.98)	85 **At** Astatine (209.99)	86 **Rn** Radon (222.02)
87 **Fr** Francium (223.02)	88 **Ra** Radium (226.02)							

101. What is the atomic number for aluminum from the figure above?
    a. 13
    b. 14
    c. 26.98
    d. 26.9815

    ANS: A   DIF: III   OBJ: 3-3.2

102. In the figure above, a neutral atom of silicon contains
    a. 14 electrons.
    b. 28.09 electrons.
    c. 16 electrons.
    d. 38 electrons.

    ANS: A   DIF: II   OBJ: 3-3.3

103. An atom of potassium has 19 protons and 20 neutrons. What is its mass number?
    a. 19
    b. 20
    c. 39
    d. 10

    ANS: C   DIF: III   OBJ: 3-3.2

104. A neutral carbon atom (atomic number 6) has
    a. 3 electrons and 3 neutrons.
    b. 6 protons.
    c. 3 protons and 3 electrons.
    d. 3 protons and 3 neutrons.

    ANS: B   DIF: III   OBJ: 3-3.3

105. Zn-66 (atomic number 30) has
    a. 30 neutrons.
    b. 33 neutrons.
    c. 36 neutrons.
    d. 96 neutrons.

    ANS: C   DIF: III   OBJ: 3-3.3

106. Ag-109 has 62 neutrons. The neutral atom has
   a. 40 electrons.   c. 53 electrons.
   b. 47 electrons.   d. 62 electrons.

   ANS: B   DIF: III   OBJ: 3-3.3

107. Chlorine has atomic number 17 and mass number 35. It has
   a. 17 protons, 17 electrons, and 18   c. 17 protons, 17 electrons, and 52
       neutrons.   neutrons.
   b. 35 protons, 35 electrons, and 17   d. 18 protons, 18 electrons, and 17
       neutrons.   neutrons.

   ANS: A   DIF: III   OBJ: 3-3.3

108. Nickel-60 (atomic number 28) has
   a. 28 neutrons.   c. 60 neutrons.
   b. 32 neutrons.   d. 88 neutrons.

   ANS: B   DIF: III   OBJ: 3-3.3

109. Carbon-14 (atomic number 6), the radioactive nuclide used in dating fossils, has
   a. 6 neutrons.   c. 10 neutrons.
   b. 8 neutrons.   d. 14 neutrons.

   ANS: B   DIF: III   OBJ: 3-3.3

110. Sulphur-34 (atomic number 16) contains
   a. 34 protons.   c. 18 neutrons.
   b. 18 protons.   d. 16 neutrons.

   ANS: C   DIF: III   OBJ: 3-3.3

111. Phosphorus-33 (atomic number 15) contains
   a. 33 protons.   c. 33 neutrons.
   b. 18 neutrons.   d. 18 protons.

   ANS: B   DIF: III   OBJ: 3-3.3

112. Silicon-30 contains 14 protons. It also contains
   a. 16 electrons.   c. 30 neutrons.
   b. 16 neutrons.   d. 44 neutrons.

   ANS: B   DIF: III   OBJ: 3-3.3

113. Neon-22 contains 12 neutrons. It also contains
   a. 12 protons.   c. 22 electrons.
   b. 22 protons.   d. 10 protons.

   ANS: D   DIF: III   OBJ: 3-3.3

114. Calcium-48 (atomic number 20) contains
    a. 20 electrons.   c. 20 neutrons.
    b. 48 protons.   d. 28 protons.

    ANS: A   DIF: III   OBJ: 3-3.3

115. Argon (atomic number 18 and mass number 40) has _____ protons in its nucleus.
    a. 22   c. 40
    b. 9   d. 18

    ANS: D   DIF: III   OBJ: 3-3.3

116. An electrically neutral atom of mercury (atomic number 80) has
    a. 80 neutrons and 80 electrons.   c. 80 protons and 80 neutrons.
    b. 40 protons and 40 electrons.   d. 80 protons and 80 electrons.

    ANS: D   DIF: III   OBJ: 3-3.3

117. The number of atoms in 1 mol of carbon is
    a. 6.022×10^{22}.   c. 5.022×10^{22}.
    b. 6.022×10^{23}.   d. 5.022×10^{23}.

    ANS: B   DIF: I   OBJ: 3-3.4

118. The number of atoms in a mole of any pure substance is called
    a. its atomic number.   c. its mass number.
    b. Avogadro's constant.   d. its gram-atomic number.

    ANS: B   DIF: I   OBJ: 3-3.4

119. As the atomic masses of the elements in the periodic table increase, the number of atoms in 1 mol of each element
    a. decreases.   c. remains the same.
    b. increases.   d. becomes a negative number.

    ANS: C   DIF: I   OBJ: 3-3.4

120. The atomic number of neon is 10. The atomic number of calcium is 20. Compared with a mole of neon, a mole of calcium contains
    a. twice as many atoms.   c. an equal number of atoms.
    b. half as many atoms.   d. 20 times as many atoms.

    ANS: C   DIF: III   OBJ: 3-3.4

121. To determine the molar mass of an element, one must know the element's
    a. Avogadro constant.   c. number of isotopes.
    b. atomic number.   d. average atomic mass.

    ANS: D   DIF: II   OBJ: 3-3.4

122. If samples of two different elements each represent one mole, then
   a. they are equal in mass.
   c. their molar masses are equal.
   b. they contain the same number of atoms.
   d. they have the same atomic mass.

   ANS: B   DIF: I   OBJ: 3-3.4

123. An Avogadro's constant amount of any element is equivalent to
   a. the atomic number of that element.
   c. 6.022×10^{23} particles.
   b. the mass number of that element.
   d. 100 g of that element.

   ANS: C   DIF: I   OBJ: 3-3.4

124. Avogadro's constant is
   a. the maximum number of electrons that all the energy levels can accommodate.
   b. the number of protons and neutrons that can fit in the shells of the nucleus.
   c. the number of particles in 1 mole of a pure substance.
   d. the number of particles in exactly 1 gram of a pure substance.

   ANS: C   DIF: I   OBJ: 3-3.4

125. Molar mass
   a. is the mass in grams of one mole of a substance.
   b. is numerically equal to the average atomic mass of the element.
   c. both a and b
   d. neither a nor b

   ANS: C   DIF: I   OBJ: 3-3.4

126. The mass of 1 mol of chromium (atomic mass 51.996 amu) is
   a. 12 g.
   c. 51.996 g.
   b. 198 g.
   d. 6.02×10^{23} g.

   ANS: C   DIF: III   OBJ: 3-3.5

127. A mass of 6.005 g of carbon (atomic mass 12.010 amu) contains
   a. 1 mol C.
   c. 0.5000 mol C.
   b. 2 atoms C.
   d. 1 atom O.

   ANS: C   DIF: III   OBJ: 3-3.5

128. A quantity of sodium (atomic mass 22.99 amu) contains 6.02×10^{23} atoms. The mass of the sodium is
   a. 6.02×10^{23} g.
   c. 22.99 g.
   b. 3.88 g.
   d. not determinable.

   ANS: C   DIF: III   OBJ: 3-3.5

129. The mass of two moles of oxygen atoms (atomic mass 16 amu) is
   a. 16 g.
   c. 48 g.
   b. 32 g.
   d. 64 g.

   ANS: B   DIF: III   OBJ: 3-3.5

130. The mass of a sample containing 3.5 mol of silicon atoms (atomic mass 28.0855 amu) is
    a. 28 g.   c. 72 g.
    b. 35 g.   d. 98 g.

    ANS: D   DIF: III   OBJ: 3-3.5

131. What is the number of moles of chemical units represented by 9.03×10^{24} units?
    a. 1.50 mol   c. 10.0 mol
    b. 9.03 mol   d. 15.0 mol

    ANS: D   DIF: III   OBJ: 3-3.5

132. The mass of 2.50 mol of calcium atoms (atomic mass 40.08 amu) is approximately
    a. 10.0 g.   c. 100 g.
    b. 42.5 g.   d. 250 g.

    ANS: C   DIF: III   OBJ: 3-3.5

133. How many moles of atoms are in 50.15 g of mercury (atomic mass 200.59 amu)?
    a. 0.1001 mol   c. 0.2500 mol
    b. 0.1504 mol   d. 0.4000 mol

    ANS: C   DIF: III   OBJ: 3-3.5

134. A prospector finds 39.39 g of gold (atomic mass 196.9665 amu). She has
    a. 1.20×10^{23} atoms.   c. 4.30×10^{23} atoms.
    b. 2.30×10^{23} atoms.   d. 6.02×10^{23} atoms.

    ANS: A   DIF: III   OBJ: 3-3.5

135. A sample of tin (atomic mass 118.69 amu) contains 3.01×10^{23} atoms. The mass of the sample is
    a. 3.01 g.   c. 72.6 g.
    b. 59.3 g.   d. 11 g.

    ANS: B   DIF: III   OBJ: 3-3.5

136. The mass of a sample of nickel (atomic mass 58.69 amu) is 176.07 g. It contains
    a. 1.7607×10^{24} atoms.   c. 5.869×10^{23} atoms.
    b. 1.806×10^{24} atoms.   d. 5.869×10^{24} atoms.

    ANS: B   DIF: III   OBJ: 3-3.5

137. The mass of a sample of nickel (atomic mass 58.69 amu) is 11.74 g. It contains
    a. 1.174×10^{23} atoms.   c. 1.869×10^{23} atoms.
    b. 1.205×10^{23} atoms.   d. 3.256×10^{23} atoms.

    ANS: B   DIF: III   OBJ: 3-3.5

138. The mass of exactly 5 mol of cesium (atomic mass 132.9 amu) is
    a. 664.5 g.   c. 6.02×10^{23} g.
    b. 132.9 g.   d. 5 g.

    ANS: A   DIF: III   OBJ: 3-3.5

SHORT ANSWER

1. How has Dalton's theory about atomic particles been modified by modern science?

   ANS:
   The discovery of subatomic particles proved that atoms are divisible. Dalton believed that atoms were indivisible.

   DIF: II   OBJ: 3-1.3

2. What is the relationship between isotopes, mass number, and neutrons?

   ANS:
   Isotopes are atoms of the same element with different numbers of neutrons and therefore different mass numbers.

   DIF: II   OBJ: 3-3.2

3. Why do chemists work with moles instead of individual atoms?

   ANS:
   A mole is a collection of atoms that is large enough to measure in the laboratory. A single atom is too small.

   DIF: II   OBJ: 3-3.4

PROBLEM

1. How many atoms are present in 8.00 mol of chlorine atoms?

   ANS:
   $$8.00 \text{ mol Cl} \times \frac{6.022 \times 10^{23} \text{ atoms Cl}}{1 \text{ mol Cl}} = 4.82 \times 10^{24} \text{ atoms Cl}$$

   4.82×10^{24} atoms Cl

   DIF: III   OBJ: 3-3.5

2. How many atoms are present in 80.0 mol of zirconium?

   ANS:
   $$80.0 \text{ mol Zr} \times \frac{6.022 \times 10^{23} \text{ atoms Zr}}{1 \text{ mol Zr}} = 4.82 \times 10^{25} \text{ atoms Zr}$$

   4.82×10^{25} atoms Zr

   DIF: III   OBJ: 3-3.5

3. How many moles of platinum are equivalent to 1.20×10^{24} atoms?

ANS:

$$1.20 \times 10^{24} \text{ atoms Pt} \times \frac{1 \text{ mol Pt}}{6.022 \times 10^{23} \text{ atoms Pt}} = 1.99 \text{ mol Pt}$$

1.99 mol Pt

DIF: III   OBJ: 3-3.5

4. How many moles of iron are equivalent to 1.11×10^{25} atoms?

ANS:

$$1.11 \times 10^{25} \text{ atoms Fe} \times \frac{1 \text{ mol Fe}}{6.022 \times 10^{23} \text{ atoms Fe}} = 18.4 \text{ mol Fe}$$

18.4 mol Fe

DIF: III   OBJ: 3-3.5

5. Determine the mass in grams of 5.00 mol of oxygen. The molar mass of oxygen is 16.00 g/mol.

ANS:

$$5.00 \text{ mol O} \times \frac{16.00 \text{ g O}}{1 \text{ mol O}} = 80.0 \text{ g O}$$

80.0 g O

DIF: III   OBJ: 3-3.5

6. Determine the mass in grams of 10.0 mol of bromine. The molar mass of bromine is 79.90 g/mol.

ANS:

$$10.0 \text{ mol Br} \times \frac{79.90 \text{ g Br}}{1 \text{ mol Br}} = 799 \text{ g Br}$$

799 g Br

DIF: III   OBJ: 3-3.5

7. Determine the number of moles of helium in 10.0 g of helium. The molar mass of helium is 4.00 g/mol.

ANS:

$$10.0 \text{ g He} \times \frac{1 \text{ mol He}}{4.00 \text{ g He}} = 2.50 \text{ mol He}$$

2.50 mol He

DIF: III   OBJ: 3-3.5

8. Determine the number of moles in 100. g of potassium. The molar mass of potassium is 39.10 g/mol.

ANS:

$$100. \text{ g K} \times \frac{1 \text{ mol K}}{39.10 \text{ g K}} = 2.56 \text{ mol K}$$

2.56 mol K

DIF: III   OBJ: 3-3.5

9. The mass of 1 mol of gold atoms is 196.97 g. Find the mass of 1 atom of gold.

ANS:

$$\frac{196.97 \text{ g Au}}{1 \text{ mol Au}} \times \frac{1 \text{ mol Au}}{6.022 \times 10^{23}} = 3.27 \times 10^{-22} \text{ g/atom Au}$$

3.27×10^{-22} g/atom Au

DIF: III   OBJ: 3-3.5

10. Calculate the mass in grams of 9.00 mol of potassium (molar mass 39.10 g/mol).

ANS:

$$9.00 \text{ mol K} \times \frac{39.10 \text{ g K}}{1 \text{ mol K}} = 352 \text{ g K}$$

352 g K

DIF: III   OBJ: 3-3.5

11. Calculate the number of atoms in 10.0 g of sulfur (molar mass 32.07 g/mol).

ANS:

$$10.0 \text{ g S} \times \frac{1 \text{ mol S}}{32.07 \text{ g S}} \times \frac{6.022 \times 10^{23} \text{ atoms S}}{1 \text{ mol S}} = 1.88 \times 10^{23} \text{ atoms S}$$

1.88×10^{23} atoms S

DIF: III   OBJ: 3-3.5

ESSAY

1. Explain what is meant by the law of definite proportions, the law of conservation of mass, and the law of multiple proportions.

   ANS:
   Definite proportions: regardless of the origin or size of samples of a particular compound, their elements are always in the same proportion. Conservation of mass: the mass of the elements combined in a compound is the same as the sum of the masses of the individual elements. Multiple proportions: when two elements combine to form two different compounds, the ratio of the masses of one element that combine with a fixed mass of the other element is a small whole number.

   DIF: III   OBJ: 3-1.1

2. Why do scientists use Dalton's theory, even though parts of it have been proven wrong? How much of Dalton's theory do scientists still accept?

   ANS:
   Dalton's theory led to the modern theory of the atom. Although scientists now know that atoms can be divided, most of his early theory is still accepted.

   DIF: II   OBJ: 3-1.3

3. What can you determine about the atomic structure of an element and one of its isotopes if you know the atomic number and mass numbers?

   ANS:
   The atomic number equals the number of protons in the nucleus of an atom and also equals the number of electrons in the neutral atom. The mass number is the sum of the number of protons and neutrons and can be used, with the atomic number, to find the number of neutrons. An isotope of an element has a different mass number but the same atomic number as the element.

   DIF: II   OBJ: 3-3.2

4. Explain the significance of Avogadro's constant, $6.022 \times 10_{23}$. What is the relationship between it and the molar mass of oxygen, 16.00 g/mol?

   ANS:
   Avogadro's constant is the number of particles in 1 mol of a substance. Because the molar mass of oxygen atoms is 16.00 g/mol, the mass of 1 mol, or $6.022 \times 10_{23}$, oxygen atoms is 16.00 g.

   DIF: II   OBJ: 3-3.4

MULTIPLE CHOICE

1. The product of the frequency and the wavelength of a wave equals the
   a. number of waves passing a point in a second.
   b. speed of the wave.
   c. distance between wave crests.
   d. time for one full wave to pass.

   ANS: B   DIF: II   OBJ: 4-1.1

2. Visible light, X rays, infrared radiation, and radio waves all have the same
   a. energy.   c. speed.
   b. wavelength.   d. frequency.

   ANS: C   DIF: I   OBJ: 4-1.1

3. For electromagnetic radiation, c (the speed of light) equals
   a. frequency minus wavelength.   c. frequency divided by wavelength.
   b. frequency plus wavelength.   d. frequency times wavelength.

   ANS: D   DIF: I   OBJ: 4-1.1

4. The speed of an electromagnetic wave is equal to the product of its wavelength and its
   a. mass.   c. velocity.
   b. color.   d. frequency.

   ANS: D   DIF: I   OBJ: 4-1.1

5. Because c, the speed of electromagnetic radiation, is a constant, the wavelength of the radiation is
   a. proportional to its frequency.   c. inversely proportional to its frequency.
   b. equal to its frequency.   d. double its frequency.

   ANS: C   DIF: I   OBJ: 4-1.1

6. In SI, the frequency of electromagnetic radiation is measured in
   a. nanometers.   c. hertz.
   b. quanta.   d. joules.

   ANS: C   DIF: I   OBJ: 4-1.1

7. Electromagnetic radiation behaves like a particle when it
   a. travels through space.   c. interacts with photons.
   b. is absorbed by matter.   d. interacts with other radiation.

   ANS: B   DIF: II   OBJ: 4-1.2

8. One of the wave properties of electromagnetic radiation, such as light, is
   a. volume.
   b. frequency.
   c. mass.
   d. weight.

   ANS: B   DIF: I   OBJ: 4-1.2

9. According to the particle model of light, certain kinds of light cannot eject electrons from metals because
   a. the mass of the light is too low.
   b. the frequency of the light is too high.
   c. the energy of the light is too low.
   d. the wavelength of the light is too short.

   ANS: C   DIF: I   OBJ: 4-1.2

10. As it travels through space, electromagnetic radiation
    a. exhibits wavelike behavior.
    b. loses energy.
    c. varies in speed.
    d. releases photons.

    ANS: A   DIF: I   OBJ: 4-1.2

11. If electromagnetic radiation A has a lower frequency than electromagnetic radiation B, then compared to B the wavelength of A is
    a. longer.
    b. shorter.
    c. equal.
    d. exactly half the length of B's wavelength.

    ANS: A   DIF: II   OBJ: 4-1.2

12. The distance between two successive peaks on a wave is its
    a. frequency.
    b. wavelength.
    c. quantum number.
    d. velocity.

    ANS: B   DIF: I   OBJ: 4-1.2

13. A quantum of electromagnetic energy is called a(n)
    a. photon.
    b. electron.
    c. excited atom.
    d. orbital.

    ANS: A   DIF: I   OBJ: 4-1.3

14. The wave model of light did not explain
    a. the frequency of light.
    b. the continuous spectrum.
    c. interference.
    d. the photoelectric effect.

    ANS: D   DIF: II   OBJ: 4-1.3

15. Max Planck proposed that a hot object radiated energy in small, specific amounts called
    a. quanta.
    b. waves.
    c. hertz.
    d. electrons.

    ANS: A   DIF: I   OBJ: 4-1.3

16. The energy of a photon, or quantum, is related to its
    a. mass.
    b. speed.
    c. frequency.
    d. size.

    ANS: C   DIF: I   OBJ: 4-1.3

17. Planck's constant
    a. depends on the frequency of the radiation.
    b. depends on the mass of the radiation.
    c. depends on the wavelength of the radiation.
    d. is the same for all forms of radiation.

    ANS: D   DIF: II   OBJ: 4-1.3

18. The emission of electrons from metals that have absorbed photons is called the
    a. interference effect.
    b. photoelectric effect.
    c. quantum effect.
    d. dual effect.

    ANS: B   DIF: I   OBJ: 4-1.3

19. In the hydrogen spectrum, the infrared spectral lines of the Paschen series, compared with the visible lines of the Balmer series, have
    a. longer wavelengths and less energy.
    b. longer wavelengths and more energy.
    c. shorter wavelengths and more energy.
    d. shorter wavelengths and less energy.

    ANS: A   DIF: II   OBJ: 4-1.3

20. A line spectrum is produced when an electron moves from one energy level
    a. to a higher energy level.
    b. to a lower energy level.
    c. into the nucleus.
    d. to another position in the same sublevel.

    ANS: B   DIF: II   OBJ: 4-1.3

21. The spectral lines of hydrogen in the ultraviolet region of the electromagnetic spectrum are called
    a. principal series.
    b. Balmer series.
    c. Lyman series.
    d. Paschen series.

    ANS: C   DIF: I   OBJ: 4-1.3

22. When the pink-colored light of glowing hydrogen gas passes through a prism, it is possible to see
    a. all the colors of the rainbow.
    b. only lavender-colored lines.
    c. four lines of different colors.
    d. black light.

    ANS: C   DIF: II   OBJ: 4-1.3

23. A bright-line spectrum of an atom is caused by the energy released when electrons
    a. jump to a higher energy level.
    b. fall to a lower energy level.
    c. absorb energy and jump to a higher energy level.
    d. absorb energy and fall to a lower energy level.

    ANS: B   DIF: II   OBJ: 4-1.3

24. Because excited hydrogen atoms always produce the same line-emission spectrum, scientists concluded that hydrogen
    a. had no electrons.
    b. did not release photons.
    c. released photons of only certain energies.
    d. could only exist in the ground state.

    ANS: C   DIF: II   OBJ: 4-1.3

25. The Bohr model of the atom was an attempt to explain hydrogen's
    a. density.
    b. flammability.
    c. mass.
    d. line-emission spectrum.

    ANS: D   DIF: I   OBJ: 4-1.4

26. For an electron in an atom to change from the ground state to an excited state,
    a. energy must be released.
    b. energy must be absorbed.
    c. radiation must be emitted.
    d. the electron must make a transition from a higher to a lower energy level.

    ANS: B   DIF: II   OBJ: 4-1.4

27. If electrons in an atom have the lowest possible energies, the atom is in the
    a. ground state.
    b. inert state.
    c. excited state.
    d. radiation-emitting state.

    ANS: A   DIF: I   OBJ: 4-1.4

28. Bohr's theory helped explain why
    a. electrons have negative charge.
    b. most of the mass of the atom is in the nucleus.
    c. excited hydrogen gas gives off certain colors of light.
    d. atoms combine to form molecules.

    ANS: C   DIF: II   OBJ: 4-1.4

29. Bohr's model of the atom works best in explaining
    a. the spectra of the first ten elements.
    b. only the spectrum of hydrogen.
    c. only the spectra of atoms with electrons in an s orbital.
    d. the entire visible spectra of atoms.

    ANS: B   DIF: II   OBJ: 4-1.4

30. According to Bohr's theory, an excited atom would
    a. collapse.
    b. absorb photons.
    c. remain stable.
    d. radiate energy.

    ANS: D   DIF: II   OBJ: 4-1.4

31. According to the Bohr model of the atom, the single electron of a hydrogen atom circles the nucleus
    a. in specific, allowed orbits.
    b. in one fixed orbit at all times.
    c. at any of an infinite number of distances, depending on its energy.
    d. counterclockwise.

    ANS: A   DIF: II   OBJ: 4-1.4

32. The drop of an electron from a high energy level to the ground state in a hydrogen atom would be most closely associated with
    a. long-wavelength radiation.   c. infrared radiation.
    b. low-frequency radiation.   d. high-frequency radiation.

    ANS: D   DIF: II   OBJ: 4-1.4

33. The electron in a hydrogen atom has its lowest total energy when the electron is in its
    a. neutral state.   c. ground state.
    b. excited state.   d. quantum state.

    ANS: C   DIF: I   OBJ: 4-1.4

34. The change of an atom from an excited state to the ground state always requires
    a. absorption of energy.   c. release of visible light.
    b. emission of electromagnetic radiation.   d. an increase in electron energy.

    ANS: B   DIF: II   OBJ: 4-1.4

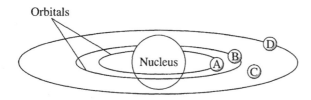

35. According to Bohr, electrons cannot reside at _____ in the figure above.
    a. point A   c. point C
    b. point B   d. point D

    ANS: C   DIF: III   OBJ: 4-1.4

36. According to the quantum theory, point D in the figure above represents
    a. the fixed position of an electron.
    b. the farthest point from the nucleus where an electron can be found.
    c. a position where an electron probably exists.
    d. a position where an electron cannot exist.

    ANS: C   DIF: II   OBJ: 4-2.2

37. The French scientist Louis de Broglie believed
    a. electrons could have a dual wave-particle nature.
    b. light waves did not have a dual wave-particle nature.
    c. the natures of light and quantized electron orbits were not similar.
    d. Bohr's model of the hydrogen atom was completely correct.

    ANS: A   DIF: I   OBJ: 4-2.1

38. All of the following are true about Louis de Broglie's research EXCEPT
    a. it led to a revolution in our basic understanding of matter.
    b. it pointed out that the behavior of quantized electron orbits was similar to the behavior of waves.
    c. it combined two well-known equations.
    d. it explained the photoelectric effect.

    ANS: D   DIF: I   OBJ: 4-2.1

39. Louis de Broglie's research suggested that
    a. frequencies of electron waves do not correspond to specific energies.
    b. electrons usually behave like particles and rarely like waves.
    c. electrons should be considered as waves confined to the space around an atomic nucleus.
    d. electron waves exist at random frequencies.

    ANS: C   DIF: I   OBJ: 4-2.1

40. The equation $E = h\nu$ helped Louis de Broglie determine
    a. how protons and neutrons behave in the nucleus.
    b. how electron wave frequencies correspond to specific energies.
    c. whether electrons behave as particles.
    d. whether electrons exist in a limited number of orbits with different energies.

    ANS: B   DIF: I   OBJ: 4-2.1

41. All of the following statements are correct EXCEPT
    a. according to the Bohr model of the atom, the single electron of a hydrogen atom circles the nucleus in specific, allowed orbits.
    b. the quantum model uses probability to locate the electron within the atom.
    c. the orbitals of the quantum model of the atom were suggested by descriptions of electrons as waves.
    d. the Bohr model of the atom explains the reactivity of hydrogen.

    ANS: D   DIF: I   OBJ: 4-2.2

42. Which model of the atom explains why excited hydrogen gas gives off certain colors of light?
    a. the Bohr model
    b. the quantum model
    c. Rutherford's model
    d. Planck's theory

    ANS: A   DIF: I   OBJ: 4-2.2

43. Which model of the atom explains the orbitals of electrons as waves?
    a. the Bohr model
    b. the quantum model
    c. Rutherford's model
    d. Planck's theory

    ANS: B   DIF: I   OBJ: 4-2.2

44. The part of the atom where the electrons CANNOT be found is the
    a. area surrounding the nucleus.
    b. nucleus.
    c. electron cloud.
    d. orbitals.

    ANS: B   DIF: I   OBJ: 4-2.2

45. The region outside the nucleus where an electron can most probably be found is the
    a. electron configuration.
    b. quantum.
    c. *s* sublevel.
    d. electron cloud.

    ANS: D   DIF: I   OBJ: 4-2.2

46. The size and shape of an electron cloud are most closely related to the electron's
    a. charge.
    b. mass.
    c. spin.
    d. energy.

    ANS: D   DIF: I   OBJ: 4-2.2

47. With the quantum model of the atom, scientists have come to believe that determining an electron's exact location around the nucleus
    a. is impossible.
    b. can be done before 2005.
    c. can be done easily.
    d. can be done only with specialized equipment.

    ANS: A   DIF: I   OBJ: 4-2.2

48. All of the following describe the Heisenberg uncertainly principle EXCEPT
    a. it states that it is impossible to determine simultaneously both the position and velocity of an electron or any other particle.
    b. it is one of the fundamental principles of our present understanding of light and matter.
    c. it helped lay the foundation for the modern quantum theory.
    d. it helps to locate an electron in an atom.

    ANS: D   DIF: I   OBJ: 4-2.3

49. All of the following describe the Schrödinger wave equation EXCEPT
   a. it is an equation that treats electrons in atoms as waves.
   b. only waves of specific energies and frequencies provide solutions to the equation.
   c. it helped lay the foundation for the modern quantum theory.
   d. it is similar to Bohr's theory.

   ANS: D   DIF: I   OBJ: 4-2.3

50. The solutions to the Schrödinger wave equation are
   a. quantum numbers.   c. energy levels.
   b. wave functions.   d. orbitals.

   ANS: B   DIF: I   OBJ: 4-2.3

51. Both the Heisenberg uncertainty principle and the Schrödinger wave equation
   a. are based on Bohr's theory.   c. led to locating an electron in an atom.
   b. treat electrons as particles.   d. led to the concept of atomic orbitals.

   ANS: D   DIF: I   OBJ: 4-2.3

52. A three-dimensional region around a nucleus where an electron may be found is called a(n)
   a. spectral line.   c. orbital.
   b. electron path.   d. orbit.

   ANS: C   DIF: I   OBJ: 4-2.3

53. Unlike in an orbit, in an orbital
   a. an electron's position cannot be known precisely.
   b. an electron has no energy.
   c. electrons cannot be found.
   d. protons cannot be found.

   ANS: A   DIF: I   OBJ: 4-2.3

54. The quantum number that indicates the position of an orbital about the three axes in space is the
   a. principal quantum number.   c. magnetic quantum number.
   b. angular momentum quantum number.   d. spin quantum number.

   ANS: C   DIF: I   OBJ: 4-2.4

55. How many quantum numbers are needed to describe the energy state of an electron in an atom?
   a. 1   c. 3
   b. 2   d. 4

   ANS: D   DIF: I   OBJ: 4-2.4

56. Quantum numbers are sets of numbers that describe the properties of
   a. the atomic nucleus.   c. atoms.
   b. atomic orbitals.   d. molecules.

   ANS: B   DIF: I   OBJ: 4-2.4

57. The main energy levels of an atom are indicated by the
   a. orbital quantum numbers.
   b. magnetic quantum numbers.
   c. spin quantum numbers.
   d. principal quantum numbers.

   ANS: D   DIF: I   OBJ: 4-2.4

58. The possible values of an electron's spin quantum number are
   a. −1, 0, or 1.
   b. $+\frac{1}{2}$ or $-\frac{1}{2}$
   c. +1 or −1.
   d. 0 or 1.

   ANS: B   DIF: I   OBJ: 4-2.4

59. The number of sublevels within each energy level of an atom is equal to the value of the
   a. principal quantum number.
   b. angular momentum quantum number.
   c. magnetic quantum number.
   d. spin quantum number.

   ANS: A   DIF: I   OBJ: 4-2.4

60. What values can the angular momentum quantum number have when $n = 2$?
   a. $+\frac{1}{2}, -\frac{1}{2}$
   b. $-\frac{1}{2}, -1, -2$
   c. 0, 1, 2
   d. 0, 1

   ANS: D   DIF: II   OBJ: 4-2.4

61. The spin quantum number indicates that the number of possible states for an electron in an orbital is
   a. 1.
   b. 2.
   c. 3.
   d. 5.

   ANS: B   DIF: II   OBJ: 4-2.4

62. The values $+\frac{1}{2}$ and $-\frac{1}{2}$ specify an electron's

   a. charge.
   b. main energy level.
   c. speed.
   d. possible state in an orbital.

   ANS: D   DIF: I   OBJ: 4-2.4

63. The spin quantum number of an electron can be thought of as describing
   a. the direction of electron spin.
   b. whether the electron's charge is positive or negative.
   c. the electron's exact location in orbit.
   d. the number of revolutions the electron makes about the nucleus per second.

   ANS: A   DIF: II   OBJ: 4-2.4

64. Because of the property described by the spin quantum number, an electron behaves as though it
   a. were positively charged.
   b. were a magnet.
   c. oscillated in one position.
   d. were spiraling toward the nucleus.

   ANS: B   DIF: II   OBJ: 4-2.4

65. An electron for which $n = 4$ has more _____ than an electron for which $n = 2$.
   a. spin
   b. stability
   c. energy
   d. wave nature

   ANS: C   DIF: II   OBJ: 4-2.4

66. The set of orbitals that are dumbbell-shaped and directed along the x, y, and z axes are called
   a. d orbitals.
   b. p orbitals.
   c. f orbitals.
   d. s orbitals.

   ANS: B   DIF: I   OBJ: 4-2.5

67. A spherical electron cloud surrounding an atomic nucleus would best represent
   a. an s orbital.
   b. a p_x orbital.
   c. a combination of p_x and p_y orbitals.
   d. a combination of an s and a p_x orbital.

   ANS: A   DIF: II   OBJ: 4-2.5

68. The major difference between a $1s$ orbital and a $2s$ orbital is that
   a. the $2s$ orbital can hold more electrons.
   b. the $2s$ orbital has a slightly different shape.
   c. the $2s$ orbital is at a higher energy level.
   d. the $1s$ orbital can have only one electron.

   ANS: C   DIF: II   OBJ: 4-2.5

69. The p orbitals are shaped like
   a. electrons.
   b. circles.
   c. dumbbells.
   d. spheres.

   ANS: C   DIF: I   OBJ: 4-2.5

70. An orbital that could never exist according to the quantum description of the atom is
   a. $3d$.
   b. $8s$.
   c. $6d$.
   d. $3f$.

   ANS: D   DIF: II   OBJ: 4-2.5

71. The letter designations for the first four sublevels with the number of electrons that can be accommodated in each sublevel are
   a. s:1, p:3, d:10, and f:14.
   b. s:1, p:3, d:5, and f:7.
   c. s:2, p:6, d:10, and f:14.
   d. s:1, p:2, d:3, and f:4.

   ANS: C   DIF: II   OBJ: 4-2.5

72. The number of possible orbital shapes for the third energy level is
   a. 1.
   b. 2.
   c. 3.
   d. 4.

   ANS: C   DIF: II   OBJ: 4-2.5

73. The number of orbitals for the *d* sublevel is
    a. 1.   c. 5.
    b. 3.   d. 7.

    ANS: C   DIF: II   OBJ: 4-2.5

74. For $n = 4$, the number of possible orbital shapes is
    a. 1.   c. 16.
    b. 4.   d. 32.

    ANS: B   DIF: II   OBJ: 4-2.5

75. For the *f* sublevel, the number of orbitals is
    a. 5.   c. 9.
    b. 7.   d. 18.

    ANS: B   DIF: II   OBJ: 4-2.5

76. The total number of orbitals that can exist at the second main energy level is
    a. 2.   c. 4.
    b. 3.   d. 8.

    ANS: C   DIF: II   OBJ: 4-2.5

77. How many orientations can an *s* orbital have about the nucleus?
    a. 1   c. 3
    b. 2   d. 5

    ANS: A   DIF: II   OBJ: 4-2.5

78. How many orbitals can exist at the third main energy level?
    a. 3   c. 9
    b. 6   d. 18

    ANS: C   DIF: II   OBJ: 4-2.5

79. How many electrons can occupy the *s* orbitals at each energy level?
    a. two, if they have opposite spins   c. one
    b. two, if they have the same spin   d. no more than eight

    ANS: A   DIF: II   OBJ: 4-2.5

80. If *n* is the principal quantum number of a main energy level, the number of electrons in that energy level is
    a. n.   c. n^2.
    b. $2n$.   d. $2n^2$.

    ANS: D   DIF: II   OBJ: 4-2.5

81. How many electrons are needed to completely fill the fourth energy level?
    a. 8
    b. 18

    c. 32
    d. 40

    ANS: C   DIF: II   OBJ: 4-3.1

82. How many electrons are needed to fill the third main energy level if it already contains 8 electrons?
    a. 0
    b. 8

    c. 10
    d. 22

    ANS: C   DIF: II   OBJ: 4-3.1

83. One main energy level can hold 18 electrons. What is n?
    a. $+\frac{1}{2}$
    b. 3

    c. 6
    d. 18

    ANS: B   DIF: III   OBJ: 4-3.1

84. At $n = 1$, the total number of electrons that could be found is
    a. 1.
    b. 2.

    c. 6.
    d. 18.

    ANS: B   DIF: II   OBJ: 4-3.1

85. If 8 electrons completely fill a main energy level, what is n?
    a. 2
    b. 4

    c. 8
    d. 32

    ANS: A   DIF: III   OBJ: 4-3.1

86. If the third main energy level contains 15 electrons, how many more could it possibly hold?
    a. 0
    b. 1

    c. 3
    d. 17

    ANS: C   DIF: III   OBJ: 4-3.1

87. What is the total number of electrons needed to fill the first two main energy levels?
    a. 2
    b. 6

    c. 10
    d. 18

    ANS: C   DIF: II   OBJ: 4-3.1

88. The main energy level that can hold only two electrons is the
    a. first.
    b. second.

    c. third.
    d. fourth.

    ANS: A   DIF: I   OBJ: 4-3.1

89. A single orbital in the 3*d* level can hold _____ electrons.
   a. 10
   b. 2
   c. 3
   d. 6

   ANS: B   DIF: II   OBJ: 4-3.1

90. The statement that an electron occupies the lowest available energy orbital is
   a. Hund's rule.
   b. the Aufbau principle.
   c. Bohr's law.
   d. the Pauli exclusion principle.

   ANS: B   DIF: I   OBJ: 4-3.2

91. "Orbitals of equal energy are each occupied by one electron before any is occupied by a second electron, and all electrons in singly occupied orbitals must have the same spin" is a statement of
   a. the Pauli exclusion principle.
   b. the Aufbau principle.
   c. the quantum effect.
   d. Hund's rule.

   ANS: D   DIF: I   OBJ: 4-3.2

92. The statement that no two electrons in the same atom can have the same four quantum numbers is
   a. the Pauli exclusion principle.
   b. Hund's rule.
   c. Bohr's law.
   d. the Aufbau principle.

   ANS: A   DIF: I   OBJ: 4-3.2

93. Which of the following rules requires that each of the *p* orbitals at a particular energy level receive one electron before any of them can have two electrons?
   a. Hund's rule
   b. the Pauli exclusion principle
   c. the Aufbau principle
   d. the quantum rule

   ANS: A   DIF: I   OBJ: 4-3.2

94. Two electrons in the 1*s* orbital must have different spin quantum numbers to satisfy
   a. Hund's rule.
   b. the magnetic rule.
   c. the Pauli exclusion principle.
   d. the Aufbau principle.

   ANS: C   DIF: II   OBJ: 4-3.2

95. The sequence in which energy sublevels are filled is specified by
   a. the Pauli exclusion principle.
   b. the orbital rule.
   c. Lyman's series.
   d. the Aufbau principle.

   ANS: D   DIF: I   OBJ: 4-3.2

96. The Aufbau principle states that an electron
   a. can have only one spin number.
   b. occupies the lowest available energy level.
   c. must be paired with another electron.
   d. must enter an s orbital.

   ANS: B   DIF: I   OBJ: 4-3.2

97. The Pauli exclusion principle states that no two electrons in the same atom can
    a. occupy the same orbital.
    b. have the same spin quantum numbers.
    c. have the same set of quantum numbers.
    d. be at the same main energy level.

    ANS: C   DIF: I   OBJ: 4-3.2

98. The atomic sublevel with the next highest energy after $4p$ is
    a. $4d$.
    b. $4f$.
    c. $5p$.
    d. $5s$.

    ANS: D   DIF: II   OBJ: 4-3.2

99. In the electron configuration for scandium (atomic number 21), what is the notation for the three highest-energy electrons?
    a. $3d^1\,4s^2$
    b. $4s^3$
    c. $3d^3$
    d. $4s^2\,4p^1$

    ANS: A   DIF: III   OBJ: 4-3.3

100. Which electron configuration is most stable?
    a. $3d^4\,4s^2$
    b. $3d^5\,4s^1$
    c. $3d^3\,4s^3$
    d. $3d^2\,4s^4$

    ANS: B   DIF: II   OBJ: 4-3.3

101. Both copper (atomic number 29) and chromium (atomic number 24) appear to break the pattern in the order of filling the $3d$ and $4s$ orbitals. This change in pattern is expressed by
    a. an increase in the number of electrons in both the $3d$ and $4s$ orbitals.
    b. a reduction in the number of electrons in both the $3d$ and $4s$ orbitals.
    c. a reduction in the number of electrons in the $3d$ orbital and an increase in the $4s$ orbital.
    d. a reduction in the number of electrons in the $4s$ orbital and an increase in the $3d$ orbital.

    ANS: D   DIF: II   OBJ: 4-3.3

102. In the ground state, the $3d$ and $4s$ sublevels of the chromium atom (atomic number 24) may be represented as
    a. $3d^6\,4s^1$.
    b. $3d^4\,4s^2$.
    c. $3d^5\,4s^1$.
    d. $4s^2\,3d^4$.

    ANS: C   DIF: II   OBJ: 4-3.3

103. The element with electron configuration $1s^2\,2s^2\,2p^6\,3s^2\,3p^2$ is
    a. Mg ($Z = 12$).
    b. C ($Z = 6$).
    c. S ($Z = 16$).
    d. Si ($Z = 14$).

    ANS: D   DIF: II   OBJ: 4-3.3

104. The electron configuration for the carbon atom (C) is $1s^2\,2s^2\,2p^2$. The atomic number of carbon is
    a. 3.
    b. 6.
    c. 11.
    d. 12.

    ANS: B   DIF: II   OBJ: 4-3.3

105. What is the electron configuration for nitrogen, atomic number 7?
    a. $1s^2 2s^2 2p^3$
    b. $1s^2 2s^3 2p^2$
    c. $1s^2 2s^3 2p^1$
    d. $1s^2 2s^2 2p^2 3s^1$

    ANS: A   DIF: II   OBJ: 4-3.3

106. The electron notation for aluminum (atomic number 13) is
    a. $1s^2 2s^2 2p^3 3s^2 3p^3 3d^1$.
    b. $1s^2 2s^2 2p^6 3s^2 2d^1$.
    c. $1s^2 2s^2 2p^6 3s^2 3p^1$.
    d. $1s^2 2s^2 2p^9$.

    ANS: C   DIF: II   OBJ: 4-3.3

107. If the s and p orbitals of the highest main energy level of an atom are filled with electrons, the atom has a(n)
    a. electron pair.
    b. octet.
    c. ellipsoid.
    d. circle.

    ANS: B   DIF: II   OBJ: 4-3.3

108. The number of electrons in the highest energy level of the argon atom (atomic number 18) is
    a. 10.
    b. 2.
    c. 6.
    d. 8.

    ANS: D   DIF: II   OBJ: 4-3.3

109. If the s and p sublevels of the highest main energy level of an atom are filled, how many electrons are in the main energy level?
    a. 2
    b. 8
    c. 16
    d. 32

    ANS: B   DIF: I   OBJ: 4-3.3

110. If an element has an octet of electrons in its highest main energy level, there are _____ electrons in this level.
    a. 2
    b. 8
    c. 10
    d. 32

    ANS: B   DIF: I   OBJ: 4-3.3

111. An element with 8 electrons in its highest main energy level is a(n)
    a. octet element.
    b. third period element.
    c. Aufbau element.
    d. noble gas.

    ANS: D   DIF: I   OBJ: 4-3.3

SHORT ANSWER

1. Explain Louis de Broglie's contribution to the quantum model of the atom.

   ANS:
   If light has a particle nature, de Broglie reasoned, could particles have a wave nature? He compared the behavior of Bohr's quantized electron orbits to the known behavior of waves. Finally, he hypothesized that electrons are confined to the space around an atomic nucleus and that electron waves exist only at specific energies.

   DIF: II   OBJ: 4-2.1

2. What do quantum numbers describe?

   ANS:
   Three of the quantum numbers describe the location of an electron, and the fourth gives its spin.

   DIF: II   OBJ: 4-2.4

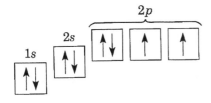

3. How does the figure above illustrate Hund's rule?

   ANS:
   According to Hund's rule, the arrangement of electrons with the maximum number of unpaired electrons is the most stable arrangement.

   DIF: III   OBJ: 4-3.2

4. How does the figure above illustrate the Pauli exclusion principle?

   ANS:
   According to the Pauli exclusion principle, no two electrons can have the same set of four quantum numbers. Therefore, no more than two electrons can occupy an orbital, and these two must have opposite spins.

   DIF: III   OBJ: 4-3.2

5. Use the electron configuration for nitrogen, $1s^2\, 2s^2\, 2p^3$, to state how many electrons are in each main energy level.

ANS:
The principal quantum number n describes the energy level that an electron occupies. In a nitrogen atom, $n = 1$ for two electrons and $n = 2$ for five electrons.

DIF: II   OBJ: 4-3.3

6. The electron configuration for nitrogen is $1s^2\, 2s^2\, 2p^3$. What does the 3 in $2p^3$ mean?

ANS:
The 3 in $2p^3$ indicates that three electrons are in the p orbitals of the second energy level.

DIF: II   OBJ: 4-3.3

PROBLEM

1. Write the electron configuration for beryllium, atomic number 4.

ANS:
$1s^2\, 2s^2$

DIF: III   OBJ: 4-3.3

2. Write the electron configuration for argon, atomic number 18.

ANS:
$1s^2\, 2s^2\, 2p^6\, 3s^2\, 3p^6$

DIF: III   OBJ: 4-3.3

3. Write the electron configuration for nitrogen, atomic number 7.

ANS:
$1s^2\, 2s^2\, 2p^3$

DIF: III   OBJ: 4-3.3

1 **H** Hydrogen 1.01																	Group 18 2 **He** Helium 4.00
Group 1	Group 2									Group 13	Group 14	Group 15	Group 16	Group 17		
3 **Li** Lithium 6.94	4 **Be** Beryllium 9.01									5 **B** Boron 10.81	6 **C** Carbon 12.01	7 **N** Nitrogen 14.01	8 **O** Oxygen 16.00	9 **F** Fluorine 19.00	10 **Ne** Neon 20.18
11 **Na** Sodium 22.99	12 **Mg** Magnesium 24.30									13 **Al** Aluminum 26.98	14 **Si** Silicon 28.08	15 **P** Phosphorus 30.97	16 **S** Sulfur 32.07	17 **Cl** Chlorine 35.45	18 **Ar** Argon 39.95
19 **K** Potassium 39.10	20 **Ca** Calcium 40.08									31 **Ga** Gallium 69.72	32 **Ge** Germanium 72.61	33 **As** Arsenic 74.92	34 **Se** Selenium 78.96	35 **Br** Bromine 79.90	36 **Kr** Krypton 83.80
37 **Rb** Rubidium 85.47	38 **Sr** Strontium 87.62									49 **In** Indium 114.82	50 **Sn** Tin 118.71	51 **Sb** Antimony 121.76	52 **Te** Tellurium 127.60	53 **I** Iodine 126.90	54 **Xe** Xenon 131.29
55 **Cs** Cesium 132.90	56 **Ba** Barium 137.33									81 **Tl** Thallium 204.38	82 **Pb** Lead 207.2	83 **Bi** Bismuth 208.98	84 **Po** Polonium (208.98)	85 **At** Astatine (209.99)	86 **Rn** Radon (222.02)
87 **Fr** Francium (223.02)	88 **Ra** Radium (226.02)														

4. Use the figure above. Which element has the following electron configuration: [Ar] $4s^2\,3d^{10}\,4p^5$?

ANS:
bromine

DIF: III   OBJ: 4-3.3

5. Use the figure above. Which element has the following electron configuration: [Kr] $5s^2\,4d^{10}\,5p^3$?

ANS:
antimony

DIF: III   OBJ: 4-3.3

6. Use the figure above and the symbols for the noble gases to write the electron configuration for silicon.

ANS:
[Ne] $3s^2\,3p^2$

DIF: III   OBJ: 4-3.3

7. Use the figure above and the symbols for the noble gases to write the electron configuration for strontium.

ANS:
[Kr] $5s^2$

DIF: III   OBJ: 4-3.3

8. Refer to the figure above and draw the orbital diagram for phosphorus.

ANS:

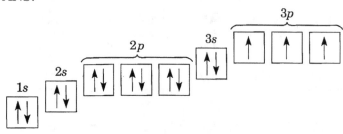

DIF: III   OBJ: 4-3.3

9. Refer to the figure above and draw the orbital diagram for calcium.

ANS:

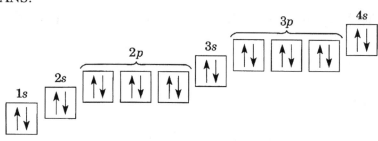

DIF: III   OBJ: 4-3.3

10. Refer to the figure above and draw the orbital diagram for carbon.

ANS:

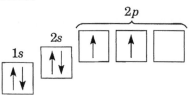

DIF: III   OBJ: 4-3.3

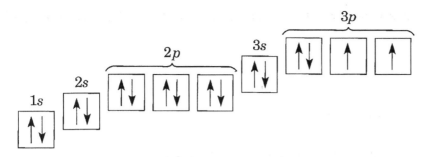

11. Use the symbols for the noble gases to write the electron configuration represented in the figure above.

ANS:
[Ne] $3s^2\,3p^4$

DIF: III   OBJ: 4-3.3

12. What element's orbital diagram is shown in the figure above?

ANS:
sulfur

DIF: III   OBJ: 4-3.3

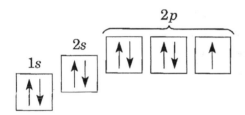

13. Write the electron configuration of the figure above.

ANS:
$1s^2\,2s^2\,2p^5$

DIF: III   OBJ: 4-3.3

14. What element's orbital diagram is shown in the figure above?

ANS:
fluorine

DIF: III   OBJ: 4-3.3

ESSAY

1. Why does the last electron of potassium enter the $4s$ level rather than the $3d$ level?

   ANS:
   The $4s$ level fills first because it is in a lower energy state than $3d$.

   DIF: III   OBJ: 4-3.2

MULTIPLE CHOICE

1. The idea of arranging the elements in the periodic table according to their chemical and physical properties is attributed to
   a. Mendeleev.   c. Bohr.
   b. Moseley.   d. Ramsay.

   ANS: A   DIF: I   OBJ: 5-1.1

2. Mendeleev left spaces in his periodic table and predicted several elements and their
   a. atomic numbers.   c. properties.
   b. colors.   d. radioactivity.

   ANS: C   DIF: I   OBJ: 5-1.1

3. Mendeleev attempted to organize the chemical elements based on their
   a. symbols.   c. atomic numbers.
   b. properties.   d. electron configurations.

   ANS: B   DIF: I   OBJ: 5-1.1

4. Mendeleev noticed that properties of elements usually repeated at regular intervals when the elements were arranged in order of increasing
   a. atomic number.   c. reactivity.
   b. density.   d. atomic mass.

   ANS: D   DIF: I   OBJ: 5-1.1

5. Mendeleev is credited with developing the first successful
   a. periodic table.   c. test for radioactivity.
   b. method for determining atomic number.   d. use of X rays.

   ANS: A   DIF: I   OBJ: 5-1.1

6. Mendeleev did not always list elements in his periodic table in order of increasing atomic mass because he grouped together elements with similar
   a. properties.   c. densities.
   b. atomic numbers.   d. colors.

   ANS: A   DIF: I   OBJ: 5-1.1

7. In developing his periodic table, Mendeleev listed on cards each element's name, atomic mass, and
   a. atomic number.   c. isotopes.
   b. electron configuration.   d. properties.

   ANS: D   DIF: I   OBJ: 5-1.1

8. Mendeleev predicted that the spaces in his periodic table represented
   a. isotopes.
   b. radioactive elements.
   c. permanent gaps.
   d. undiscovered elements.

   ANS: D   DIF: I   OBJ: 5-1.1

9. Mendeleev's table was called periodic because the properties of the elements
   a. showed no pattern.
   b. occurred at repeated intervals called periods.
   c. occurred at regular time intervals called periods.
   d. were identical.

   ANS: B   DIF: I   OBJ: 5-1.1

10. The person whose work led to a periodic table based on increasing atomic number was
    a. Moseley.
    b. Mendeleev.
    c. Rutherford.
    d. Cannizzaro.

    ANS: A   DIF: I   OBJ: 5-1.1

11. Moseley's work led to the realization that elements with similar properties occurred at regular intervals when the elements were arranged in order of increasing
    a. atomic mass.
    b. density.
    c. radioactivity.
    d. atomic number.

    ANS: D   DIF: I   OBJ: 5-1.1

12. Who used his experimental evidence to determine the order of the elements according to atomic number?
    a. Meyer
    b. Ramsay
    c. Stas
    d. Moseley

    ANS: D   DIF: I   OBJ: 5-1.1

13. The most useful source of general information about the elements for anyone associated with chemistry is a
    a. calculator.
    b. table of metric equivalents.
    c. periodic table.
    d. table of isotopes.

    ANS: C   DIF: I   OBJ: 5-1.2

14. The periodic table
    a. permits the properties of an element to be predicted before the element is discovered.
    b. will be completed with element 118.
    c. has been of little use to chemists since the early 1900s.
    d. was completed with the discovery of the noble gases.

    ANS: A   DIF: I   OBJ: 5-1.2

15. Evidence gathered since Mendeleev's time indicates that a better arrangement than atomic mass for elements in the periodic table is an arrangement by
    a. mass number.
    c. group number.
    b. atomic number.
    d. series number.

    ANS: B   DIF: I   OBJ: 5-1.2

16. What are the elements whose discovery added an entirely new row to Mendeleev's periodic table?
    a. noble gases
    c. transition elements
    b. radioactive elements
    d. metalloids

    ANS: A   DIF: I   OBJ: 5-1.2

17. What are the radioactive elements with atomic numbers from 90 to 103 in the periodic table called?
    a. the noble gases
    c. the actinides
    b. the lanthanides
    d. the rare-earth elements

    ANS: C   DIF: I   OBJ: 5-1.2

18. What are the elements with atomic numbers from 58 to 71 in the periodic table called?
    a. the lanthanide elements
    c. the actinide elements
    b. the noble gases
    d. the alkali metals

    ANS: A   DIF: I   OBJ: 5-1.2

19. Argon, krypton, and xenon are
    a. alkaline earth metals.
    c. actinides.
    b. noble gases.
    d. lanthanides.

    ANS: B   DIF: I   OBJ: 5-1.2

20. Which two periods have the same number of elements?
    a. 2 and 4
    c. 4 and 5
    b. 3 and 4
    d. 5 and 6

    ANS: C   DIF: I   OBJ: 5-1.2

21. The discovery of the noble gases changed Mendeleev's periodic table by adding a new
    a. period.
    c. group.
    b. series.
    d. sublevel block.

    ANS: C   DIF: I   OBJ: 5-1.2

22. In the modern periodic table, elements are ordered according to
    a. decreasing atomic mass.
    c. increasing atomic number.
    b. Mendeleev's original design.
    d. the date of their discovery.

    ANS: C   DIF: I   OBJ: 5-1.2

23. The periodic law states that the physical and chemical properties of elements are periodic functions of their atomic
    a. masses.
    b. numbers.
    c. radii.
    d. structures.

    ANS: B   DIF: I   OBJ: 5-1.3

24. The periodic law states that the properties of elements are periodic functions of their atomic numbers. This means that the _____ determines the position of each element in the periodic table.
    a. mass number
    b. number of neutrons
    c. number of protons
    d. number of nucleons

    ANS: C   DIF: I   OBJ: 5-1.3

25. The principle that states that the physical and chemical properties of the elements are periodic functions of their atomic numbers is
    a. the periodic table.
    b. the periodic law.
    c. the law of properties.
    d. Mendeleev's law.

    ANS: B   DIF: I   OBJ: 5-1.3

26. The periodic law allows some properties of an element to be predicted based on its
    a. position in the periodic table.
    b. number of isotopes.
    c. symbol.
    d. color.

    ANS: A   DIF: I   OBJ: 5-1.3

27. The periodic law states that
    a. no two electrons with the same spin can be found in the same place in an atom.
    b. the physical and chemical properties of the elements are functions of their atomic numbers.
    c. electrons exhibit properties of both particles and waves.
    d. the chemical properties of elements can be grouped according to periodicity but physical properties cannot.

    ANS: B   DIF: I   OBJ: 5-1.3

28. Elements in a group or column in the periodic table can be expected to have similar
    a. atomic masses.
    b. atomic numbers.
    c. numbers of neutrons.
    d. properties.

    ANS: D   DIF: I   OBJ: 5-1.4

29. A horizontal row of blocks in the periodic table is called a(n)
    a. group.
    b. period.
    c. family.
    d. octet.

    ANS: B   DIF: I   OBJ: 5-1.4

30. The atomic number of lithium, the first element in Group 1, is 3. The atomic number of the second element in this group is
    a. 4.   c. 11.
    b. 10.   d. 18.

    ANS: C   DIF: II   OBJ: 5-1.4

31. For groups 1, 2, and 18, the atomic number of the fourth element in the group is _____ more than the preceding element.
    a. 3   c. 18
    b. 4   d. 20

    ANS: C   DIF: II   OBJ: 5-1.4

32. Krypton, atomic number 36, is the fourth element in Group 18. What is the atomic number of xenon, the fifth element in Group 18?
    a. 54   c. 72
    b. 68   d. 90

    ANS: A   DIF: II   OBJ: 5-1.4

33. Barium, atomic number 56, is the fifth element in Group 2. What is the atomic number of radium, the next element in Group 2?
    a. 64   c. 88
    b. 74   d. 103

    ANS: C   DIF: II   OBJ: 5-1.4

34. For elements in Groups 1, 2, and 18, the increase in atomic number for successive elements follows the pattern 8, 8, 18, _____, 32.
    a. 18   c. 24
    b. 20   d. 26

    ANS: A   DIF: II   OBJ: 5-1.4

	1					Group 18

| | 1
H
Hydrogen
1.01 | | | | | 2
He
Helium
4.00 |

Group 18

	Group 1	Group 2		Group 17		
2	3 **Li** Lithium 6.94	4 **Be** Beryllium 9.01		9 **F** Fluorine 19.00	10 **Ne** Neon 20.18	
3	11 **Na** Sodium 22.99	12 **Mg** Magnesium 24.30		17 **Cl** Chlorine 35.45	18 **Ar** Argon 39.95	
4	19 **K** Potassium 39.10	20 **Ca** Calcium 40.08		35 **Br** Bromine 79.90	36 **Kr** Krypton 83.80	
5	37 **Rb** Rubidium 85.47	38 **Sr** Strontium 87.62		53 **I** Iodine 126.90	54 **Xe** Xenon 131.29	
6	55 **Cs** Cesium 132.90	56 **Ba** Barium 137.33		85 **At** Astatine (209.99)	86 **Rn** Radon (222.02)	
7	87 **Fr** Francium (223.02)	88 **Ra** Radium (226.02)				

35. To which group do lithium and potassium belong? Refer to the figure above.
    a. alkali metals
    b. transition metals
    c. halogens
    d. noble gases

    ANS: A   DIF: I   OBJ: 5-1.4

36. Refer to the figure above. To which group do fluorine and chlorine belong?
    a. alkaline-earth metals
    b. transition elements
    c. halogens
    d. actinides

    ANS: C   DIF: I   OBJ: 5-1.4

37. The electron configuration of aluminum, atomic number 13, is [Ne] $3s^2 3p^1$. Aluminum is in Period
    a. 2.
    b. 3.
    c. 6.
    d. 13.

    ANS: B   DIF: I   OBJ: 5-2.1

38. Identify the sublevels in a period that contains 32 elements.
    a. s, f
    b. s, p
    c. s, p, d
    d. s, p, d, f

    ANS: D   DIF: II   OBJ: 5-2.1

39. How many elements are in a period in which only the *s* and *p* sublevels are filled?
    a. 2
    b. 8
    c. 18
    d. 32

    ANS: B   DIF: II   OBJ: 5-2.1

40. The electron configuration of cesium, atomic number 55, is [Xe] $6s^1$. In what period is cesium?
    a. Period 1
    b. Period 6
    c. Period 8
    d. Period 55

    ANS: B   DIF: I   OBJ: 5-2.1

41. The length of each period in the periodic table is determined by the
    a. atomic masses of the elements.
    b. atomic numbers of the elements.
    c. sublevels being filled with electrons.
    d. number of isotopes of the elements in the period.

    ANS: C   DIF: I   OBJ: 5-2.1

42. Because the first energy level contains only the $1s$ sublevel, the number of elements in this period is
    a. 1.
    b. 2.
    c. 4.
    d. 8.

    ANS: B   DIF: III   OBJ: 5-2.1

43. In Period 3 there are 8 elements. What sublevel(s) is (are) being filled?
    a. *s*
    b. *s* and *d*
    c. *s* and *p*
    d. *d* and *f*

    ANS: C   DIF: II   OBJ: 5-2.1

44. Period 4 contains 18 elements. How many of these elements have electrons in the *d* sublevel?
    a. 8
    b. 10
    c. 16
    d. 18

    ANS: C   DIF: II   OBJ: 5-2.1

45. The period of an element can be determined from its
    a. reactivity.
    b. density.
    c. symbol.
    d. electron configuration.

    ANS: D   DIF: I   OBJ: 5-2.1

46. Calcium, atomic number 20, has the electron configuration [Ar] $4s^2$. In what period is calcium?
    a. Period 2
    b. Period 4
    c. Period 8
    d. Period 20

    ANS: B   DIF: I   OBJ: 5-2.1

Group 3	Group 4	Group 5	Group 6
21 **Sc** Scandium 44.96	22 **Ti** Titanium 47.88	23 **V** Vanadium 50.94	24 **Cr** Chromium 52.00
39 **Y** Yttrium 88.90	40 **Zr** Zirconium 91.22	41 **Nb** Niobium 92.91	42 **Mo** Molybdenum 95.94

47. In the elements shown in the figure above, the *s* sublevel of the highest occupied energy level
    a. always contains one electron.
    b. always contains two electrons.
    c. varies in the number of electrons it contains.
    d. is always empty.

    ANS: C   DIF: II   OBJ: 5-2.1

48. Elements to the right side of the periodic table (*p*-block elements) have properties most associated with
    a. gases.   c. metals.
    b. nonmetals.   d. metalloids.

    ANS: B   DIF: I   OBJ: 5-2.2

49. Neutral atoms with an s^2p^6 electron configuration in the highest energy level are best classified as
    a. metalloids.   c. nonmetals.
    b. metals.   d. gases.

    ANS: D   DIF: II   OBJ: 5-2.2

50. Elements in which the *d*-sublevel is being filled have the properties of
    a. metals.   c. metalloids.
    b. nonmetals.   d. gases.

    ANS: A   DIF: I   OBJ: 5-2.2

51. The elements that border the zigzag line in the periodic table are
    a. inactive.   c. metalloids.
    b. metals.   d. nonmetals.

    ANS: C   DIF: I   OBJ: 5-2.2

52. The group of 14 elements in the sixth period that have occupied 4*f* orbitals is the
    a. actinides.   c. transition elements.
    b. lanthanides.   d. metalloids.

    ANS: B   DIF: I   OBJ: 5-2.2

53. Within the *p*-block elements, the elements at the top of the table, compared with those at the bottom,
    a. have larger radii.
    b. are more metallic.
    c. have lower ionization energies.
    d. are less metallic.

    ANS: D   DIF: I   OBJ: 5-2.2

54. The electron configurations of the noble gases from neon to radon in the periodic table end with filled
    a. *f* orbitals.
    b. *d* orbitals.
    c. *s* orbitals.
    d. *p* orbitals.

    ANS: D   DIF: I   OBJ: 5-2.2

55. Hydrogen is placed separately from other elements in the periodic table because it
    a. is a gas.
    b. has one electron.
    c. has atomic number one.
    d. has many unique properties.

    ANS: D   DIF: I   OBJ: 5-2.2

56. Which orbitals are characteristic of the lanthanide elements?
    a. *d* orbitals
    b. *s* orbitals
    c. *f* orbitals
    d. *p* orbitals

    ANS: C   DIF: I   OBJ: 5-2.2

57. The elements whose electron configurations end with $s^2 p^5$ in the highest occupied energy level belong to Group
    a. 3.
    b. 7.
    c. 10.
    d. 17.

    ANS: D   DIF: II   OBJ: 5-2.3

58. For Groups 13 through 18, the total number of electrons in the highest occupied level equals the group number
    a. plus 1.
    b. minus 1.
    c. plus 5.
    d. minus 10.

    ANS: D   DIF: II   OBJ: 5-2.3

59. Strontium's highest occupied energy level is $5s^2$. To what group does strontium belong?
    a. Group 2
    b. Group 4
    c. Group 6
    d. Group 8

    ANS: A   DIF: II   OBJ: 5-2.3

60. If n stands for the highest occupied energy level, the outer configuration for all Group 1 elements is
    a. ns^1.
    b. $2n$.
    c. $n - s$.
    d. np^1.

    ANS: A   DIF: II   OBJ: 5-2.3

61. Titanium, atomic number 22, has the configuration [Ar] $3d^2\,4s^2$. To what group does titanium belong?
    a. Group 2
    b. Group 3
    c. Group 4
    d. Group 5

    ANS: C   DIF: II   OBJ: 5-2.3

62. Nitrogen's electron configuration is $1s^2\,2s^2\,2p^3$. To what group does nitrogen belong?
    a. Group 2
    b. Group 7
    c. Group 15
    d. Group 17

    ANS: C   DIF: II   OBJ: 5-2.3

63. Periods with occupied f sublevels
    a. have only Group 1 and 2 elements.
    b. are not assigned group numbers.
    c. have 32 groups.
    d. contain only Group 18 elements.

    ANS: B   DIF: I   OBJ: 5-2.3

64. The electron configuration of an element is [Kr] $4d^6\,5s^1$. To what group does this element belong?
    a. Group 4
    b. Group 5
    c. Group 7
    d. Group 9

    ANS: C   DIF: II   OBJ: 5-2.3

65. Bromine, atomic number 35, belongs to Group 17. How many electrons does bromine have in its outermost energy level?
    a. 7
    b. 17
    c. 18
    d. 35

    ANS: A   DIF: II   OBJ: 5-2.3

66. Magnesium, atomic number 12, has the electron configuration [Ne] $3s^2$. To what group does magnesium belong?
    a. Group 2
    b. Group 3
    c. Group 5
    d. Group 12

    ANS: A   DIF: II   OBJ: 5-2.3

67. In nature, the alkali metals occur as
    a. elements.
    b. compounds.
    c. complex ions.
    d. gases.

    ANS: B   DIF: I   OBJ: 5-2.4

68. The elements in Group 1 are also known as the
    a. alkali metals.
    b. rare-earth series.
    c. Period 1 elements.
    d. actinide series.

    ANS: A   DIF: I   OBJ: 5-2.4

69. The alkali metals belong to the _____-block in the periodic table.
    a. *s*   c. *d*
    b. *p*   d. *f*

    ANS: A   DIF: I   OBJ: 5-2.4

70. The most reactive group of the nonmetals are the
    a. lanthanides.   c. halogens.
    b. transition elements.   d. rare-earth elements.

    ANS: C   DIF: I   OBJ: 5-2.4

71. The group of soft, silvery active metals, all of which have one electron in an *s* orbital, is known as the
    a. alkaline-earth metals.   c. alkali metals.
    b. transition metals.   d. metalloids.

    ANS: C   DIF: I   OBJ: 5-2.4

72. The first member of the noble gas family, whose highest energy level consists of an octet of electrons, is
    a. helium.   c. neon.
    b. argon.   d. krypton.

    ANS: C   DIF: I   OBJ: 5-2.4

73. Among the alkali metals below, which has the lowest melting point?
    a. sodium (atomic number 11)   c. rubidium (atomic number 37)
    b. potassium (atomic number 19)   d. cesium (atomic number 55)

    ANS: D   DIF: I   OBJ: 5-2.4

74. The most characteristic property of the noble gases is that they
    a. have low boiling points.   c. are gases at ordinary temperatures.
    b. are radioactive.   d. are largely unreactive.

    ANS: D   DIF: I   OBJ: 5-2.4

75. Compared to the alkali metals, the alkaline-earth metals
    a. are less reactive.   c. are less dense.
    b. have lower melting points.   d. combine more readily with nonmetals.

    ANS: A   DIF: II   OBJ: 5-2.4

76. When determining the size of an atom by measuring the distance between identical adjacent nuclei, the radius of an atom is
    a. equal to the distance between nuclei.   c. twice the distance between nuclei.
    b. one-half the distance between nuclei.   d. one-fourth the distance between nuclei.

    ANS: B   DIF: I   OBJ: 5-3.1

77. When an electron is added to a neutral atom, a certain amount of energy is
    a. always absorbed.
    b. always released.
    c. either released or absorbed.
    d. burned away.

    ANS: C   DIF: I   OBJ: 5-3.1

78. Atomic size is determined by measuring the
    a. radius of an individual atom.
    b. distance between nuclei of adjacent atoms.
    c. diameter of an individual atom.
    d. volume of the electron cloud of adjacent atoms.

    ANS: B   DIF: I   OBJ: 5-3.1

79. Which represents a neutral atom acquiring an electron in an exothermic process?
    a. $A + e^- + energy \rightarrow A^-$
    b. $A + e^- \rightarrow A^- - energy$
    c. $A + e^- \rightarrow A^- + energy$
    d. $A^- + energy \rightarrow A + e^-$

    ANS: C   DIF: II   OBJ: 5-3.1

80. The energy required to remove an electron from an atom is the atom's
    a. electron affinity.
    b. electron energy.
    c. electronegativity.
    d. ionization energy.

    ANS: D   DIF: I   OBJ: 5-3.1

81. A measure of the ability of an atom in a chemical compound to attract electrons is called
    a. electron affinity.
    b. electron configuration.
    c. electronegativity.
    d. ionization potential.

    ANS: C   DIF: I   OBJ: 5-3.1

82. The element that has the greatest electronegativity is
    a. oxygen.
    b. sodium.
    c. chlorine.
    d. fluorine.

    ANS: D   DIF: II   OBJ: 5-3.1

83. One-half the distance between the nuclei of identical atoms that are bonded together is called the
    a. atomic radius.
    b. atomic diameter.
    c. atomic volume.
    d. electron cloud.

    ANS: A   DIF: I   OBJ: 5-3.1

84. Ionization energy is the energy required to remove _____ from an atom of an element.
    a. the electron cloud
    b. the nucleus
    c. an electron
    d. an ion

    ANS: C   DIF: I   OBJ: 5-3.1

85. When an electron is acquired by a neutral atom, the energy change is called
    a. electron affinity.   c. ionization energy.
    b. electronegativity.   d. electron configuration.

    ANS: A   DIF: I   OBJ: 5-3.1

86. A positive ion is known as a(n)
    a. ionic radius.   c. cation.
    b. valence electron.   d. anion

    ANS: C   DIF: I   OBJ: 5-3.2

87. A negative ion is known as a(n)
    a. ionic radius.   c. cation.
    b. valence electron.   d. anion.

    ANS: D   DIF: None   OBJ: None

88. In a row in the periodic table, as the atomic number increases, the atomic radius generally
    a. decreases.   c. increases.
    b. remains constant.   d. becomes unmeasurable.

    ANS: A   DIF: II   OBJ: 5-3.2

89. Within a group of elements, as the atomic number increases, the atomic radius
    a. increases.   c. decreases regularly.
    b. remains approximately constant.   d. decreases, but not regularly.

    ANS: A   DIF: II   OBJ: 5-3.2

90. In the alkaline-earth group, atoms with the smallest radii
    a. are the most reactive.   c. are all gases.
    b. have the largest volume.   d. have the highest ionization energies.

    ANS: D   DIF: II   OBJ: 5-3.2

91. As the atomic number of the metals of Group 1 increases, the ionic radius
    a. increases.   c. remains the same.
    b. decreases.   d. cannot be determined.

    ANS: A   DIF: II   OBJ: 5-3.2

92. Across a period in the periodic table, atomic radii
    a. gradually decrease.   c. gradually increase.
    b. gradually decrease, then sharply increase. d. gradually increase, then sharply decrease.

    ANS: A   DIF: II   OBJ: 5-3.2

93. The ionization energies for removing successive electrons from sodium are 496 kJ/mol, 4562 kJ/mol, 6912 kJ/mol, and 9544 kJ/mol. The great jump in ionization energy after the first electron is removed indicates that
    a. sodium has four or five electrons.
    b. the atomic radius has increased.
    c. a d-electron has been removed.
    d. the noble gas configuration has been reached.

    ANS: D   DIF: II   OBJ: 5-3.2

94. Which is the best reason that the atomic radius generally increases with atomic number in each group of elements?
    a. The nuclear charge increases.   c. The number of energy levels increases.
    b. The number of neutrons increases.   d. A new octet forms.

    ANS: C   DIF: II   OBJ: 5-3.2

95. The ionization energies required to remove successive electrons from one mole of calcium atoms are 590 kJ/mol, 1145 kJ/mol, 4912 kJ/mol, and 6474 kJ/mol. The most common ion of calcium is probably
    a. Ca^+.   c. Ca^{3+}.
    b. Ca^{2+}.   d. Ca^{4+}.

    ANS: B   DIF: III   OBJ: 5-3.2

96. For each successive electron removed from an atom, the ionization energy
    a. increases.   c. remains the same.
    b. decreases.   d. shows no pattern.

    ANS: A   DIF: II   OBJ: 5-3.2

97. As you move down the periodic table from carbon through lead, atomic radii
    a. generally increase.   c. do not change.
    b. generally decrease.   d. vary unpredictably.

    ANS: A   DIF: II   OBJ: 5-3.2

98. As you move left to right from gallium through bromine, atomic radii
    a. generally increase.   c. do not change.
    b. generally decrease.   d. vary unpredictably.

    ANS: B   DIF: II   OBJ: 5-3.2

99. The energy required to remove an electron from an atom _____ as you move left to right from potassium through iron.
    a. generally increases   c. does not change
    b. generally decreases   d. varies unpredictably

    ANS: A   DIF: II   OBJ: 5-3.2

100. The force of attraction by Group 1 metals for their valence electrons is
    a. weak.
    b. zero.
    c. strong.
    d. greater than that for inner shell electrons.

    ANS: A   DIF: I   OBJ: 5-3.3

101. The electrons available to be lost, gained, or shared when atoms form molecules are called
    a. ions.
    b. valence electrons.
    c. *d* electrons.
    d. electron clouds.

    ANS: B   DIF: I   OBJ: 5-3.3

102. When chemical compounds form, valence electrons are those that may be
    a. lost only.
    b. gained only.
    c. shared only.
    d. lost, gained, or shared.

    ANS: D   DIF: I   OBJ: 5-3.3

103. Valence electrons are those
    a. closest to the nucleus.
    b. in the lowest energy level.
    c. in the highest energy level.
    d. combined with protons.

    ANS: C   DIF: I   OBJ: 5-3.3

104. The number of valence electrons in Group 1 elements is
    a. 1.
    b. 2.
    c. 8.
    d. equal to the period number.

    ANS: A   DIF: I   OBJ: 5-3.3

105. In Group 2 elements, the valence electrons are in sublevel
    a. *d.*
    b. *p.*
    c. *s.*
    d. *f.*

    ANS: C   DIF: I   OBJ: 5-3.3

106. The number of valence electrons in Group 17 elements is
    a. 7.
    b. 8.
    c. 17.
    d. equal to the period number.

    ANS: A   DIF: I   OBJ: 5-3.3

107. For Groups 13 through 18, the number of valence electrons is equal to the group number
    a. plus 1.
    b. plus the period number.
    c. minus the period number.
    d. minus 10.

    ANS: D   DIF: I   OBJ: 5-3.3

108. The number of valence electrons in Group 2 elements is
    a. 2.
    b. 8.
    c. 18.
    d. equal to the period number.

    ANS: A   DIF: I   OBJ: 5-3.3

109. In Groups 13 through 18, valence electrons may be in sublevels
  a. *s* and *d*.
  b. *s* and *p*.
  c. *d* and *f*.
  d. *p* and *d*.

  ANS: B   DIF: II   OBJ: 5-3.3

110. Across a period, the atomic radii of *d*-block elements generally
  a. increase.
  b. decrease.
  c. remain constant.
  d. increase and then decrease.

  ANS: B   DIF: II   OBJ: 5-3.4

111. As with main-group elements, ionization energies of *d*-block elements generally _____
  across a period.
  a. increase
  b. decrease
  c. remain constant
  d. drop to zero

  ANS: A   DIF: II   OBJ: 5-3.4

112. In contrast to elements in the main group, the first ionization energies of *d*-block elements
  _____ as one proceeds down each group.
  a. remain constant
  b. decrease
  c. are identical
  d. increase

  ANS: D   DIF: II   OBJ: 5-3.4

113. The first electrons to be removed when *d*-block elements form ions are the
  a. *d* electrons.
  b. *p* electrons.
  c. *s* electrons.
  d. *f* electrons.

  ANS: C   DIF: I   OBJ: 5-3.4

114. Which groups in the main group have lower electronegativity than d-block elements?
  a. Groups 1 and 2
  b. Groups 13 through 18
  c. Groups 17 and 18
  d. Groups 13 through 17

  ANS: A   DIF: II   OBJ: 5-3.4

115. Among the *d*-block elements, as atomic radii decrease, electronegativity values
  a. remain constant.
  b. increase.
  c. decrease.
  d. drop to zero.

  ANS: B   DIF: II   OBJ: 5-3.4

116. In compounds, *d*-block elements most often form ions with charge
  a. 2–.
  b. 1–.
  c. 1+.
  d. 2+.

  ANS: D   DIF: II   OBJ: 5-3.4

SHORT ANSWER

1. In terms of the periodic law, explain which two of these elements are most similar: sodium (element 11), phosphorus (element 15), and sulfur (element 16).

   ANS:
   Their locations in the periodic table indicate that phosphorus and sulfur are nonmetals and sodium is a metal. Nonmetals are a group with characteristic properties, so phosphorus and sulfur are the most similar elements of the three.

   DIF: III   OBJ: 5-1.3

2. What can you predict about the properties of xenon and helium, both in Group 18 in the periodic table? Why?

   ANS:
   In the periodic table, elements in the same column or group have similar properties. Because helium and xenon are in the same group, they have similar properties.

   DIF: I   OBJ: 5-1.4

ESSAY

1. Describe the differences between Mendeleev's periodic table and the modern periodic table.

   ANS:
   Mendeleev developed a table of periodicity based on atomic mass. It had some inconsistencies because the physical and chemical characteristics of the elements were not functions of atomic mass and some gaps because some elements had not yet been discovered. Moseley refined the table by organizing the elements according to increasing atomic number. The periods in Mendeleev's table were columns. In the modern periodic table, the periods are rows.

   DIF: II   OBJ: 5-1.2

MULTIPLE CHOICE

1. A mutual electrical attraction between the nuclei and valence electrons of different atoms that binds the atoms together is called a(n)
    a. dipole.
    b. Lewis structure.
    c. chemical bond.
    d. London force.

    ANS: C   DIF: I   OBJ: 6-1.1

2. A chemical bond results from the mutual attraction of the nuclei of atoms and
    a. electrons.
    b. protons.
    c. neutrons.
    d. dipoles.

    ANS: A   DIF: I   OBJ: 6-1.1

3. The electrons involved in the formation of a chemical bond are called
    a. dipoles.
    b. *s* electrons.
    c. Lewis electrons.
    d. valence electrons.

    ANS: D   DIF: I   OBJ: 6-1.1

4. The electrostatic attraction between positively charged nuclei and negatively charged electrons permits two atoms to be held together by a(n)
    a. chemical bond.
    b. London force.
    c. neutron.
    d. ion.

    ANS: A   DIF: I   OBJ: 6-1.1

5. In a chemical bond, the link between atoms results from the attraction between electrons and
    a. Lewis structures.
    b. nuclei.
    c. van der Waals forces.
    d. isotopes.

    ANS: B   DIF: I   OBJ: 6-1.1

6. As independent particles, atoms are
    a. at relatively high potential energy.
    b. at relatively low potential energy.
    c. very stable.
    d. part of a chemical bond.

    ANS: A   DIF: I   OBJ: 6-1.2

7. Atoms are _____ when they are combined.
    a. more stable
    b. less stable
    c. not bound together
    d. at a high potential energy

    ANS: A   DIF: I   OBJ: 6-1.2

8. Atoms naturally move
    a. toward high potential energy.
    b. toward low potential energy.
    c. toward less stability.
    d. away from each other.

    ANS: B   DIF: I   OBJ: 6-1.2

9. As atoms bond with each other, they
   a. increase their potential energy, thus creating less-stable arrangements of matter.
   b. decrease their potential energy, thus creating less-stable arrangements of matter.
   c. increase their potential energy, thus creating more-stable arrangements of matter.
   d. decrease their potential energy, thus creating more-stable arrangements of matter.

   ANS: D   DIF: I   OBJ: 6-1.2

10. A chemical bond resulting from the electrostatic attraction between positive and negative ions is called a(n)
    a. covalent bond.   c. charged bond.
    b. ionic bond.   d. dipole bond.

    ANS: B   DIF: I   OBJ: 6-1.3

11. The chemical bond formed when two atoms share electrons is called a(n)
    a. ionic bond.   c. Lewis structure.
    b. orbital bond.   d. covalent bond.

    ANS: D   DIF: I   OBJ: 6-1.3

12. If two covalently bonded atoms are identical, the bond is
    a. nonpolar covalent.   c. nonionic.
    b. polar covalent.   d. coordinate covalent.

    ANS: A   DIF: I   OBJ: 6-1.3

13. When atoms share electrons, the electrical attraction of an atom for the electrons is called the atom's
    a. electron affinity.   c. resonance.
    b. electronegativity.   d. hybridization.

    ANS: B   DIF: I   OBJ: 6-1.3

14. If the atoms that share electrons have an unequal attraction for the electrons, the bond is called
    a. nonpolar.   c. ionic.
    b. polar.   d. dipolar.

    ANS: B   DIF: I   OBJ: 6-1.3

15. The electrostatic attraction between _____ forms an ionic bond.
    a. ions   c. electrons
    b. dipoles   d. orbitals

    ANS: A   DIF: I   OBJ: 6-1.3

16. A covalent bond results when _____ are shared.
    a. ions   c. electrons
    b. Lewis structures   d. dipoles

    ANS: C   DIF: I   OBJ: 6-1.3

17. Most chemical bonds are
    a. purely ionic.   c. partly ionic and partly covalent.
    b. purely covalent.   d. metallic.

    ANS: C   DIF: I   OBJ: 6-1.4

18. Nonpolar covalent bonds are not common because
    a. one atom usually attracts electrons more strongly than the other.
    b. ions always form when atoms join.
    c. the electrons usually remain equally distant from both atoms.
    d. dipoles are rare in nature.

    ANS: A   DIF: I   OBJ: 6-1.4

19. Purely ionic bonds do not occur because the atom that gives up an electron in such a bond
    a. is not an ion.
    b. still has some attraction for the electron.
    c. has a negative charge.
    d. shares the electron equally with the other atom in the bond.

    ANS: B   DIF: I   OBJ: 6-1.4

20. The greater the electronegativity difference between two bonded atoms, the greater the
    percentage of
    a. ionic character.   c. metallic character.
    b. covalent character.   d. electron sharing.

    ANS: A   DIF: I   OBJ: 6-1.4

21. Bonds that are between 5% and 50% ionic are considered
    a. ionic.   c. polar covalent.
    b. pure covalent.   d. nonpolar covalent.

    ANS: C   DIF: I   OBJ: 6-1.4

22. Bonds that are more than 50% ionic are considered
    a. polyatomic.   c. ionic.
    b. polar covalent.   d. nonpolar covalent.

    ANS: C   DIF: I   OBJ: 6-1.4

23. A bond that is less than 5% ionic is considered
    a. polar covalent.   c. nonpolar covalent.
    b. ionic.   d. metallic.

    ANS: C   DIF: I   OBJ: 6-1.4

24. The pair of elements that forms a bond with the least ionic character is
    a. Na and Cl.   c. O and Cl.
    b. H and Cl.   d. Br and Cl.

    ANS: D   DIF: III   OBJ: 6-1.5

25. The B—F bond in BF_3 (electronegativity for B is 2.0; electronegativity for F is 4.0) is
    a. polar covalent.
    c. nonpolar covalent.
    b. ionic.
    d. pure covalent.

    ANS: B   DIF: III   OBJ: 6-1.5

26. The percentage ionic character for the C—F bond (electronegativity for C is 2.5; electronegativity for F is 4.0) in CF_4 is
    a. 9%.
    c. 38%.
    b. 15%.
    d. 62%.

    ANS: C   DIF: III   OBJ: 6-1.5

27. The percentage ionic character and the type of bond in Br_2 (electronegativity for Br is 2.8) is
    a. 0%; nonpolar covalent.
    c. 0%; pure ionic.
    b. 100%; polar covalent.
    d. 100%; pure ionic.

    ANS: A   DIF: III   OBJ: 6-1.5

28. The percentage ionic character and the type of bond in LiCl (electronegativity for Li is 1.0; electronegativity for Cl is 3.0) is
    a. 50%; ionic.
    c. 67%; polar covalent.
    b. 50%; polar covalent.
    d. 67%; ionic.

    ANS: D   DIF: III   OBJ: 6-1.5

29. In which of these compounds is the bond between the atoms NOT a nonpolar covalent bond?
    a. Cl_2
    c. HCl
    b. H_2
    d. O_2

    ANS: C   DIF: II   OBJ: 6-1.5

30. A neutral group of atoms held together by covalent bonds is a
    a. molecular formula.
    c. compound.
    b. chemical formula.
    d. molecule.

    ANS: D   DIF: I   OBJ: 6-2.1

31. A molecule is a
    a. negatively charged group of atoms held together by covalent bonds.
    b. positively charged group of atoms held together by covalent bonds.
    c. neutral group of atoms held together by covalent bonds.
    d. neutral group of atoms held together by ionic bonds.

    ANS: C   DIF: I   OBJ: 6-2.1

32. All of the following are true statements about a molecule EXCEPT
    a. it is capable of existing on its own.
    b. it may consist of two or more atoms of the same type.
    c. it must be part of a molecular compound.
    d. it may consist of two or more atoms of different types.

    ANS: C   DIF: I   OBJ: 6-2.1

33. A _____ shows the types and numbers of atoms joined in a single molecule of a molecular compound.
   a. molecular formula
   b. chemical formula
   c. covalent bond
   d. ionic bond

   ANS: A   DIF: I   OBJ: 6-2.1

34. Which of the following is NOT an example of a molecular formula?
   a. H_2O
   b. B
   c. NH_3
   d. O_2

   ANS: B   DIF: II   OBJ: 6-2.1

35. When a covalent bond forms, as the distance between two atoms decreases, the potential energy
   a. increases.
   b. decreases.
   c. remains constant.
   d. becomes zero.

   ANS: B   DIF: I   OBJ: 6-2.2

36. A covalent bond forms when the attraction between two atoms is balanced by repulsion and the potential energy is
   a. at a maximum.
   b. zero.
   c. at a minimum.
   d. equal to the kinetic energy.

   ANS: C   DIF: I   OBJ: 6-2.2

37. Bond length is
   a. the separation for which potential energy is at a minimum.
   b. the separation for which kinetic energy is at a maximum.
   c. the separation for which potential energy is at a maximum.
   d. one-half the diameter of the electron cloud.

   ANS: A   DIF: I   OBJ: 6-2.2

38. Bond energy is the energy
   a. required to break a chemical bond.
   b. released when a chemical bond breaks.
   c. required to form a chemical bond.
   d. absorbed when a chemical bond forms.

   ANS: A   DIF: I   OBJ: 6-2.2

39. When two atoms are separated and do not attract each other, the potential energy is
   a. at a minimum.
   b. zero.
   c. at a maximum.
   d. equal to the sum of the potential energy of each atom divided by two.

   ANS: B   DIF: I   OBJ: 6-2.2

40. The energy released when a covalent bond forms is the difference between zero and the
   a. maximum potential energy.
   b. kinetic energy of the atom.
   c. minimum potential energy.
   d. bond length expressed in nanometers.

   ANS: C   DIF: I   OBJ: 6-2.2

41. In a molecule of fluorine, the two shared electrons give each fluorine atom _____ electron(s) in the outer energy level.
   a. 1
   b. 2
   c. 8
   d. 32

   ANS: C   DIF: II   OBJ: 6-2.3

42. In many compounds, atoms of main-group elements form bonds so that the number of electrons in the outermost energy levels of each atom is
   a. 2.
   b. 6.
   c. 8.
   d. 10.

   ANS: C   DIF: I   OBJ: 6-2.3

43. The octet rule states that chemical compounds tend to form so that each atom has an octet of electrons in
   a. its highest occupied energy level.
   b. the $1s$ orbital.
   c. its d orbitals.
   d. its p orbitals.

   ANS: A   DIF: I   OBJ: 6-2.3

44. An octet is equal to
   a. 2.
   b. 4.
   c. 5.
   d. 8.

   ANS: D   DIF: I   OBJ: 6-2.3

45. What principle states that atoms tend to form compounds so that each atom can have eight electrons in its outermost energy level?
   a. rule of eights
   b. Avogadro principle
   c. configuration rule
   d. octet rule

   ANS: D   DIF: I   OBJ: 6-2.3

46. The electron configuration of nitrogen is $1s^2\,2s^2\,2p^3$. How many more electrons does nitrogen need to satisfy the octet rule?
   a. 1
   b. 3
   c. 5
   d. 8

   ANS: B   DIF: II   OBJ: 6-2.3

47. The elements of the _____ group satisfy the octet rule without forming compounds.
   a. main
   b. noble gas
   c. alkali metal
   d. alkaline-earth metal

   ANS: B   DIF: I   OBJ: 6-2.3

48. When the octet rule is satisfied, the outermost _____ are filled.
   a. d and f orbitals
   b. s and p orbitals
   c. s and d orbitals
   d. d and p orbitals

   ANS: B   DIF: II   OBJ: 6-2.3

49. In drawing a Lewis structure, each nonmetal atom except hydrogen should be surrounded by
    a. 2 electrons.   c. 8 electrons.
    b. 4 electrons.   d. 10 electrons.

    ANS: C   DIF: I   OBJ: 6-2.4

50. In drawing a Lewis structure, the central atom is the
    a. atom with the greatest mass.   c. atom with the fewest electrons.
    b. atom with the highest atomic number.   d. least electronegative atom.

    ANS: D   DIF: I   OBJ: 6-2.4

51. To draw a Lewis structure, one must know the
    a. number of valence electrons in each atom.
    b. atomic mass of each atom.
    c. bond length of each atom.
    d. ionization energy of each atom.

    ANS: A   DIF: I   OBJ: 6-2.4

52. After drawing a Lewis structure, one should
    a. determine the number of each type of atom in the molecule.
    b. add unshared pairs of electrons around nonmetal atoms.
    c. determine the total number of valence electrons in each atom.
    d. determine the electronegativity of each atom.

    ANS: C   DIF: I   OBJ: 6-2.4

53. To draw a Lewis structure, it is NOT necessary to know
    a. bond energies.
    b. the types of atoms in the molecule.
    c. the number of valence electrons for each atom.
    d. the number of atoms in the molecule.

    ANS: A   DIF: I   OBJ: 6-2.4

54. The Lewis structure for the ammonium ion has
    a. 2 valence electrons.   c. 8 valence electrons.
    b. 4 valence electrons.   d. 12 valence electrons.

    ANS: C   DIF: I   OBJ: 6-2.4

55. If, after drawing a Lewis structure, too many valence electrons have been used, the molecule probably contains
    a. too many atoms.   c. too many unshared pairs of electrons.
    b. one or more multiple covalent bonds.   d. an ionic bond.

    ANS: B   DIF: I   OBJ: 6-2.5

56. Multiple covalent bonds may occur in atoms that contain carbon, nitrogen, or
    a. chlorine.
    b. hydrogen.
    c. oxygen.
    d. helium.

    ANS: C   DIF: II   OBJ: 6-2.5

57. The substance whose Lewis structure shows three covalent bonds is
    a. H_2O.
    b. CH_2Cl_2.
    c. NH_3.
    d. CCl_4.

    ANS: C   DIF: III   OBJ: 6-2.5

58. How many double bonds are in the Lewis structure for hydrogen fluoride, HF?
    a. none
    b. one
    c. two
    d. three

    ANS: A   DIF: II   OBJ: 6-2.5

59. How many extra electrons are in the Lewis structure of the phosphate ion, PO_4^{3-}?
    a. 0
    b. 2
    c. 3
    d. 4

    ANS: C   DIF: III   OBJ: 6-2.5

60. How many electrons must be shown in the Lewis structure of the hydroxide ion, OH^-?
    a. 1
    b. 8
    c. 9
    d. 10

    ANS: B   DIF: III   OBJ: 6-2.5

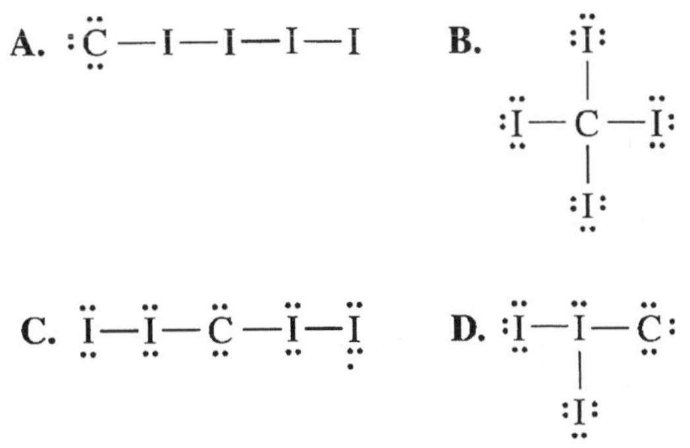

A. :C̈—I—I—I—I   B.   :Ï:
                         |
                    :Ï—C—Ï:
                         |
                        :Ï:

C. Ï—Ï—C̈—Ï—Ï   D. :Ï—Ï—C̈:
                         |
                        :Ï:

61. What is the Lewis structure for hydrogen chloride, HCl?
    a. A
    b. B
    c. C
    d. D

    ANS: D   DIF: III   OBJ: 6-2.5

A. Cl—H: **B.** :H—Cl: **C.** :H—C̈l: **D.** H—C̈l:

62. What is the Lewis structure for carbon tetraiodide, which contains one carbon atom and four iodine atoms?

    a. A
    b. B

    c. C
    d. D

    ANS: B   DIF: III   OBJ: 6-2.5

63. Bonding in molecules or ions that cannot be correctly represented by a single Lewis structure is
    a. covalent bonding.
    b. resonance.

    c. single bonding.
    d. double bonding.

    ANS: B   DIF: I   OBJ: 6-2.6

64. Resonance structures are also called
    a. dichotomous structures.
    b. Lewis structures.

    c. molecular structures.
    d. resonance hybrids.

    ANS: D   DIF: I   OBJ: 6-2.6

65. To indicate resonance, a _____ is placed between a molecule's resonance structures.
    a. double-headed arrow
    b. single-headed arrow

    c. series of dots
    d. Lewis structure

    ANS: A   DIF: I   OBJ: 6-2.6

66. Chemists once believed that a molecule that contains a single bond and a double bond split its time existing as one of these two structures. This effect became known as
    a. alternation.
    b. resonance.

    c. Lewis structure.
    d. single-double bonding.

    ANS: B   DIF: I   OBJ: 6-2.6

67. The chemical formula for an ionic compound represents the
    a. number of atoms in each molecule.
    b. number of ions in each molecule.
    c. simplest ratio of the combined ions that balances total charges.
    d. total number of ions in the crystal lattice.

    ANS: C   DIF: I   OBJ: 6-3.1

68. A shorthand representation of the composition of a substance using atomic symbols and numerical subscripts is called a(n)
    a. Lewis structure.
    b. chemical formula.

    c. polyatomic ion.
    d. multiple bond.

    ANS: B   DIF: I   OBJ: 6-3.1

69. A formula that shows the types and numbers of atoms combined in a single molecule is called a(n)
    a. molecular formula.
    b. ionic formula.
    c. Lewis structure.
    d. covalent formula.

    ANS: A   DIF: I   OBJ: 6-3.1

70. A formula unit of an ionic compound
    a. is an independent unit that can be isolated and studied.
    b. is the simplest ratio of ions that balances total charge.
    c. describes the crystal lattice.
    d. all of the above

    ANS: B   DIF: I   OBJ: 6-3.1

71. The chemical formula for water, a covalent compound, is H_2O. This formula is an example of a(n)
    a. formula unit.
    b. Lewis structure.
    c. ionic formula.
    d. molecular formula.

    ANS: D   DIF: II   OBJ: 6-3.1

72. An ionic compound is not represented by a molecular formula because an ionic compound
    a. does not contain bonds.
    b. does not have charged particles.
    c. lacks molecules.
    d. always has a positive charge.

    ANS: C   DIF: I   OBJ: 6-3.1

73. In the NaCl crystal, each Na^+ and Cl^- ion has clustered around it _____ of the oppositely charged ions.
    a. 1
    b. 2
    c. 4
    d. 6

    ANS: D   DIF: I   OBJ: 6-3.2

74. In an ionic compound, the orderly arrangement of ions in a crystal is the state of
    a. maximum potential energy.
    b. minimum potential energy.
    c. average potential energy.
    d. zero potential energy.

    ANS: B   DIF: I   OBJ: 6-3.2

75. The ions in an ionic compound are organized into a(n)
    a. molecule.
    b. Lewis structure.
    c. polyatomic ion.
    d. crystal.

    ANS: D   DIF: I   OBJ: 6-3.2

76. In a crystal of an ionic compound, each cation is surrounded by
    a. molecules.
    b. positive ions.
    c. dipoles.
    d. anions.

    ANS: D   DIF: I   OBJ: 6-3.2

77. In a crystal, the valence electrons of adjacent ions
   a. repel each other.   c. neutralize each other.
   b. attract each other.   d. have no effect on each other.

   ANS: A   DIF: I   OBJ: 6-3.2

78. The simplest repeating unit of a crystal is known as a(n)
   a. unit cell.   c. ionic compound.
   b. crystal lattice.   d. salt.

   ANS: A   DIF: I   OBJ: 6-3.2

79. The energy released when 1 mol of an ionic crystalline compound is formed from gaseous ions is
   called the
   a. bond energy.   c. lattice energy.
   b. potential energy.   d. energy of crystallization.

   ANS: C   DIF: I   OBJ: 6-3.3

80. The lattice energy is a measure of the
   a. strength of an ionic bond.   c. strength of a covalent bond.
   b. strength of a metallic bond.   d. number of ions in a crystal.

   ANS: A   DIF: I   OBJ: 6-3.3

81. Lattice energy is the energy released in the formation of a(n)
   a. Lewis structure.   c. nonpolar covalent compound.
   b. polar covalent compound.   d. ionic compound.

   ANS: D   DIF: I   OBJ: 6-3.3

82. Compared with energies of neutral atoms, a crystal lattice has
   a. higher potential energy.   c. equal potential energy.
   b. lower potential energy.   d. less stability.

   ANS: B   DIF: II   OBJ: 6-3.3

83. If the lattice energy of compound A is greater than that of compound B,
   a. compound A is not an ionic compound.
   b. the bonds in compound A are stronger than the bonds in compound B.
   c. compound B is probably a gas.
   d. compound A has larger crystals than compound B.

   ANS: B   DIF: II   OBJ: 6-3.3

84. Which of the following is NOT a property of an ionic compound?
   a. low boiling point   c. hardness
   b. brittleness   d. molten compound conducts electricity

   ANS: A   DIF: I   OBJ: 6-3.4

85. Compared with ionic compounds, molecular compounds
   a. have higher boiling points.   c. have lower melting points.
   b. are brittle.   d. are harder.

   ANS: C   DIF: I   OBJ: 6-3.4

86. The forces of attraction between molecules in a molecular compound are
   a. stronger than the forces of ionic bonding.
   b. weaker than the forces of ionic bonding.
   c. approximately equal to the forces of ionic bonding.
   d. zero.

   ANS: B   DIF: I   OBJ: 6-3.4

87. A compound that vaporizes readily at room temperature is most likely to be a(n)
   a. molecular compound.   c. metal.
   b. ionic compound.   d. brittle compound.

   ANS: A   DIF: I   OBJ: 6-3.4

88. Because the particles in ionic compounds are more strongly attracted than in molecular
   compounds, the melting points of ionic compounds are
   a. equal for all ionic compounds.
   b. lower than melting points of molecular compounds.
   c. higher than melting points of molecular compounds.
   d. approximately equal to room temperature.

   ANS: C   DIF: II   OBJ: 6-3.4

89. Ionic compounds are brittle because the strong attractive forces
   a. allow the layers to shift easily.
   b. cause the compound to vaporize easily.
   c. keep the surface dull.
   d. hold the layers in relatively fixed positions.

   ANS: D   DIF: II   OBJ: 6-3.4

90. The properties of both ionic and molecular compounds are related to the
   a. lattice energies of the compounds.
   b. strengths of attraction between the particles in the compounds.
   c. number of covalent bonds each contains.
   d. mobile electrons that they contain.

   ANS: B   DIF: I   OBJ: 6-3.4

91. A chemical bond formed by the attraction between positive ions and surrounding mobile
   electrons is a(n)
   a. nonpolar covalent bond.   c. polar covalent bond.
   b. ionic bond.   d. metallic bond.

   ANS: D   DIF: I   OBJ: 6-4.1

92. Compared with nonmetals, the number of valence electrons in metals is generally
   a. smaller.
   c. about the same.
   b. greater.
   d. almost triple that of nonmetals.

   ANS: A   DIF: I   OBJ: 6-4.1

93. In metals, the valence electrons
   a. are attached to particular positive ions.
   c. are immobile.
   b. are shared by all of the atoms.
   d. form covalent bonds.

   ANS: B   DIF: I   OBJ: 6-4.1

94. In the electron-sea model of a metallic bond,
   a. electrons are stationary.
   b. electrons are bonded to particular positive ions.
   c. some electrons are valence electrons and some are not.
   d. mobile electrons are shared by all the atoms.

   ANS: D   DIF: I   OBJ: 6-4.1

95. A metallic bond forms when positive ions attract
   a. stationary electrons.
   c. cations.
   b. nonvalence electrons.
   d. mobile electrons.

   ANS: D   DIF: I   OBJ: 6-4.1

96. In metallic bonds, the mobile electrons surrounding the positive ions are called a(n)
   a. Lewis structure.
   c. electron cloud.
   b. electron sea.
   d. dipole.

   ANS: B   DIF: I   OBJ: 6-4.1

97. The electron-sea model of bonding represents
   a. covalent bonding.
   c. ionic bonding.
   b. metallic bonding.
   d. hydrogen bonding.

   ANS: B   DIF: I   OBJ: 6-4.1

98. To appear shiny, a material must be able to
   a. form crystals.
   b. absorb and re-emit light of many wavelengths.
   c. absorb light and change it all to heat.
   d. change light to electricity.

   ANS: B   DIF: I   OBJ: 6-4.2

99. The shiny appearance of a metal is most closely related to the metal's
   a. electron sea.
   c. brittle crystalline structure.
   b. covalent bonds.
   d. positive ions.

   ANS: A   DIF: I   OBJ: 6-4.2

100. As light strikes the surface of a metal, the electrons in the electron sea
    a. allow the light to pass through.
    b. become attached to particular positive ions.
    c. fall to lower energy levels.
    d. absorb and re-emit the light.

    ANS: D   DIF: I   OBJ: 6-4.2

101. The electron-sea model of the metallic bond helps to explain why metals are
    a. brittle.   c. shiny.
    b. dull.   d. malleable.

    ANS: C   DIF: I   OBJ: 6-4.2

102. If a material can be shaped or extended by physical pressure, such as hammering, which property
     does the material have?
    a. conductivity   c. ductility
    b. malleability   d. luster

    ANS: B   DIF: I   OBJ: 6-4.3

103. Metals are malleable because the metallic bonding
    a. holds the layers of ions in rigid positions.
    b. does not produce ions.
    c. allows one plane of ions to slide past another.
    d. is easily broken.

    ANS: C   DIF: I   OBJ: 6-4.3

104. Which best explains the observation that metals are malleable and ionic crystals are brittle?
    a. their chemical bonds   c. their heats of vaporization
    b. their London forces   d. their polarity

    ANS: A   DIF: I   OBJ: 6-4.3

105. Because metallic bonds permit one plane of ions to slide past another without breaking bonds,
     metals are
    a. brittle.   c. nonreflective.
    b. malleable.   d. poor conductors of electricity.

    ANS: B   DIF: I   OBJ: 6-4.3

106. Malleability and ductility are characteristic of substances with
    a. covalent bonds.   c. Lewis structures.
    b. ionic bonds.   d. metallic bonds.

    ANS: D   DIF: I   OBJ: 6-4.3

107. Compared with metals, ionic crystals are
    a. ductile.   c. brittle.
    b. malleable.   d. lustrous.

    ANS: C   DIF: I   OBJ: 6-4.3

108. Planes of ions can slide past one another without breaking bonds in substances with
   a. metallic bonds.
   b. ionic bonds.
   c. covalent bonds.
   d. no bonds.

   ANS: A   DIF: I   OBJ: 6-4.3

109. Shifting the layers of an ionic crystal causes the crystal to
   a. be drawn into a wire.
   b. shatter.
   c. become metallic.
   d. freeze.

   ANS: B   DIF: I   OBJ: 6-4.3

110. The model for predicting the shape of a molecule that is based on the repulsion of electrons for each other is called
   a. hybridization.
   b. Lewis structure.
   c. London force model.
   d. VSEPR theory.

   ANS: D   DIF: I   OBJ: 6-5.1

111. According to VSEPR theory, an AB_2 molecule is
   a. trigonal planar.
   b. tetrahedral.
   c. linear.
   d. octahedral.

   ANS: C   DIF: II   OBJ: 6-5.1

112. VSEPR theory is a model for predicting
   a. the strength of metallic bonds.
   b. the shape of molecules.
   c. lattice energy values.
   d. ionization energy.

   ANS: B   DIF: I   OBJ: 6-5.1

113. The concept that electrostatic repulsion between electron pairs surrounding an atom causes these pairs to be separated as far as possible is the foundation of
   a. VSEPR theory.
   b. the hybridization model.
   c. the electron-sea model.
   d. Lewis theory.

   ANS: A   DIF: I   OBJ: 6-5.1

114. According to VSEPR theory, the electrostatic repulsion between electron pairs surrounding an atom causes
   a. an electron sea to form.
   b. positive ions to form.
   c. these pairs to be separated as far as possible.
   d. light to reflect.

   ANS: C   DIF: I   OBJ: 6-5.1

115. According to VSEPR theory, the shape of an AB_3 molecule is
   a. trigonal planar.
   b. tetrahedral.
   c. linear.
   d. bent.

   ANS: A   DIF: I   OBJ: 6-5.1

116. According to VSEPR theory, the structure of the ammonia molecule, NH_3, is
    a. linear.
    b. bent.
    c. pyramidal.
    d. tetrahedral.

    ANS: C   DIF: II   OBJ: 6-5.2

117. Use VSEPR theory to predict the shape of the hydrogen chloride molecule, HCl.
    a. tetrahedral
    b. linear
    c. bent
    d. trigonal planar

    ANS: B   DIF: III   OBJ: 6-5.2

118. Use VSEPR theory to predict the shape of the magnesium hydride molecule, MgH_2.
    a. tetrahedral
    b. linear
    c. bent
    d. octahedral

    ANS: B   DIF: III   OBJ: 6-5.2

119. Use VSEPR theory to predict the shape of the carbon tetraiodide molecule, CI_4.
    a. tetrahedral
    b. linear
    c. bent
    d. trigonal planar

    ANS: A   DIF: III   OBJ: 6-5.2

120. Use VSEPR theory to predict the shape of the chlorate ion, ClO_3^-.
    a. trigonal planar
    b. octahedral
    c. trigonal pyramidal
    d. bent

    ANS: C   DIF: III   OBJ: 6-5.2

121. Use VSEPR theory to predict the shape of the hydrogen sulfide molecule, H_2S.
    a. tetrahedral
    b. linear
    c. bent
    d. octahedral

    ANS: C   DIF: III   OBJ: 6-5.2

122. Use VSEPR theory to predict the shape of carbon dioxide, CO_2.
    a. tetrahedral
    b. linear
    c. bent
    d. octahedral

    ANS: B   DIF: III   OBJ: 6-5.2

123. The hybridized orbitals responsible for the bent shape of the water molecule are
    a. $1s^2 \, 2s^2$.
    b. ps^1.
    c. sp^3.
    d. $2s^2 \, sp^2$.

    ANS: C   DIF: II   OBJ: 6-5.3

124. The hybridized orbitals responsible for the shape of the CH_4 molecule are
    a. $1s^1 \, 1p^3$.
    b. sp^2.
    c. $2s^2 \, 2p^2$.
    d. sp^3.

    ANS: D   DIF: II   OBJ: 6-5.3

125. The mixing of two or more atomic orbitals of similar energies on the same atom to produce new orbitals of equal energies is called
    a. VSEPR theory.
    b. malleability.
    c. hybridization.
    d. dipole-dipole interaction.

    ANS: C   DIF: I   OBJ: 6-5.3

126. Orbitals of equal energy produced by the combination of two or more orbitals in the same atom are called
    a. q-orbitals.
    b. hybrid orbitals.
    c. Lewis orbitals.
    d. n-orbitals.

    ANS: B   DIF: I   OBJ: 6-5.3

127. Which hybrid orbitals help explain how methane bonds?
    a. sp^3 orbitals
    b. sp orbitals
    c. pd^3 orbitals
    d. df^3 orbitals

    ANS: A   DIF: I   OBJ: 6-5.3

128. Hybridization helps explain molecular bonding when the valence electrons on the uncombined atoms
    a. total more than 8.
    b. are in orbitals with different shapes.
    c. total less than 3.
    d. are not available for bonding.

    ANS: B   DIF: I   OBJ: 6-5.3

129. Four hybrid sp^3 orbitals are formed from
    a. two s orbitals and two p orbitals.
    b. an s orbital and a p orbital.
    c. three s orbitals and one p orbital.
    d. one s orbital and three p orbitals.

    ANS: D   DIF: II   OBJ: 6-5.3

130. Dipole-dipole forces are considered the most important forces in polar substances because the London dispersion forces
    a. act only in nonpolar substances.
    b. are usually much weaker than the dipole-dipole forces.
    c. are too unpredictable.
    d. act only in solids.

    ANS: B   DIF: II   OBJ: 6-5.4

131. The strength of London dispersion forces between molecules depends on
    a. only the number of electrons in the molecule.
    b. only the number of protons in the molecule.
    c. both the number of electrons in the molecule and the mass of the molecule.
    d. both the number of electrons and the number of neutrons in the molecule.

    ANS: C   DIF: II   OBJ: 6-5.4

132. The strong forces of attraction between the positive and negative regions of molecules are called
    a. dipole-dipole forces.
    b. London forces.
    c. lattice forces.
    d. orbital forces.

    ANS: A   DIF: I   OBJ: 6-5.4

133. The intermolecular attraction between a hydrogen atom bonded to a strongly electronegative atom and the unshared pair of electrons on another strongly electronegative atom is called
    a. electron affinity.
    b. covalent bonding.
    c. hydrogen bonding.
    d. electronegativity.

    ANS: C   DIF: I   OBJ: 6-5.4

134. The weak intermolecular forces resulting from instantaneous and induced dipoles are called
    a. London dispersion forces.
    b. dipole-dipole forces.
    c. hydrogen forces.
    d. polar covalent bonding.

    ANS: A   DIF: I   OBJ: 6-5.4

135. Compared with molecular bonds, the strength of intermolecular forces is
    a. weaker.
    b. stronger.
    c. about the same.
    d. too variable to compare.

    ANS: A   DIF: I   OBJ: 6-5.4

136. The equal but opposite charges present in the two regions of a polar molecule create a(n)
    a. electron sea.
    b. dipole.
    c. crystal lattice.
    d. ionic bond.

    ANS: B   DIF: I   OBJ: 6-5.4

137. That the boiling point of water (H_2O) is higher than the boiling point of hydrogen sulfide (H_2S) is partially explained by
    a. London forces.
    b. covalent bonding.
    c. ionic bonding.
    d. hydrogen bonding.

    ANS: D   DIF: II   OBJ: 6-5.4

138. The following molecules contain polar bonds. The only polar molecule is
    a. CCl_4.
    b. CO_2.
    c. NH_3.
    d. CH_4.

    ANS: C   DIF: III   OBJ: 6-5.5

139. The following molecules contain polar bonds. The only nonpolar molecule is
    a. HCl.
    b. H_2O.
    c. CO_2.
    d. NH_3.

    ANS: C   DIF: III   OBJ: 6-5.5

140. A polar molecule contains
    a. ions.
    b. a region of positive charge and a region of negative charge.
    c. only London forces.
    d. no bonds.

    ANS: B   DIF: I   OBJ: 6-5.5

141. A molecule of hydrogen chloride is polar because
    a. it is composed of ions.
    b. it is magnetic.
    c. it contains metallic bonds.
    d. the chlorine attracts the shared electrons more strongly than does the hydrogen atom.

    ANS: D   DIF: II   OBJ: 6-5.5

142. When a polar molecule attracts the electron in a nonpolar molecule,
    a. a dipole is induced.   c. an ionic bond forms.
    b. a crystal lattice forms.   d. a Lewis structure forms.

    ANS: A   DIF: I   OBJ: 6-5.5

143. Iodine monochloride (ICl) has a higher boiling point than bromine (Br_2) partly because iodine monochloride is a(n)
    a. nonpolar molecule.   c. crystal.
    b. ion.   d. polar molecule.

    ANS: D   DIF: II   OBJ: 6-5.5

SHORT ANSWER

1. Why do most atoms form chemical bonds?

    ANS:
    Atoms form chemical bonds to establish a more-stable arrangement. As independent particles, they are at high potential energy. By bonding, they decrease their potential energy, thus becoming more stable.

    DIF: II   OBJ: 6-1.2

2. Explain why scientists use resonance structures to represent some molecules.

    ANS:
    Resonance structures picture the bonding in molecules that cannot be correctly pictured with a single Lewis structure.

    DIF: II   OBJ: 6-2.6

3. Differentiate between an ionic compound and a molecular compound.

ANS:
Atoms in a molecular compound share electrons to achieve stability. Atoms in an ionic compound gain or lose electrons to form ions.

DIF: II   OBJ: 6-3.1

4. Explain what is meant by the lattice energy of BeF_2.

ANS:
The lattice energy of BeF_2 is the energy released when 1 mol of solid BeF_2 is formed from beryllium ions and fluoride ions.

DIF: II   OBJ: 6-3.3

PROBLEM

1. Draw a Lewis structure for the oxalate ion, $C_2O_4^{2-}$.

ANS:
2 C atoms with 4 electrons $\Longrightarrow$ $2 \times 4 = 8$
4 O atoms with 6 electrons $\Longrightarrow$ $4 \times 6 = 24$
$8 + 24 = 32$ valence electrons
$2-$ charge $\Longrightarrow$ 2 extra electrons
$32 + 2 = 34$ electrons total

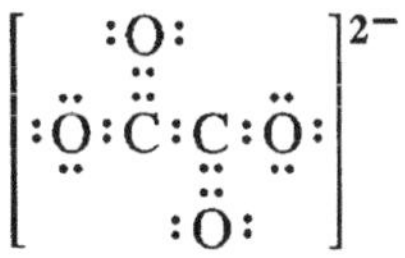

DIF: III   OBJ: 6-3.5

2. Draw a Lewis structure for the ammonium ion, NH_4^+.

ANS:

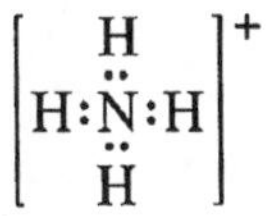

Ammonium ion

DIF: III   OBJ: 6-3.5

3. Draw a Lewis structure for the nitrate ion, NO_3^-.

ANS:

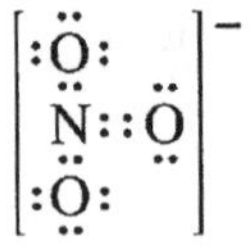

$$\left[\begin{array}{c} :\ddot{O}: \\ \ddot{N}::\ddot{O} \\ :\ddot{O}: \end{array}\right]^{-}$$

Nitrate ion

DIF: III   OBJ: 6-3.5

4. Draw a Lewis structure for the sulfate ion, SO_4^{2-}.

ANS:

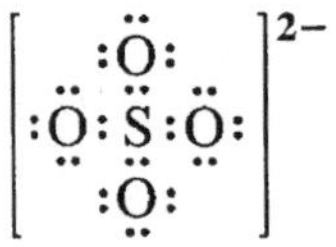

$$\left[\begin{array}{c} :\ddot{O}: \\ :\ddot{O}:\ddot{S}:\ddot{O}: \\ :\ddot{O}: \end{array}\right]^{2-}$$

Sulfate ion

DIF: III   OBJ: 6-3.5

5. Draw a Lewis structure for the phosphate ion, PO_4^{3-}.

ANS:

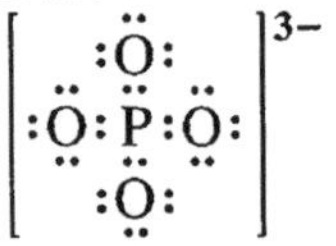

$$\left[\begin{array}{c} :\ddot{O}: \\ :\ddot{O}:\ddot{P}:\ddot{O}: \\ :\ddot{O}: \end{array}\right]^{3-}$$

Phosphate ion

DIF: III   OBJ: 6-3.5

ESSAY

1. List the six basic steps used in drawing Lewis structures.

ANS:
1. Determine the type and number of atoms in the molecule.

2. Write the electron-dot notation for each type of atom in the molecule.

3. Determine the total number of valence electrons in the atoms to be combined.

4. Arrange the atoms to form a skeleton structure for the molecule. If carbon is present, it is the central atom. If not, the least-electronegative atom is central.

5. Add unshared pairs of electrons so that each hydrogen atom shares a pair of electrons and each other nonmetal is surrounded by eight electrons.

6. Count the electrons in the structure to check that the number of valence electrons used equals the number available.

DIF: II   OBJ: 6-2.4

MULTIPLE CHOICE

1. A chemical formula includes the symbols of the elements in the compound and subscripts that indicate
   a. the number of moles in each element.
   b. how many atoms or ions of each type are combined in the simplest unit.
   c. the formula mass.
   d. the charges on the elements or ions.

   ANS: B   DIF: I   OBJ: 7-1.1

2. A chemical formula for a molecular compound represents the composition of
   a. a molecule.   c. the ions that make up the compound.
   b. an atom.   d. the crystal lattice.

   ANS: A   DIF: I   OBJ: 7-1.1

3. How many atoms of fluorine are present in a molecule of carbon tetrafluoride, CF_4?
   a. 1   c. 4
   b. 2   d. 5

   ANS: C   DIF: II   OBJ: 7-1.1

4. Changing a subscript in a correctly written chemical formula
   a. changes the number of moles represented by the formula.
   b. changes the charges on the other ions in the compound.
   c. changes the formula so that it no longer represents that compound.
   d. has no effect on the formula.

   ANS: C   DIF: I   OBJ: 7-1.1

5. The formula for carbon dioxide, CO_2, can represent
   a. one molecule of carbon dioxide.   c. one molar mass of carbon dioxide.
   b. 1 mol of carbon dioxide molecules.   d. all of the above.

   ANS: D   DIF: II   OBJ: 7-1.1

6. Which formula does NOT represent a molecule?
   a. H_2O (water)   c. CO_2 (carbon dioxide)
   b. NH_3 (ammonia)   d. NaCl (table salt)

   ANS: D   DIF: II   OBJ: 7-1.1

7. What is the formula for zinc fluoride?
   a. ZnF   c. Zn_2F
   b. ZnF_2 d. Zn_2F_3

   ANS: B   DIF: III   OBJ: 7-1.2

8. What is the formula for the compound formed by calcium ions and chloride ions?
   a. $CaCl$
   b. Ca_2Cl
   c. $CaCl_3$
   d. $CaCl_2$

   ANS: D   DIF: III   OBJ: 7-1.2

9. What is the formula for the compound formed by lead(II) ions and chromate ions?
   a. $PbCrO_4$
   b. Pb_2CrO_4
   c. $Pb_2(CrO_4)_3$
   d. $Pb(CrO_4)_2$

   ANS: A   DIF: III   OBJ: 7-1.2

10. What is the formula for aluminum sulfate?
    a. $AlSO_4$
    b. Al_2SO_4
    c. $Al_2(SO_4)_3$
    d. $Al(SO_4)_3$

    ANS: C   DIF: III   OBJ: 7-1.2

11. What is the formula for tin(IV) chromate?
    a. $Sn(CrO_4)_4$
    b. $Sn_2(CrO_4)_2$
    c. $Sn_2(CrO_4)_4$
    d. $Sn(CrO_4)_2$

    ANS: D   DIF: III   OBJ: 7-1.2

12. What is the formula for barium hydroxide?
    a. $BaOH$
    b. $BaOH_2$
    c. $Ba(OH)_2$
    d. $Ba(OH)$

    ANS: C   DIF: III   OBJ: 7-1.2

13. Name the compound $Ni(ClO_3)_2$.
    a. nickel chlorate
    b. nickel chloride
    c. nickel chlorite
    d. nickel peroxide

    ANS: A   DIF: III   OBJ: 7-1.3

14. Name the compound $Zn_3(PO_4)_2$.
    a. zinc potassium oxide
    b. trizinc polyoxide
    c. zinc phosphate
    d. zinc phosphite

    ANS: C   DIF: III   OBJ: 7-1.3

15. Name the compound $Hg_2(NO_3)_2$.
    a. mercury(II) nitrate
    b. dimercury dinitrate
    c. mercury(I) nitrate
    d. mercuric nitrate

    ANS: C   DIF: III   OBJ: 7-1.3

16. Name the compound $KClO_3$.
    a. potassium chloride
    b. potassium trioxychlorite
    c. potassium chlorate
    d. hypochlorite

    ANS: C   DIF: III   OBJ: 7-1.3

17. Name the compound $Fe(NO_2)_2$.
    a. iron(II) nitrate
    b. iron(II) nitrite
    c. ferric nitrate
    d. ferrous nitride

    ANS: B   DIF: III   OBJ: 7-1.3

18. Name the compound $CuCO_3$.
    a. copper(I) carbonate
    b. cupric trioxycarbide
    c. cuprous carbide
    d. copper(II) carbonate

    ANS: D   DIF: III   OBJ: 7-1.3

19. What is the name of $Sn_3(PO_4)_4$ under the Stock system of nomenclature?
    a. stannous phosphate
    b. tin(IV) phosphate
    c. tin(III) phosphate
    d. tin(II) phosphate

    ANS: B   DIF: II   OBJ: 7-1.4

20. What is the name of $Cr_2(SO_4)_3$ under the Stock system of nomenclature?
    a. chromium(II) sulfate
    b. chromic sulfate
    c. chromium(III) sulfate
    d. chromous sulfate

    ANS: C   DIF: II   OBJ: 7-1.4

21. What is the metallic ion in copper(II) chloride?
    a. Co^{2+}
    b. Cl^{2-}
    c. Cu^{2+}
    d. Cl^-

    ANS: C   DIF: II   OBJ: 7-1.4

22. Using the Stock system, name the compound PbO.
    a. plumbous oxide
    b. lead oxide
    c. potassium oxide
    d. lead(II) oxide

    ANS: D   DIF: II   OBJ: 7-1.4

23. Name the compound CF_4.
    a. calcium fluoride
    b. carbon fluoride
    c. carbon tetrafluoride
    d. monocarbon quadrafluoride

    ANS: C   DIF: II   OBJ: 7-1.5

24. Name the compound SiO_2.
    a. silver oxide
    b. silicon oxide
    c. silicon dioxide
    d. monosilicon dioxide

    ANS: C   DIF: II   OBJ: 7-1.5

25. Name the compound N_2O_4.
    a. sodium tetroxide
    b. dinitrogen tetroxide
    c. nitrous oxide
    d. binitrogen oxide

    ANS: B   DIF: II   OBJ: 7-1.5

26. Name the compound SO_3.
    a. sulfur trioxide
    b. silver trioxide
    c. selenium trioxide
    d. sodium trioxide

    ANS: A   DIF: II   OBJ: 7-1.5

27. Name the compound N_2O_5.
    a. dinickel pentoxide
    b. dinitrogen pentoxide
    c. neon oxide
    d. nitric oxide

    ANS: B   DIF: II   OBJ: 7-1.5

28. Which compound's name includes the Greek numerical prefixes di- and tri-?
    a. Fe_2O_3
    b. $Ca_3(PO_4)_2$
    c. N_2O_3
    d. Al_2S_3

    ANS: C   DIF: II   OBJ: 7-1.5

29. Name the compound N_2O_3.
    a. dinitrogen oxide
    b. nitrogen trioxide
    c. nitric oxide
    d. dinitrogen trioxide

    ANS: D   DIF: II   OBJ: 7-1.5

30. What is the formula for nitrogen monoxide?
    a. N_2O
    b. NOO
    c. NO
    d. N_2O_2

    ANS: C   DIF: II   OBJ: 7-1.6

31. What is the formula for silicon dioxide?
    a. SO_2
    b. SiO_2
    c. Si_2O
    d. S_2O

    ANS: B   DIF: II   OBJ: 7-1.6

32. What is the formula for nitrogen trifluoride?
    a. NiF_3
    b. NF_3
    c. N_3F
    d. Ni_3F

    ANS: B   DIF: II   OBJ: 7-1.6

33. What is the formula for dinitrogen trioxide?
    a. Ni_2O_3
    b. NO_3
    c. N_2O_6
    d. N_2O_3

    ANS: D   DIF: II   OBJ: 7-1.6

34. What is the formula for sulfur dichloride?
    a. $NaCl_2$
    b. SCl_2
    c. S_2Cl
    d. S_2Cl_2

    ANS: B   DIF: II   OBJ: 7-1.6

35. What is the formula for diphosphorous pentoxide?
    a. P_2PeO_5
    b. PO_5
    c. P_2O_4
    d. P_2O_5

    ANS: D   DIF: II   OBJ: 7-1.6

36. What is the formula for carbon disulfide?
    a. CaS_2
    b. CS_2
    c. S_2C
    d. SC_2

    ANS: B   DIF: II   OBJ: 7-1.6

37. The oxidation number of fluorine is
    a. always 0.
    b. −1 in all compounds
    c. +1 in all compounds.
    d. equal to the positive charge of all the metal ions in a compound.

    ANS: D   DIF: II   OBJ: 7-2.1

38. What is the oxidation number of oxygen in most compounds?
    a. −8
    b. −2
    c. 0
    d. +1

    ANS: B   DIF: II   OBJ: 7-2.1

39. What is the oxidation number of an uncombined element?
    a. −1
    b. 0
    c. +1
    d. 8

    ANS: B   DIF: II   OBJ: 7-2.1

40. In a compound, the algebraic sum of the oxidation numbers of all atoms equals
    a. 0.
    b. 1.
    c. 8.
    d. the charge on the compound.

    ANS: A   DIF: II   OBJ: 7-2.1

41. What is the oxidation number of hydrogen in compounds containing metals?
    a. −1
    b. 0
    c. +1
    d. It is equal to the charge on the metal ion.

    ANS: A   DIF: II   OBJ: 7-2.1

42. What is the oxidation number of oxygen in peroxides?
    a. −2
    b. −1
    c. 0
    d. +2

    ANS: B   DIF: II   OBJ: 7-2.1

43. In a polyatomic ion, the algebraic sum of the oxidation numbers of all atoms is equal to
   a.   0.
   b.   the number of atoms in the ion.
   c.   10.
   d.   the charge of the ion.

   ANS: D   DIF: II   OBJ: 7-2.1

44. What is the oxidation number of hydrogen in most compounds?
   a.   −1
   b.   0
   c.   +1
   d.   It is equal to the algebraic sum of the oxidation numbers of the nonmetals.

   ANS: C   DIF: II   OBJ: 7-2.1

45. What is the oxidation number of oxygen in H_2O_2?
   a.   −2
   b.   −1
   c.   +2
   d.   +4

   ANS: B   DIF: III   OBJ: 7-2.2

46. What is the oxidation number of carbon in CI_4?
   a.   −4
   b.   +1
   c.   +4
   d.   +5

   ANS: C   DIF: III   OBJ: 7-2.2

47. What is the oxidation number of hydrogen in KH?
   a.   −1
   b.   0
   c.   +1
   d.   +2

   ANS: A   DIF: III   OBJ: 7-2.2

48. What is the oxidation number of hydrogen in H_2O?
   a.   0
   b.   +1
   c.   +2
   d.   +3

   ANS: B   DIF: III   OBJ: 7-2.2

49. What is the oxidation number of magnesium in MgO?
   a.   −1
   b.   0
   c.   +1
   d.   +2

   ANS: D   DIF: III   OBJ: 7-2.2

50. What is the oxidation number of sulfur in SO_2?
   a.   0
   b.   +1
   c.   +2
   d.   +4

   ANS: D   DIF: III   OBJ: 7-2.2

51. What is the oxidation number of sulfur in H_2SO_4?
    a. −2   c. +4
    b. 0   d. +6

    ANS: D   DIF: III   OBJ: 7-2.2

52. What is the oxidation number of oxygen in CO_2?
    a. −4   c. 0
    b. −2   d. +4

    ANS: B   DIF: III   OBJ: 7-2.2

53. Name the compound N_2O_2 using the Stock system.
    a. dinitrogen monoxide   c. nitrogen(II) oxide
    b. nitrous oxide   d. nitrogen oxide(II)

    ANS: C   DIF: III   OBJ: 7-2.3

54. Name the compound SO_2 using the Stock system.
    a. sulfur(II) oxide   c. sulfur dioxide
    b. sulfur(IV) oxide   d. sulfuric oxide

    ANS: B   DIF: III   OBJ: 7-2.3

55. Name the compound CCl_4 using the Stock system.
    a. carbon(IV) chloride   c. carbon chloride
    b. carbon tetrachloride   d. carbon hypochlorite

    ANS: A   DIF: III   OBJ: 7-2.3

56. Name the compound H_2O using the Stock system.
    a. water   c. hydrogen(I) oxide
    b. hydrogen dioxide   d. hydrogen(II) oxide

    ANS: C   DIF: III   OBJ: 7-2.3

57. Name the compound CO_2 using the Stock system.
    a. carbon(IV) oxide   c. monocarbon dioxide
    b. carbon dioxide   d. carbon oxide

    ANS: A   DIF: III   OBJ: 7-2.3

58. Name the compound N_2O_5 using the Stock system.
    a. nitrogen(II) oxide   c. nitrogen(VII) oxide
    b. nitrogen(V) oxide   d. nitrogen pentoxide

    ANS: B   DIF: III   OBJ: 7-2.3

59. Name the compound SiO_2 using the Stock system.
    a. silica   c. silicon dioxide
    b. silicon oxide   d. silicon(IV) oxide

    ANS: D   DIF: III   OBJ: 7-2.3

60. Name the compound PBr_5 using the Stock system.
   a. potassium hexabromide
   b. phosphorus(V) pentabromide
   c. phosphorus(V) bromide
   d. phosphoric acid

   ANS: C   DIF: III   OBJ: 7-2.3

61. The molar mass of an element is the mass of one
   a. atom of the element.
   b. liter of the element.
   c. gram of the element.
   d. mole of the element.

   ANS: D   DIF: I   OBJ: 7-3.1

62. What is the sum of the atomic masses of all the atoms in a formula for a compound?
   a. molecular mass
   b. formula mass
   c. atomic mass
   d. actual mass

   ANS: B   DIF: I   OBJ: 7-3.1

63. What is the formula mass of magnesium chloride, $MgCl_2$?
   a. 46 amu
   b. 59.763 amu
   c. 95.211 amu
   d. 106.354 amu

   ANS: C   DIF: III   OBJ: 7-3.1

64. What is the formula mass of ethyl alcohol, C_2H_5OH?
   a. 30.328 amu
   b. 33.271 amu
   c. 45.061 amu
   d. 46.069 amu

   ANS: D   DIF: III   OBJ: 7-3.1

65. What is the formula mass of $(NH_4)_2SO_4$?
   a. 114.09 amu
   b. 118.34 amu
   c. 128.06 amu
   d. 132.13 amu

   ANS: D   DIF: III   OBJ: 7-3.1

66. The molar mass of MgI_2 is
   a. the sum of the masses of 1 mol of Mg and 2 mol of I.
   b. the sum of the masses of 1 mol of Mg and 1 mol of I.
   c. the sum of the masses of 2 mol of Mg and 2 mol of I.
   d. impossible to calculate.

   ANS: A   DIF: II   OBJ: 7-3.1

67. The molar mass of NO_2 is 46.01 g/mol. How many moles of NO_2 are present in 114.95 g?
   a. 0.4003 mol
   b. 1.000 mol
   c. 2.498 mol
   d. 114.95 mol

   ANS: C   DIF: III   OBJ: 7-3.2

68. The molar mass of CCl_4 is 153.81 g/mol. How many grams of CCl_4 are needed to have 5.000 mol?
   a. 5 g
   b. 30.76 g
   c. 769.0 g
   d. 796.05 g

   ANS: C   DIF: III   OBJ: 7-3.2

69. The molar mass of H_2O is 18.015 g/mol. How many grams of H_2O are present in 0.20 mol?
   a. 0.2 g
   b. 3.6 g
   c. 35.9 g
   d. 89.9 g

   ANS: B   DIF: III   OBJ: 7-3.2

70. The molar mass of LiF is 25.94 g/mol. How many moles of LiF are present in 10.37 g?
   a. 0.3998 mol
   b. 1.333 mol
   c. 2.500 mol
   d. 36.32 mol

   ANS: A   DIF: III   OBJ: 7-3.2

71. The molar mass of CS_2 is 76.14 g/mol. How many grams of CS_2 are present in 10.00 mol?
   a. 0.13 g
   b. 7.614 g
   c. 10.00 g
   d. 761.4 g

   ANS: D   DIF: III   OBJ: 7-3.2

72. The molar mass of NH_3 is 17.03 g/mol. How many moles of NH_3 are present in 107.1 g?
   a. 0.1623 mol
   b. 3.614 mol
   c. 6.289 mol
   d. 107.1 mol

   ANS: C   DIF: III   OBJ: 7-3.2

73. How many Cl^- ions are present in 2.00 mol of KCl?
   a. 12.04×10^{23}
   b. 6.02×10^{24}
   c. 2.00
   d. 0.5

   ANS: A   DIF: III   OBJ: 7-3.3

74. How many OH^- ions are present in 3.00 mol of $Ca(OH)_2$?
   a. 3.00
   b. 6.00
   c. 3.61×10^{24}
   d. 2.06×10^{23}

   ANS: C   DIF: III   OBJ: 7-3.3

75. How many oxygen atoms are there in 0.5 mol of CO_2?
   a. 6.02×10^{23}
   b. 3.01×10^{23}
   c. 15.9994
   d. 11

   ANS: A   DIF: III   OBJ: 7-3.3

76. How many Mg^{2+} ions are found in 1.00 mol of MgO?
    a. 3.01×10^{23}
    b. 6.02×10^{23}
    c. 12.04×10^{23}
    d. 6.02×10^{25}

    ANS: B   DIF: III   OBJ: 7-3.3

77. If 0.500 mol of Na^+ combines with 0.500 mol of Cl^- to form NaCl, how many formula units of NaCl are present?
    a. 3.01×10^{23}
    b. 6.02×10^{23}
    c. 6.02×10^{24}
    d. 1

    ANS: A   DIF: III   OBJ: 7-3.3

78. What is the percentage composition of CF_4?
    a. 20% C, 80% F
    b. 13.6% C, 86.4% F
    c. 16.8% C, 83.2% F
    d. 81% C, 19% F

    ANS: B   DIF: III   OBJ: 7-3.4

79. What is the percentage composition of CO?
    a. 50% C, 50% O
    b. 12% C, 88% O
    c. 25% C, 75% O
    d. 43% C, 57% O

    ANS: D   DIF: III   OBJ: 7-3.4

80. What is the percentage composition of $CuCl_2$?
    a. 33% Cu, 66% Cl
    b. 50% Cu, 50% Cl
    c. 65.50% Cu, 34.50% Cl
    d. 47.263% Cu, 52.737% Cl

    ANS: D   DIF: III   OBJ: 7-3.4

81. The percentage composition of sulfur in SO_2 is about 50%. What is the percentage of oxygen in this compound?
    a. 25%
    b. 50%
    c. 75%
    d. 90%

    ANS: B   DIF: II   OBJ: 7-3.4

82. What is the percentage composition of OH^- in $Ca(OH)_2$?
    a. 45.9%
    b. 66.6%
    c. 75%
    d. 90.1%

    ANS: A   DIF: III   OBJ: 7-3.4

83. What is the percentage composition of chlorine in NaCl?
    a. 35.45%
    b. 50%
    c. 60.7%
    d. 64.5%

    ANS: C   DIF: III   OBJ: 7-3.4

84. What is the percentage composition of oxygen in H_2O?
    a. 15.99%   c. 88.8%
    b. 33%   d. 99.8%

    ANS: C   DIF: III   OBJ: 7-3.4

85. A formula that shows the simplest whole-number ratio of the atoms in a compound is the
    a. molecular formula.   c. structural formula.
    b. ideal formula.   d. empirical formula.

    ANS: D   DIF: I   OBJ: 7-4.1

86. The empirical formula is always the accepted formula for a(n)
    a. atom.   c. molecular compound.
    b. molecule.   d. ionic compound.

    ANS: D   DIF: I   OBJ: 7-4.1

87. The empirical formula for a compound shows the symbols of the elements with subscripts indicating the
    a. actual numbers of atoms in a molecule.
    b. number of moles of the compound in 100 g.
    c. smallest whole-number ratio of the atoms.
    d. atomic masses of each element.

    ANS: C   DIF: I   OBJ: 7-4.1

88. The empirical formula may not represent the actual composition of a unit of a(n)
    a. ionic compound.   c. atom.
    b. molecular compound.   d. crystal.

    ANS: B   DIF: I   OBJ: 7-4.1

89. What is the empirical formula for a compound that is 31.9% potassium, 28.9% chlorine, and 39.2% oxygen?
    a. $KClO_2$ c. $K_2Cl_2O_3$
    b. $KClO_3$ d. $K_2Cl_2O_5$

    ANS: B   DIF: III   OBJ: 7-4.2

90. What is the empirical formula for a compound that is 43.6% phosphorus and 56.4% oxygen?
    a. P_3O_7 c. P_2O_3
    b. PO_3 d. P_2O_5

    ANS: D   DIF: III   OBJ: 7-4.2

91. What is the empirical formula for a compound that is 53.3% O and 46.7% Si?
    a. SiO c. Si_2O
    b. SiO_2 d. Si_2O_3

    ANS: B   DIF: III   OBJ: 7-4.2

92. What is the empirical formula for a compound that is 7.9% Li and 92.1% Br?
    a.   LiBr
    b. $LiBr_2$
    c. $LiBr_3$
    d. $LiBr_4$

    ANS: A   DIF: III   OBJ: 7-4.2

93. A compound contains 259.2 g of F and 40.8 g of C. What is the empirical formula for this compound?
    a. CF_4
    b. C_4F
    c.   CF
    d. CF_2

    ANS: A   DIF: III   OBJ: 7-4.2

94. A compound contains 64 g of O and 4 g of H. What is the empirical formula for this compound?
    a. H_2O
    b. H_2O_2
    c. HO_2
    d.   HO

    ANS: D   DIF: III   OBJ: 7-4.2

95. What is the empirical formula for a compound that is 36.1% Ca and 63.9% Cl?
    a.   CaCl
    b. Ca_2Cl
    c. $CaCl_2$
    d. Ca_2Cl_2

    ANS: C   DIF: III   OBJ: 7-4.2

96. A compound contains 27.3 g of C and 72.7 g of O. What is the empirical formula for this compound?
    a.   CO
    b. CO_2
    c. C_2O
    d. C_2O_4

    ANS: B   DIF: III   OBJ: 7-4.2

97. To find the molecular formula from the empirical formula, one must determine the compound's
    a.   density.
    b.   formula mass.
    c.   structural formula.
    d.   crystal lattice.

    ANS: B   DIF: II   OBJ: 7-4.3

98. The empirical formula and the formula mass of a compound are needed to determine the compound's
    a.   molecular formula.
    b.   bond energy.
    c.   lattice structure.
    d.   toxicity.

    ANS: A   DIF: I   OBJ: 7-4.3

99. A molecular compound has the empirical formula XY_3. Which of the following is a possible molecular formula?
    a. X_2Y_3
    b. XY_4
    c. X_2Y_5
    d. X_2Y_6

    ANS: D   DIF: II   OBJ: 7-4.3

100. The molecular formula for vitamin C is $C_6H_8O_6$. What is the empirical formula?
   a. CHO
   b. CH_2O
   c. $C_3H_4O_3$
   d. $C_2H_4O_2$

   ANS: C   DIF: III   OBJ: 7-4.3

101. Of the following molecular formulas for hydrocarbons, which is an empirical formula?
   a. CH_4
   b. C_2H_2
   c. C_3H_6
   d. C_4H_{10}

   ANS: A   DIF: III   OBJ: 7-4.3

102. A compound's empirical formula is C_2H_5. If the formula mass is 58 amu, what is the molecular formula?
   a. C_3H_6
   b. C_4H_{10}
   c. C_5H_8
   d. C_5H_{15}

   ANS: B   DIF: III   OBJ: 7-4.4

103. A compound's empirical formula is N_2O_5. If the formula mass is 108 amu, what is the molecular formula?
   a. N_2O_5
   b. N_4O_{10}
   c. NO_3
   d. N_2O_4

   ANS: A   DIF: III   OBJ: 7-4.4

104. A compound's empirical formula is CH. If the formula mass is 26 amu, what is the molecular formula?
   a. C_2H_2
   b. CH_3
   c. CH_4
   d. C_4H

   ANS: A   DIF: III   OBJ: 7-4.4

105. A compound's empirical formula is NO_2. If the formula mass is 92 amu, what is the molecular formula?
   a. NO
   b. N_2O_2
   c. NO_4
   d. N_2O_4

   ANS: D   DIF: III   OBJ: 7-4.4

106. A compound's empirical formula is CH_3. If the formula mass is 30 amu, what is the molecular formula?
   a. CH_3
   b. CH_4
   c. C_2H_6
   d. C_3H_9

   ANS: C   DIF: III   OBJ: 7-4.4

107. A compound's empirical formula is HO. If the formula mass is 34 amu, what is the molecular formula?
a. H_2O
b. H_2O_2
c. HO_3
d. H_2O_3

ANS: B   DIF: III   OBJ: 7-4.4

SHORT ANSWER

1. Explain the term empirical formula. What is the empirical formula of strontium bromide, $SrBr_2$?

ANS:
The empirical formula shows the simplest whole-number ratio of the atoms that form a compound. The simplest ratio in which strontium and bromine combine is 1:2, so the empirical formula is $SrBr_2$.

DIF: II   OBJ: 7-4.1

PROBLEM

1. The molar mass of aluminum is 26.98 g/mol and the molar mass of fluorine is 19.00 g/mol. Calculate the molar mass of aluminum trifluoride, AlF_3.

ANS:
83.98 g/mol AlF_3

26.98 g/mol Al + (3 × 19.00 g/mol F) = 89.3 g/mol AlF_3

DIF: III   OBJ: 7-3.1

2. The molar mass of copper is 63.55 g/mol, the molar mass of sulfur is 32.07 g/mol, and the molar mass of oxygen is 16.00 g/mol. Calculate the molar mass of copper(II) sulfate, $CuSO_4$.

ANS:
159.62 g/mol $CuSO_4$

63.55 g/mol Cu + 32.07 g/mol S + (4 × 16.00 g/mol O) = 159.62 g/mol $CuSO_4$

DIF: III   OBJ: 7-3.1

3. The molar mass of iron is 55.85 g/mol, the molar mass of silicon is 28.09 g/mol, and the molar mass of oxygen is 16.00 g/mol. Calculate the molar mass of iron(II) silicate, Fe_2SiO_4.

ANS:
203.79 g/mol Fe_2SiO_4

(2 × 55.85 g/mol Fe) + (28.09 g/mol Si) + (4 × 16.00 g/mol O) = 203.79 g/mol Fe_2SiO_4

DIF: III   OBJ: 7-3.1

4. The molar mass of aluminum is 26.98 g/mol and the molar mass of oxygen is 16.00 g/mol. Determine the molar mass of Al_2O_3.

ANS:
101.96 g/mol Al_2O_3

$(2 \times 26.98 \text{ g/mol Al}) + (3 \times 16.00 \text{ g/mol O}) = 101.96 \text{ g/mol Al}_2O_3$

DIF: III   OBJ: 7-3.1

MULTIPLE CHOICE

1. Knowledge about what products are produced in a chemical reaction is obtained by
   a. inspecting the chemical equation.   c. laboratory analysis.
   b. balancing the chemical equation.   d. writing a word equation.

   ANS: C   DIF: I   OBJ: 8-1.1

2. A chemical reaction has NOT occurred if the products have
   a. the same mass as the reactants.
   b. less total bond energy than the reactants.
   c. more total bond energy than the reactants.
   d. the same chemical properties as the reactants.

   ANS: D   DIF: I   OBJ: 8-1.1

3. Which observation does NOT indicate that a chemical reaction has occurred?
   a. formation of a precipitate   c. evolution of heat and light
   b. production of a gas   d. change in total mass of substances

   ANS: D   DIF: I   OBJ: 8-1.1

4. A solid produced by a chemical reaction in solution that separates from the solution is called
   a. a precipitate.   c. a molecule.
   b. a reactant.   d. the mass of the product.

   ANS: A   DIF: I   OBJ: 8-1.1

5. After the correct formula for a reactant in an equation has been written, the
   a. subscripts are adjusted to balance the equation.
   b. formula should not be changed.
   c. same formula must appear as the product.
   d. symbols in the formula must not appear on the product side of the equation.

   ANS: B   DIF: I   OBJ: 8-1.2

6. In writing an equation that produces hydrogen gas, the correct representation of hydrogen gas is
   a. H.   c. H_2.
   b. 2H.   d. OH.

   ANS: C   DIF: I   OBJ: 8-1.2

7. What is the small whole number that appears in front of a formula in a chemical equation?
   a. a subscript   c. a ratio
   b. a superscript   d. a coefficient

   ANS: D   DIF: I   OBJ: 8-1.2

8. To balance a chemical equation, it may be necessary to adjust the
    a. coefficients.   c. formulas of the products.
    b. subscripts.   d. number of products.

    ANS: A   DIF: I   OBJ: 8-1.2

9. According to the law of conservation of mass, the total mass of the reacting substances is
    a. always more than the total mass of the products.
    b. always less than the total mass of the products.
    c. sometimes more and sometimes less than the total mass of the products.
    d. always equal to the total mass of the products.

    ANS: D   DIF: I   OBJ: 8-1.2

10. A chemical equation is balanced when the
    a. coefficients of the reactants equal the coefficients of the products.
    b. same number of each kind of atom appears in the reactants and in the products.
    c. products and reactants are the same chemicals.
    d. subscripts of the reactants equal the subscripts of the products.

    ANS: B   DIF: I   OBJ: 8-1.2

11. Which word equation represents the reaction that produces water from hydrogen and oxygen?
    a. Water is produced from hydrogen and oxygen.
    b. Hydrogen plus oxygen yields water.
    c. $H_2 + O_2 \rightarrow$ water.
    d. Water can be separated into hydrogen and oxygen.

    ANS: B   DIF: II   OBJ: 8-1.3

12. How would oxygen be represented in the formula equation for the reaction of methane and oxygen to yield carbon dioxide and water?
    a. oxygen   c. O_2
    b. O   d. O_3

    ANS: C   DIF: II   OBJ: 8-1.3

13. Which of the following is a formula equation for the formation of carbon dioxide from carbon and oxygen?
    a. Carbon plus oxygen yields carbon   c. $CO_2 \rightarrow C + O_2$
        dioxide.
    b. $C + O_2 \rightarrow CO_2$ d. $2C + O \rightarrow CO_2$

    ANS: B   DIF: II   OBJ: 8-1.3

14. For the formula equation $2Mg + O_2 \rightarrow 2MgO$, the word equation would begin
    a. Manganese plus oxygen . . .   c. Magnesium plus oxygen . . .
    b. Molybdenum plus oxygen . . .   d. Heat plus oxygen . . .

    ANS: C   DIF: II   OBJ: 8-1.3

15. In an equation, the symbol for a substance in water solution is followed by
    a. (l).
    b. (g).
    c. (aq).
    d. (s).

    ANS: C   DIF: I   OBJ: 8-1.3

16. A chemical formula written over the arrow in a chemical equation signifies
    a. a byproduct.
    b. the formation of a gas.
    c. a catalyst for the reaction.
    d. an impurity.

    ANS: C   DIF: I   OBJ: 8-1.3

17. When the equation $Fe_3O_4 + Al \rightarrow Al_2O_3 + Fe$ is correctly balanced, what is the coefficient of Fe?
    a. 3
    b. 4
    c. 6
    d. 9

    ANS: D   DIF: III   OBJ: 8-1.4

18. Which coefficients correctly balance the formula equation $NH_4NO_2(s) \rightarrow N_2(g) + H_2O(l)$?
    a. 1, 2, 2
    b. 1, 1, 2
    c. 2, 1, 1
    d. 2, 2, 2

    ANS: B   DIF: III   OBJ: 8-1.4

19. Which coefficients correctly balance the formula equation $CaO + H_2O \rightarrow Ca(OH)_2$?
    a. 2, 1, 2
    b. 1, 2, 3
    c. 1, 2, 1
    d. 1, 1, 1

    ANS: D   DIF: III   OBJ: 8-1.4

20. After the first steps in writing an equation, the equation is balanced by
    a. adjusting subscripts to the formula(s).
    b. adjusting coefficients to the smallest whole-number ratio.
    c. changing the products formed.
    d. making the number of reactants equal to the number of products.

    ANS: B   DIF: II   OBJ: 8-1.4

21. The complete balanced equation for the reaction between zinc hydroxide and acetic acid is
    a. $ZnOH + CH_3COOH \rightarrow ZnCH_3COO + H_2O$.
    b. $Zn(OH)_2 + CH_3COOH \rightarrow Zn + 2CO_2 + 3H_2O$.
    c. $Zn(OH)_2 + 2CH_3COOH \rightarrow Zn(CH_3COO)_2 + 2H_2O$.
    d. $Zn(OH)_2 + 2CH_3COOH \rightarrow Zn(CH_3COO)_2 + H_2 + O_2$.

    ANS: C   DIF: III   OBJ: 8-1.4

22. What is the balanced equation for the combustion of sulfur?
    a. $S(s) + O_2(g) \rightarrow SO(g)$
    b. $S(s) + O_2(g) \rightarrow SO_2(g)$
    c. $2S(s) + 3O_2(g) \rightarrow SO_3(s)$
    d. $S(s) + 2O_2(g) \rightarrow SO_4^{2-}(aq)$

    ANS: B   DIF: III   OBJ: 8-1.4

23. Which equation is NOT balanced?
    a. $2H_2 + O_2 \rightarrow 2H_2O$
    b. $4H_2 + 2O_2 \rightarrow 4H_2O$
    c. $H_2 + H_2 + O_2 \rightarrow H_2O + H_2O$
    d. $2H_2 + O_2 \rightarrow H_2O$

    ANS: D   DIF: III   OBJ: 8-1.4

24. In what kind of reaction do two or more substances combine to form a new compound?
    a. decomposition reaction   c. double-replacement reaction
    b. ionic reaction   d. synthesis reaction

    ANS: D   DIF: I   OBJ: 8-2.1

25. The equation $AX \rightarrow A + X$ is the general equation for a
    a. synthesis reaction.   c. combustion reaction.
    b. decomposition reaction.   d. single-replacement reaction.

    ANS: B   DIF: II   OBJ: 8-2.1

26. In what kind of reaction does one element replace a similar element in a compound?
    a. single-replacement reaction   c. decomposition reaction
    b. double-replacement reaction   d. ionic reaction

    ANS: A   DIF: I   OBJ: 8-2.1

27. The equation $AX + BY \rightarrow AY + BX$ is the general equation for a
    a. synthesis reaction.   c. single-replacement reaction.
    b. decomposition reaction.   d. double-replacement reaction.

    ANS: D   DIF: II   OBJ: 8-2.1

28. The equation $A + X \rightarrow AX$ is the general equation for a(n)
    a. combustion reaction.   c. synthesis reaction.
    b. ionic reaction.   d. double-replacement reaction.

    ANS: C   DIF: II   OBJ: 8-2.1

29. In what kind of reaction does a single compound produce two or more simpler substances?
    a. decomposition reaction   c. single-replacement reaction
    b. synthesis reaction   d. ionic reaction

    ANS: A   DIF: I   OBJ: 8-2.1

30. The equation $A + BX \rightarrow AX + B$ is the general equation for a
    a. double-replacement reaction.   c. single-replacement reaction.
    b. decomposition reaction.   d. combustion reaction.

    ANS: C   DIF: II   OBJ: 8-2.1

31. In what kind of reaction do the ions of two compounds exchange places in aqueous solution to form two new compounds?
   a. synthesis reaction
   b. double-replacement reaction
   c. decomposition reaction
   d. combustion reaction

   ANS: B   DIF: I   OBJ: 8-2.1

32. The reaction $2Mg(s) + O_2(g) \rightarrow 2MgO(s)$ is a
   a. synthesis reaction.
   b. decomposition reaction.
   c. single-replacement reaction.
   d. double-replacement reaction.

   ANS: A   DIF: II   OBJ: 8-2.2

33. The reaction $Mg(s) + 2HCl(aq) \rightarrow H_2(g) + MgCl_2(aq)$ is a
   a. composition reaction.
   b. decomposition reaction.
   c. single-replacement reaction.
   d. double-replacement reaction.

   ANS: C   DIF: II   OBJ: 8-2.2

34. The reaction $2HgO(s) \rightarrow 2Hg(l) + O_2(g)$ is a(n)
   a. single-replacement reaction.
   b. synthesis reaction.
   c. ionic reaction.
   d. decomposition reaction.

   ANS: D   DIF: II   OBJ: 8-2.2

35. The reaction $Pb(NO_3)_2(aq) + 2KI(aq) \rightarrow PbI_2(s) + 2KNO_3(aq)$ is a
   a. double-replacement reaction.
   b. synthesis reaction.
   c. decomposition reaction.
   d. combustion reaction.

   ANS: A   DIF: II   OBJ: 8-2.2

36. The reaction $2KClO_3(s) \rightarrow 2KCl(s) + 3O_2(g)$ is a(n)
   a. synthesis reaction.
   b. decomposition reaction.
   c. combustion reaction.
   d. ionic reaction.

   ANS: B   DIF: II   OBJ: 8-2.2

37. The reaction $Cl_2(g) + 2KBr(aq) \rightarrow 2KCl(aq) + Br_2(l)$ is a(n)
   a. synthesis reaction.
   b. ionic reaction.
   c. single-replacement reaction.
   d. combustion reaction.

   ANS: C   DIF: II   OBJ: 8-2.2

38. In one type of synthesis reaction, an element combines with oxygen to yield a(n)
   a. acid.
   b. hydroxide.
   c. oxide.
   d. metal.

   ANS: C   DIF: II   OBJ: 8-2.3

39. The decomposition of a substance by an electric current is called
    a. electrolysis.   c. ionization.
    b. conduction.   d. transformation.

    ANS: A   DIF: I   OBJ: 8-2.3

40. When heated, a metal carbonate decomposes into a metal oxide and
    a. carbon.   c. oxygen.
    b. carbon dioxide.   d. hydrogen.

    ANS: B   DIF: II   OBJ: 8-2.3

41. Oxides of active metals, such as CaO, react with water to produce
    a. metal carbonates.   c. acids.
    b. metal hydrides.   d. metal hydroxides.

    ANS: D   DIF: II   OBJ: 8-2.3

42. An active metal and a halogen react to form a(n)
    a. salt.   c. acid.
    b. hydroxide.   d. oxide.

    ANS: A   DIF: II   OBJ: 8-2.3

43. When a binary compound decomposes, what is produced?
    a. an oxide   c. a tertiary compound
    b. an acid   d. two elements

    ANS: D   DIF: II   OBJ: 8-2.3

44. Many metal hydroxides decompose when heated to yield metal oxides and
    a. metal hydrides.   c. carbon dioxide.
    b. water.   d. an acid.

    ANS: B   DIF: II   OBJ: 8-2.3

45. When a metal chlorate is heated, it decomposes to yield a metal chloride and
    a. a metal oxide.   c. hydrogen.
    b. a metal hydroxide.   d. oxygen.

    ANS: D   DIF: II   OBJ: 8-2.3

46. Some acids, such as carbonic acid, decompose to nonmetal oxides and
    a. water.   c. oxygen.
    b. a salt.   d. peroxide.

    ANS: A   DIF: II   OBJ: 8-2.3

47. When heated, metallic chlorates decompose into
    a. metallic oxides and chlorine.
    b. metallic chlorides and oxygen.
    c. a metal and a compound of chlorine and oxygen.
    d. a metal, chlorine, and oxygen.

    ANS: B   DIF: II   OBJ: 8-2.3

48. In the equation $2Al(s) + 3Fe(NO_3)_2(aq) \rightarrow 3Fe(s) + 2Al(NO_3)_3(aq)$, iron has been replaced by
    a. nitrate.   c. aluminum.
    b. water.   d. nitrogen.

    ANS: C   DIF: II   OBJ: 8-2.4

49. Group 1 metals react with water to produce metal hydroxides and
    a. metal hydroxides.   c. oxygen.
    b. hydrochloric acid.   d. hydrogen.

    ANS: D   DIF: I   OBJ: 8-2.4

50. The replacement of bromine by chlorine in a salt is an example of a single-replacement reaction by
    a. halogens.   c. water.
    b. sodium.   d. electrolysis.

    ANS: A   DIF: II   OBJ: 8-2.4

51. When a slightly soluble solid compound is produced in a double-replacement reaction, a
    a. gas bubbles off.   c. combustion reaction takes place.
    b. precipitate is formed.   d. halogen is produced.

    ANS: B   DIF: I   OBJ: 8-2.4

52. An insoluble gas that forms in a double-replacement reaction in aqueous solution
    a. bubbles out of solution.   c. disassociates into ions.
    b. forms a precipitate.   d. reacts with the water.

    ANS: A   DIF: I   OBJ: 8-2.4

53. In a double-replacement reaction, hydrogen chloride and sodium hydroxide react to produce sodium chloride. Another product is
    a. sodium hydride.   c. water.
    b. potassium chloride.   d. hydrogen gas.

    ANS: C   DIF: II   OBJ: 8-2.4

54. Active metals react with certain acids, such as hydrochloric acid, to yield a metal compound and
    a. oxygen.   c. chlorine.
    b. hydrogen.   d. sodium.

    ANS: B   DIF: I   OBJ: 8-2.4

55. Some metals, such as iron, react with steam to produce hydrogen gas and a
    a. metal hydroxide.   c. metallic acid.
    b. metal hydride.   d. metal oxide.

    ANS: D   DIF: I   OBJ: 8-2.4

56. When potassium reacts with water, one product formed is
    a. hydrogen gas.   c. potassium oxide.
    b. oxygen gas.   d. salt.

    ANS: A   DIF: I   OBJ: 8-2.4

57. A precipitate forms in a double-replacement reaction when
    a. hydrogen gas reacts with a metal.   c. water boils out of the solution.
    b. positive ions combine with negative ions. d. a gas escapes.

    ANS: B   DIF: I   OBJ: 8-2.4

58. The reaction of calcium oxide (CaO) with water yields
    a. calcium and oxygen gas.   c. calcium and a salt.
    b. calcium hydroxide.   d. carbon dioxide and water.

    ANS: B   DIF: III   OBJ: 8-2.5

59. Predict the product of the following reaction: $MgO + CO_2 \rightarrow$
    a. $MgCO_3$ c. $MgC + O_3$
    b. $Mg + CO_3$ d. $MgCO_2 + O$

    ANS: A   DIF: III   OBJ: 8-2.5

60. Magnesium hydroxide decomposes to yield magnesium oxide and
    a. hydrogen.   c. water.
    b. oxygen.   d. salt.

    ANS: C   DIF: III   OBJ: 8-2.5

61. When sodium chlorate ($NaClO_3$) decomposes, the products are
    a. sodium hydroxide and water.   c. sodium and chlorine oxide.
    b. sodium oxide and chlorine.   d. sodium chloride and oxygen.

    ANS: D   DIF: III   OBJ: 8-2.5

62. If chlorine gas is produced by halogen replacement, the other halogen in the reaction must be
    a. bromine.   c. astatine.
    b. iodine.   d. fluorine.

    ANS: D   DIF: III   OBJ: 8-2.5

63. The formulas for the products of the reaction between sodium hydroxide and sulfuric acid are
    a. Na_2SO_4 and H_2O.   c. SI_4 and Na_2O.
    b. $NaSO_4$ and H_2O.   d. $S + O_2$ and Na.

    ANS: A   DIF: III   OBJ: 8-2.5

64. What is the balanced equation when aluminum reacts with copper(II) sulfate?
    a. $Al + Cu_2S \rightarrow Al_2S + Cu$
    c. $Al + CuSO_4 \rightarrow AlSO_4 + Cu$
    b. $2Al + 3CuSO_4 \rightarrow Al_2(SO_4)_3 + 3Cu$
    d. $2Al + Cu_2SO_4 \rightarrow Al_2SO_4 + 2Cu$

    ANS: B   DIF: III   OBJ: 8-2.5

65. The ability of an element to react is the element's
    a. valence.
    c. stability.
    b. activity.
    d. electronegativity.

    ANS: B   DIF: I   OBJ: 8-3.1

66. What is the name of a list of elements arranged according to the ease with which they undergo certain chemical reactions?
    a. reactivity list
    c. activity series
    b. reaction sequence
    d. periodic list

    ANS: C   DIF: I   OBJ: 8-3.1

67. An element in the activity series can replace any element
    a. in the periodic table.
    c. above it on the list.
    b. below it on the list.
    d. in its group.

    ANS: B   DIF: I   OBJ: 8-3.1

68. What can be predicted by using an activity series?
    a. whether a certain chemical reaction will occur
    b. the amount of energy released by a chemical reaction
    c. the electronegativity values of elements
    d. the melting points of elements

    ANS: A   DIF: II   OBJ: 8-3.1

69. The activity series of metals indicates the ease with which metal
    a. atoms gain neutrons.
    c. atoms form covalent bonds.
    b. nuclei fuse.
    d. atoms lose electrons.

    ANS: D   DIF: I   OBJ: 8-3.1

70. If metal X is lower than metal Y in the activity series, then metal X
    a. replaces ions of metal Y in a solution.
    b. loses electrons more readily than does metal Y.
    c. loses electrons less readily than does metal Y.
    d. forms positive ions more readily than does metal Y.

    ANS: C   DIF: II   OBJ: 8-3.1

71. If a certain metal is placed in an ionic solution containing another metal and no reaction occurs, then the metal originally in the solution is
    a. a halogen.
    c. not on the activity series.
    b. higher on the activity series.
    d. unreactive.

    ANS: B   DIF: II   OBJ: 8-3.1

72. Predict what happens when calcium metal is added to a solution of magnesium chloride.
    a. No reaction occurs.
    b. Calcium chloride forms.
    c. Magnesium calcite forms.
    d. Gaseous calcium is produced.

    ANS: B   DIF: III   OBJ: 8-3.2

73. Predict what happens when zinc is added to water.
    a. No reaction occurs.
    b. Steam is produced.
    c. Zinc oxide forms.
    d. Hydrogen is released.

    ANS: A   DIF: III   OBJ: 8-3.2

74. Predict what happens when lead is added to nitric acid.
    a. No reaction occurs.
    b. Oxygen is released.
    c. Lead oxide forms.
    d. Hydrogen is released.

    ANS: D   DIF: III   OBJ: 8-3.2

75. Predict what happens when nickel is added to a solution of potassium chloride.
    a. No reaction occurs.
    b. Nickel chloride forms.
    c. Potassium nickel chloride forms.
    d. Hydrochloric acid forms.

    ANS: A   DIF: III   OBJ: 8-3.2

76. Magnesium bromide + chlorine yield
    a. Mg and BrCl.
    b. MgCl and Br_2.
    c. MgBrCl.
    d. $Mg(Cl)_2$ and Br_2.

    ANS: D   DIF: III   OBJ: 8-3.2

77. Which reaction does NOT occur?
    a. $2HF + Cl_2 \rightarrow F_2 + 2HCl$
    b. $2Na + ZnF_2 \rightarrow 2NaF + Zn$
    c. $Fe + CuCl_2 \rightarrow FeCl_2 + Cu$
    d. $2HCl + Mg \rightarrow MgCl_2 + H_2$

    ANS: A   DIF: III   OBJ: 8-3.2

78. Which reaction can be predicted from the activity series?
    a. $2Cl \rightarrow Cl_2$
    b. $HCl + NaOH \rightarrow NaCl + H_2O$
    c. $2HCl + 2Na \rightarrow 2NaCl + H_2$
    d. $Cl_2 \rightarrow 2Cl$

    ANS: C   DIF: III   OBJ: 8-3.2

SHORT ANSWER

1. When a glass blower shapes molten glass into an ornament, does a chemical reaction occur? Explain.

    ANS:
    A chemical reaction does not occur. Analysis of the molten glass and the glass in the ornament would reveal that both substances have the same chemical properties.

    DIF: II   OBJ: 8-1.1

MULTIPLE CHOICE

1. Which branch of chemistry deals with the mass relationships of elements in compounds and the mass relationships among reactants and products in chemical reactions?
   a. qualitative analysis
   b. entropy
   c. chemical kinetics
   d. stoichiometry

   ANS: D   DIF: I   OBJ: 9-1.1

2. What is the study of the mass relationships of elements in compounds?
   a. reaction stoichiometry
   b. composition stoichiometry
   c. percent yield
   d. Avogadro's principle

   ANS: B   DIF: I   OBJ: 9-1.1

3. What is the study of the mass relationships among reactants and products in a chemical reaction?
   a. reaction stoichiometry
   b. composition stoichiometry
   c. electron configuration
   d. periodic law

   ANS: A   DIF: I   OBJ: 9-1.1

4. Which of the following would be investigated in reaction stoichiometry?
   a. the masses of hydrogen and oxygen in water
   b. the amount of energy released in chemical reactions
   c. the mass of potassium required to produce a known mass of potassium chloride
   d. the types of bonds that break and form when acids react with metals

   ANS: C   DIF: II   OBJ: 9-1.1

5. A determination of the masses and number of moles of sulfur and oxygen in the compound sulfur dioxide would be studied in
   a. reaction stoichiometry.
   b. chemical kinetics.
   c. chemical equilibrium.
   d. composition stoichiometry.

   ANS: D   DIF: I   OBJ: 9-1.1

6. Which of the following would NOT be studied in the branch of chemistry called stoichiometry?
   a. the mole ratio of aluminum and chlorine in aluminum chloride
   b. the amount of energy required to break the ionic bonds in calcium fluoride
   c. the mass of carbon produced when a known mass of sucrose decomposes
   d. the number of moles of hydrogen that react completely with a known quantity of oxygen

   ANS: B   DIF: II   OBJ: 9-1.1

7. A balanced chemical equation allows one to determine the
   a. mole ratio of any two substances in the reaction.
   b. energy released in the reaction.
   c. electron configuration of all elements in the reaction.
   d. mechanism involved in the reaction.

   ANS: A   DIF: I   OBJ: 9-1.2

8. The coefficients in a chemical equation represent the
   a. masses, in grams, of all reactants and products.
   b. relative numbers of moles of reactants and products.
   c. number of atoms in each compound in a reaction.
   d. number of valence electrons involved in the reaction.

   ANS: B   DIF: I   OBJ: 9-1.2

9. Each of the four types of reaction stoichiometry problems requires using a
   a. table of bond energies.   c. Lewis structure.
   b. chart of electron configurations.   d. mole ratio.

   ANS: D   DIF: I   OBJ: 9-1.2

10. If one knows the mole ratio of a reactant and product in a chemical reaction, one can
    a. estimate the energy released in the reaction.
    b. calculate the speed of the reaction.
    c. calculate the mass of the product produced from a known mass of reactant.
    d. decide whether the reaction is reversible.

    ANS: C   DIF: I   OBJ: 9-1.2

11. If one knows the mass and molar mass of reactant A and the molar mass of product D in a
    chemical reaction, one can determine the mass of product D produced by using the
    a. mole ratio of D to A from the chemical equation.
    b. group numbers of the elements of A and D in the periodic table.
    c. estimating bond energies involved in the reaction.
    d. electron configurations of the atoms in A and D.

    ANS: A   DIF: I   OBJ: 9-1.2

12. In the chemical reaction $wA + xB \rightarrow yC + zD$, a comparison of the number of moles of A to the
    number of moles of C would be a(n)
    a. mass ratio.   c. electron ratio.
    b. mole ratio.   d. energy proportion.

    ANS: B   DIF: II   OBJ: 9-1.2

13. In the chemical equation $wA + xB \rightarrow yC + zD$, if one knows the mass of A and the molar masses
    of A, B, C, and D, one can determine
    a. the mass of any of the reactants or products.
    b. the mass of B only.
    c. the total mass of C and D only.
    d. the total mass of A and B only.

    ANS: A   DIF: II   OBJ: 9-1.2

14. In the reaction $N_2 + 3H_2 \rightarrow 2NH_3$, what is the mole ratio of nitrogen to ammonia?
    a. 1:1   c. 1:3
    b. 1:2   d. 2:3

    ANS: B   DIF: III   OBJ: 9-1.3

15. In the reaction $2Al_2O_3 \rightarrow 4Al + 3O_2$, what is the mole ratio of aluminum to oxygen?
    a. 10:6   c. 2:3
    b. 3:4   d. 4:3

    ANS: D   DIF: III   OBJ: 9-1.3

16. In the reaction $2H_2 + O_2 \rightarrow 2H_2O$, what is the mole ratio of oxygen to water?
    a. 1:2   c. 8:1
    b. 2:1   d. 1:4

    ANS: A   DIF: III   OBJ: 9-1.3

17. In the reaction $Ca + Cl_2 \rightarrow CaCl_2$, what is the mole ratio of chlorine to calcium chloride?
    a. 2:3   c. 1:2
    b. 2:1   d. 1:1

    ANS: D   DIF: III   OBJ: 9-1.3

18. In the reaction $Zn + H_2SO_4 \rightarrow ZnSO_4 + H_2$, what is the mole ratio of zinc to sulfuric acid?
    a. 1:6   c. 1:2
    b. 1:1   d. 3:1

    ANS: B   DIF: III   OBJ: 9-1.3

19. In the reaction $C + 2H_2 \rightarrow CH_4$, what is the mole ratio of hydrogen to methane?
    a. 1:1   c. 1:2
    b. 2:1   d. 2:4

    ANS: B   DIF: III   OBJ: 9-1.3

20. In the reaction $N_2 + 3H_2 \rightarrow 2NH_3$, what is the mole ratio of hydrogen to ammonia?
    a. 1:1   c. 3:2
    b. 2:1   d. 6:8

    ANS: C   DIF: III   OBJ: 9-1.3

21. The Haber process for producing ammonia commercially is represented by the equation $N_2(g) + 3H_2(g) \rightarrow 2NH_3(g)$. To completely convert 9.0 mol hydrogen gas to ammonia gas, how many moles of nitrogen gas are required?
    a. 1.0 mol   c. 3.0 mol
    b. 2.0 mol   d. 6.0 mol

    ANS: C   DIF: III   OBJ: 9-2.1

22. In the equation $2KClO_3 \rightarrow 2KCl + 3O_2$, how many moles of oxygen are produced when 3.0 mol of $KClO_3$ decompose completely?
    a. 1.0 mol   c. 3.0 mol
    b. 2.5 mol   d. 4.5 mol

    ANS: D   DIF: III   OBJ: 9-2.1

23. For the reaction $C + 2H_2 \rightarrow CH_4$, how many moles of hydrogen are required to produce 10 mol of methane, CH_4?
   a. 2 mol
   b. 4 mol
   c. 10 mol
   d. 20 mol

   ANS: D   DIF: III   OBJ: 9-2.1

24. For the reaction $2H_2 + O_2 \rightarrow 2H_2O$, how many moles of water can be produced from 6.0 mol of oxygen?
   a. 2.0 mol
   b. 6.0 mol
   c. 12 mol
   d. 18 mol

   ANS: C   DIF: III   OBJ: 9-2.1

25. For the reaction $N_2 + 3H_2 \rightarrow 2NH_3$, how many moles of nitrogen are required to produce 18 mol of ammonia?
   a. 9.0 mol
   b. 18 mol
   c. 27 mol
   d. 36 mol

   ANS: A   DIF: III   OBJ: 9-2.1

26. For the reaction $AgNO_3 + NaCl \rightarrow NaNO_3 + AgCl$, how many moles of silver chloride, $AgCl$, are produced from 7 mol of silver nitrate $AgNO_3$?
   a. 1.0 mol
   b. 2.3 mol
   c. 7.0 mol
   d. 21 mol

   ANS: C   DIF: III   OBJ: 9-2.1

Element	Symbol	Atomic mass
Bromine	Br	79.904
Calcium	Ca	40.078
Carbon	C	12.011
Chlorine	Cl	35.4527
Cobalt	Co	58.933 20
Copper	Cu	63.546
Fluorine	F	18.998 4032
Hydrogen	H	1.007 94
Iodine	I	126.904
Iron	Fe	55.847
Lead	Pb	207.2
Magnesium	Mg	24.3050
Mercury	Hg	200.59
Nitrogen	N	14.006 74
Oxygen	O	15.9994
Potassium	K	39.0983
Sodium	Na	22.989 768
Sulfur	S	32.066

27. For the reaction $2H_2 + O_2 \rightarrow 2H_2O$, how many grams of water are produced from 6.00 mol of hydrogen?
    a. 2.00 g
    b. 6.00 g
    c. 54.0 g
    d. 108 g

    ANS: D   DIF: III   OBJ: 9-2.2

28. For the reaction $2Na + 2H_2O \rightarrow 2NaOH + H_2$, how many grams of sodium hydroxide are produced from 3.0 mol of water?
    a. 40. g
    b. 80. g
    c. 120 g
    d. 240 g

    ANS: C   DIF: III   OBJ: 9-2.2

29. For the reaction $SO_3 + H_2O \rightarrow H_2SO_4$, how many grams of sulfur trioxide are required to produce 4.00 mol of sulfuric acid?
    a. 80.0 g
    b. 160. g
    c. 240. g
    d. 320. g

    ANS: D   DIF: III   OBJ: 9-2.2

30. For the reaction $2Fe + O_2 \rightarrow 2FeO$, how many grams of iron oxide are produced from 8.00 mol of iron?
    a. 71.8 g
    b. 574 g
    c. 712 g
    d. 1310 g

    ANS: B   DIF: III   OBJ: 9-2.2

31. For the reaction $2Na + Cl_2 \rightarrow 2NaCl$, how many grams of chlorine gas are required to react completely with 2.00 mol of sodium?
    a. 35.5 g
    b. 70.9 g
    c. 141.8 g
    d. 212.7 g

    ANS: B   DIF: III   OBJ: 9-2.2

32. For the reaction $2HNO_3 + Mg(OH)_2 \rightarrow Mg(NO_3)_2 + 2H_2O$, how many grams of magnesium nitrate are produced from 8.00 mol of nitric acid, HNO_3?
    a. 148 g
    b. 445 g
    c. 592 g
    d. 818 g

    ANS: C   DIF: III   OBJ: 9-2.2

33. For the reaction $CH_4 + 2O_2 \rightarrow CO_2 + 2H_2O$, how many moles of carbon dioxide are produced from the combustion of 100. g of methane?
    a. 6.23 mol
    b. 10.8 mol
    c. 12.5 mol
    d. 25 mol

    ANS: A   DIF: III   OBJ: 9-2.3

34. For the reaction $HCl + NaOH \rightarrow NaCl + H_2O$, how many moles of hydrochloric acid are required to produce 150. g of water?
    a. 1.50 mol
    b. 4.16 mol
    c. 8.32 mol
    d. 12.2 mol

    ANS: C   DIF: III   OBJ: 9-2.3

35. For the reaction $Pb(NO_3)_2 + 2KI \rightarrow PbI_2 + 2KNO_3$, how many moles of lead iodide are produced from 300. g of potassium iodide?
    a. 0.903 mol
    b. 1.81 mol
    c. 3.61 mol
    d. 11.0 mol

    ANS: A   DIF: III   OBJ: 9-2.3

36. For the reaction $Cl_2 + 2KBr \rightarrow 2KCl + Br_2$, how many moles of potassium chloride are produced from 119 g of potassium bromide?
    a. 0.119 mol
    b. 0.236 mol
    c. 0.581 mol
    d. 1.00 mol

    ANS: D   DIF: III   OBJ: 9-2.3

37. For the reaction $3Fe + 4H_2O \rightarrow Fe_3O_4 + 4H_2$, how many moles of iron oxide are produced from 500 g of iron?
    a. 1 mol
    b. 3 mol
    c. 9 mol
    d. 12 mol

    ANS: B   DIF: III   OBJ: 9-2.3

38. For the reaction $2KlO_3 \rightarrow 2KCl + 3O_2$, how many moles of potassium chlorate are required to produce 250 g of oxygen?
    a.   2.0 mol
    c.   4.9 mol
    b.   4.3 mol
    d.   5.2 mol

    ANS: D   DIF: III   OBJ: 9-2.3

39. For the reaction $Cl_2 + 2KBr \rightarrow 2KCl + Br_2$, how many grams of potassium chloride can be produced from 300. g each of chlorine and potassium bromide?
    a.   98.7 g
    c.   188 g
    b.   111 g
    d.   451 g

    ANS: C   DIF: III   OBJ: 9-2.4

40. For the reaction $2Na + 2H_2O \rightarrow 2NaOH + H_2$, how many grams of hydrogen are produced if 120. g of sodium and 80. g of water are available?
    a.   4.5 g
    c.   80. g
    b.   44 g
    d.   200. g

    ANS: A   DIF: III   OBJ: 9-2.4

41. For the reaction $2Na + Cl_2 \rightarrow 2NaCl$, how many grams of sodium chloride can be produced from 500. g each of sodium and chlorine?
    a.   112 g
    c.   409 g
    b.   319 g
    d.   825 g

    ANS: D   DIF: III   OBJ: 9-2.4

42. For the reaction $SO_3 + H_2O \rightarrow H_2SO_4$, how many grams of sulfuric acid can be produced from 200. g of sulfur trioxide and 100. g of water?
    a.   100. g
    c.   245 g
    b.   200. g
    d.   285 g

    ANS: C   DIF: III   OBJ: 9-2.4

43. For the reaction $2Zn + O_2 \rightarrow 2ZnO$, how many grams of zinc oxide can be produced from 100. g each of zinc and oxygen?
    a.   100. g
    c.   189 g
    b.   124 g
    d.   200. g

    ANS: B   DIF: III   OBJ: 9-2.4

44. For the reaction $SO_3 + H_2O \rightarrow H_2SO_4$, calculate the percent yield if 500. g of sulfur trioxide react with excess water to produce 575 g of sulfuric acid.
    a.   82.7%
    c.   91.2%
    b.   88.3%
    d.   93.9%

    ANS: D   DIF: III   OBJ: 9-3.4

45. For the reaction $Cl_2 + 2KBr \rightarrow 2KCl + Br_2$, calculate the percent yield if 200. g of chlorine react with excess potassium bromide to produce 410. g of bromine.
   a.   73.4%   c.   91.0%
   b.   82.1%   d.   98.9%

   ANS: C   DIF: III   OBJ: 9-3.4

46. For the reaction $CH_4 + 2O_2 \rightarrow 2H_2O + CO_2$, calculate the percent yield of carbon dioxide if 1000. g of methane react with excess oxygen to produce 2300. g of carbon dioxide.
   a.   83.88%   c.   92.76%
   b.   89.14%   d.   96.78%

   ANS: A   DIF: III   OBJ: 9-3.4

47. For the reaction $Mg + 2HCl \rightarrow H_2 + MgCl_2$, calculate the percent yield of magnesium chloride if 100. g of magnesium react with excess hydrochloric acid to yield 330. g of magnesium chloride.
   a.   71.8%   c.   81.6%
   b.   74.3%   d.   84.2%

   ANS: D   DIF: III   OBJ: 9-3.4

48. For the reaction $2Na + 2H_2O \rightarrow 2NaOH + H_2$, calculate the percent yield if 80. g of water react with excess sodium to produce 4.14 g of hydrogen.
   a.   87%   c.   92%
   b.   89%   d.   98%

   ANS: C   DIF: III   OBJ: 9-3.4

49. For the reaction $2Na + Cl_2 \rightarrow 2NaCl$, calculate the percent yield if 200. g of chlorine react with excess sodium to produce 240. g of sodium chloride.
   a.   61.2%   c.   83.4%
   b.   72.8%   d.   88.4%

   ANS: B   DIF: III   OBJ: 9-3.4

50. Which reactant controls the amount of product formed in a chemical reaction?
   a.   excess reactant   c.   composition reactant
   b.   mole ratio   d.   limiting reactant

   ANS: D   DIF: I   OBJ: 9-3.1

51. A chemical reaction involving substances A and B stops when B is completely used. B is the
   a.   excess reactant.   c.   primary reactant.
   b.   limiting reactant.   d.   primary product.

   ANS: B   DIF: II   OBJ: 9-3.1

52. When the limiting reactant in a chemical reaction is completely used, the
   a.   excess reactants begin combining.   c.   reaction speeds up.
   b.   reaction slows down.   d.   reaction stops.

   ANS: D   DIF: I   OBJ: 9-3.1

53. In the reaction A + B → C + D, if the quantity of B is insufficient to react with all of A,
    a. A is the limiting reactant.
    b. B is the limiting reactant.
    c. there is no limiting reactant.
    d. no product can be formed.

    ANS: B   DIF: II   OBJ: 9-3.1

54. To determine the limiting reactant in a chemical reaction, one must know the
    a. available amount of one of the reactants.
    b. amount of product formed.
    c. available amount of each reactant.
    d. speed of the reaction.

    ANS: C   DIF: II   OBJ: 9-3.1

55. To determine the limiting reactant in a chemical reaction involving known masses of A and B, one could first calculate
    a. the mass of 100 mol of A and B.
    b. the masses of all products.
    c. the bond energies of A and B.
    d. the number of moles of B and the number of moles of A available.

    ANS: D   DIF: I   OBJ: 9-3.1

56. After calculating the amount of reactant B required to completely react with A, then comparing that amount with the amount of B available, one can determine the
    a. limiting reactant.
    b. rate of the reaction.
    c. energy released in the reaction.
    d. pathway of the reaction.

    ANS: A   DIF: I   OBJ: 9-3.1

57. What is the ratio of the actual yield to the theoretical yield, multiplied by 100%?
    a. mole ratio
    b. percent yield
    c. Avogadro yield
    d. excess yield

    ANS: B   DIF: I   OBJ: 9-3.3

58. What is the measured amount of a product obtained from a chemical reaction?
    a. mole ratio
    b. percent yield
    c. theoretical yield
    d. actual yield

    ANS: D   DIF: I   OBJ: 9-3.3

59. In most chemical reactions the amount of product obtained is
    a. equal to the theoretical yield.
    b. less than the theoretical yield.
    c. more than the theoretical yield.
    d. more than the percent yield.

    ANS: B   DIF: I   OBJ: 9-3.3

60. What is the maximum possible amount of product obtained in a chemical reaction?
    a. theoretical yield
    b. percent yield
    c. mole ratio
    d. actual yield

    ANS: A   DIF: I   OBJ: 9-3.3

61. A chemist interested in the efficiency of a chemical reaction would calculate the
    a. mole ratio.   c. percent yield.
    b. energy released.   d. rate of reaction.

    ANS: C   DIF: I   OBJ: 9-3.3

62. If the percent yield is equal to 100%, then
    a. the actual yield is greater than the theoretical yield.
    b. the actual yield is equal to the theoretical yield.
    c. the actual yield is less than the theoretical yield.
    d. there was no limiting reactant.

    ANS: B   DIF: I   OBJ: 9-3.3

PROBLEM

Element	Symbol	Atomic mass
Bromine	Br	79.904
Calcium	Ca	40.078
Carbon	C	12.011
Chlorine	Cl	35.4527
Cobalt	Co	58.933 20
Copper	Cu	63.546
Fluorine	F	18.998 4032
Hydrogen	H	1.007 94
Iodine	I	126.904
Iron	Fe	55.847
Lead	Pb	207.2
Magnesium	Mg	24.3050
Mercury	Hg	200.59
Nitrogen	N	14.006 74
Oxygen	O	15.9994
Potassium	K	39.0983
Sodium	Na	22.989 768
Sulfur	S	32.066

1. What mass in grams of sodium hydroxide is produced if 20.0 g of sodium metal reacts with excess water according to the chemical equation $2Na(s) + 2H_2O(l) \rightarrow 2NaOH(aq) + H_2(g)$?

    ANS:
    34.8 g NaOH

$$20.0 \text{ g Na} \times \frac{1 \text{ mol Na}}{22.99 \text{ g Na}} \times \frac{2 \text{ mol NaOH}}{2 \text{ mol Na}} \times \frac{40.00 \text{ g NaOH}}{1 \text{ mol NaOH}} = 34.8 \text{ g NaOH}$$

    DIF: III   OBJ: 9-3.2

2. What mass in grams of 1-chloropropane (C_3H_7Cl) is produced if 400. g of propane react with excess chlorine gas according to the equation $C_3H_8 + Cl_2 \rightarrow C_3H_7Cl + HCl$?

ANS:
712 g C_3H_7Cl

$$400. \text{ g } C_3H_8 \times \frac{1 \text{ mol } C_3H_8}{44.11 \text{ g } C_3H_8} \times \frac{1 \text{ mol } C_3H_7Cl}{1 \text{ mol } C_3H_8} \times \frac{78.55 \text{ g } C_3H_7Cl}{1 \text{ mol } C_3H_7Cl} = 712 \text{ g } C_3H_7Cl$$

DIF: III   OBJ: 9-3.2

3. What mass in grams of hydrogen gas is produced if 20.0 mol of Zn are added to excess hydrochloric acid according to the equation $Zn(s) + 2HCl(aq) \rightarrow ZnCl_2(aq) + H_2(\ )$?

ANS:
40.4 g H_2

$$20.0 \text{ mol } Zn \times \frac{1 \text{ mol } H_2}{1 \text{ mol } Zn} \times \frac{2.02 \text{ g } H_2}{1 \text{ mol } H_2} = 40.4 \text{ g } H_2$$

DIF: III   OBJ: 9-3.2

4. How many grams of ammonium sulfate can be produced if 30.0 mol of H_2SO_4 react with excess NH_3 according to the equation $2NH_3(aq) + H_2SO_4(aq) \rightarrow (NH_4)_2SO_4(aq)$?

ANS:
3960 g $(NH_4)_2SO_4$

$$30.0 \text{ mol } H_2SO_4 \times \frac{1 \text{ mol } (NH_4)_2SO_4}{1 \text{ mol } H_2SO_4} \times \frac{132.17 \text{ g } (NH_4)_2SO_4}{1 \text{ mol } (NH_4)_2SO_4} = 3960 \text{ g } (NH_4)_2SO_4$$

DIF: III   OBJ: 9-3.2

5. How many moles of Ag can be produced if 350. g of Cu are reacted with excess $AgNO_3$ according to the equation $Cu(s) + 2AgNO_3(aq) \rightarrow 2Ag(s) + Cu(NO_3)_2(aq)$?

ANS:
11.0 mol Ag

$$350. \text{ g } Cu \times \frac{1 \text{ mol } Cu}{63.55 \text{ g } Cu} \times \frac{2 \text{ mol } Ag}{1 \text{ mol } Cu} = 11.0 \text{ mol Ag}$$

DIF: III   OBJ: 9-3.2

MULTIPLE CHOICE

1. According to the kinetic-molecular theory, particles of matter
    a. are in constant motion.
    b. have different shapes.
    c. have different colors.
    d. are always fluid.

    ANS: A   DIF: I   OBJ: 10-1.1

2. According to the kinetic-molecular theory, gases condense into liquids because of
    a. gravity.
    b. atmospheric pressure.
    c. forces between molecules.
    d. elastic collisions.

    ANS: C   DIF: I   OBJ: 10-1.1

3. The kinetic-molecular theory explains the behavior of
    a. gases only.
    b. solids and liquids.
    c. liquids and gases.
    d. solids, liquids, and gases.

    ANS: D   DIF: I   OBJ: 10-1.1

4. Which process can be explained by the kinetic-molecular theory?
    a. combustion
    b. oxidation
    c. condensation
    d. replacement reactions

    ANS: C   DIF: II   OBJ: 10-1.1

5. According to the kinetic-molecular theory, which substances are made of particles?
    a. ideal gases only
    b. all gases
    c. all matter
    d. all matter except solids

    ANS: C   DIF: I   OBJ: 10-1.1

6. The kinetic-molecular theory explains the properties of solids, liquids, and gases in terms of the energy of the particles and
    a. gravitational forces.
    b. the forces that act between the particles.
    c. diffusion.
    d. the mass of the particles.

    ANS: B   DIF: I   OBJ: 10-1.1

7. According to the kinetic-molecular theory, particles of matter are in motion in
    a. gases only.
    b. gases and liquids.
    c. solids, liquids, and gases.
    d. solids only.

    ANS: C   DIF: I   OBJ: 10-1.1

8. An ideal gas is an imaginary gas
   a. not made of particles.
   b. that conforms to all of the assumptions of the kinetic theory.
   c. whose particles have zero mass.
   d. made of motionless particles.

   ANS: B   DIF: I   OBJ: 10-1.2

9. Unlike in an ideal gas, in a real gas
   a. all particles move in the same direction.
   b. all particles have the same kinetic energy.
   c. the particles cannot diffuse.
   d. the particles exert attractive forces on each other.

   ANS: D   DIF: I   OBJ: 10-1.2

10. A real gas
    a. does not obey all the assumptions of the kinetic-molecular theory.
    b. consists of particles that do not occupy space.
    c. cannot be condensed.
    d. cannot be produced in scientific laboratories.

    ANS: A   DIF: I   OBJ: 10-1.2

11. If two moving steel balls collide, their total energy after the collision is the same as before. This
    is an example of
    a. Boyle's law.   c. an elastic collision.
    b. the law of gravity.   d. both Boyle's law and Charles's law.

    ANS: C   DIF: II   OBJ: 10-1.2

12. Which is NOT an assumption of the kinetic-molecular theory?
    a. Matter is composed of tiny particles.
    b. The particles of matter are in continual motion.
    c. The total kinetic energy of colliding particles remains constant.
    d. When individual particles collide, energy is transferred.

    ANS: D   DIF: I   OBJ: 10-1.2

13. According to the kinetic-molecular theory, what is the most significant difference between gases
    and liquids?
    a. the shapes of the particles
    b. the mass of each particle
    c. the distance between the particles
    d. the type of collision that occurs between particles

    ANS: C   DIF: I   OBJ: 10-1.2

14. According to the kinetic-molecular theory, particles of a gas
    a. attract each other but do not collide.
    b. repel each other and collide.
    c. neither attract nor repel each other but collide.
    d. neither attract nor repel each other and do not collide.

    ANS: C   DIF: I   OBJ: 10-1.2

15. Which is an example of gas diffusion?
    a. inflating a flat tire
    b. the odor of perfume spreading throughout a room
    c. a cylinder of oxygen stored under high pressure
    d. All of the above

    ANS: B   DIF: II   OBJ: 10-1.3

16. By which process do gases take the shape of their container?
    a. evaporation   c. adhesion
    b. expansion   d. diffusion

    ANS: B   DIF: II   OBJ: 10-1.3

17. If a gas with an odor is released in a room, it quickly can be detected across the room because it
    a. diffuses.   c. is compressed.
    b. is dense.   d. condenses.

    ANS: A   DIF: II   OBJ: 10-1.3

18. Which substance has the lowest density?
    a. $H_2O(g)$ c. $Hg(l)$
    b. $H_2O(l)$ d. $Hg(g)$

    ANS: A   DIF: II   OBJ: 10-1.3

19. The density of a substance undergoes the greatest change when the substance changes from a
    a. liquid to a gas.   c. solid to a liquid.
    b. liquid to a solid.   d. a molecular solid to an ionic solid.

    ANS: A   DIF: I   OBJ: 10-1.3

20. According to the kinetic-molecular theory, how does a gas expand?
    a. Its particles become larger.
    b. Collisions between particles become elastic.
    c. Its temperature rises.
    d. Its particles move greater distances.

    ANS: D   DIF: I   OBJ: 10-1.3

21. Diffusion between two gases occurs most rapidly if the gases are at a
    a. high temperature and the molecules are small.
    b. low temperature and the molecules are large.
    c. low temperature and the molecules are small.
    d. high temperature and the molecules are large.

    ANS: A   DIF: I   OBJ: 10-1.3

22. Which is an example of effusion?
    a. air slowly escaping from a pinhole in a tire
    b. the aroma of a cooling pie spreading across a room
    c. helium dispersing into a room after a balloon pops
    d. oxygen and gasoline fumes mixing in an automobile carburetor

    ANS: A   DIF: II   OBJ: 10-1.3

23. What happens to the volume of a gas during compression?
    a. The volume increases.
    b. The volume decreases.
    c. The volume remains constant.
    d. It is impossible to tell because all gases are different.

    ANS: B   DIF: I   OBJ: 10-1.3

24. Under which conditions do real gases most resemble ideal gases?
    a. low pressure and low temperature   c. high pressure and high temperature
    b. low pressure and high temperature   d. high pressure and low temperature

    ANS: B   DIF: I   OBJ: 10-1.4

25. When does a real gas behave like an ideal gas?
    a. when the particles are far apart
    b. when the kinetic energy of the particles is low
    c. when the pressure is high
    d. when the gas is liquefied

    ANS: A   DIF: I   OBJ: 10-1.4

26. Why doesn't a gas at a low temperature behave like an ideal gas?
    a. There is too much space between the particles.
    b. The attractive forces are too weak.
    c. The kinetic energy of the particles is too low.
    d. The particles undergo chemical reactions.

    ANS: C   DIF: I   OBJ: 10-1.4

27. Which gases behave most like an ideal gas?
    a. gases composed of highly polar molecules
    b. gases composed of monatomic, nonpolar molecules
    c. gases composed of diatomic, polar molecules
    d. gases near their condensation temperatures

    ANS: B   DIF: I   OBJ: 10-1.4

28. As a real gas deviates from ideal gas behavior, the particles
    a. move farther apart.
    b. gain kinetic energy.
    c. collide more energetically.
    d. experience stronger attractive forces.

    ANS: D   DIF: I   OBJ: 10-1.4

29. The behavior of a gas under very high pressure is likely to
    a. conform to the assumptions of the kinetic-molecular theory.
    b. deviate from ideal gas behavior.
    c. show ideal gas behavior.
    d. be nonpolar.

    ANS: B   DIF: I   OBJ: 10-1.4

30. Pressure is the force per unit
    a. volume.
    b. surface area.
    c. length.
    d. depth.

    ANS: B   DIF: I   OBJ: 10-2.1

31. What is the SI unit of force?
    a. torr
    b. pascal
    c. pound
    d. newton

    ANS: D   DIF: I   OBJ: 10-2.1

32. If force is held constant as surface area decreases, pressure
    a. remains constant.
    b. decreases.
    c. increases.
    d. increases or decreases, depending on the volume change.

    ANS: C   DIF: I   OBJ: 10-2.1

33. What does the constant bombardment of gas molecules against the inside walls of a container produce?
    a. temperature
    b. density
    c. pressure
    d. diffusion

    ANS: C   DIF: I   OBJ: 10-2.1

34. What is the definition of pressure?
    a. $\text{pressure} = \dfrac{\text{force}}{\text{area}}$
    c. $\text{pressure} = \text{force} \times \text{area}$
    b. $\text{pressure} = \dfrac{\text{area}}{\text{force}}$
    d. $\text{pressure} = \dfrac{\text{force}}{\text{area}^2}$

    ANS: A   DIF: I   OBJ: 10-2.1

35. Why does a can collapse when a vacuum pump removes air from the can?
    a. The inside and outside forces balance out and crush the can.
    b. The unbalanced outside force from atmospheric pressure crushes the can.
    c. The atmosphere exerts pressure on the inside of the can and crushes it.
    d. The vacuum pump creates a force that crushes the can.

    ANS: B   DIF: II   OBJ: 10-2.1

36. What instrument measures atmospheric pressure?
    a. barometer   c. vacuum pump
    b. manometer   d. torrometer

    ANS: A   DIF: I   OBJ: 10-2.2

37. What instrument measures the pressure of an enclosed gas?
    a. barometer   c. vacuum pump
    b. manometer   d. torrometer

    ANS: B   DIF: I   OBJ: 10-2.2

38. A pressure of 745 mm Hg equals
    a. 745 torr.   c. 1 pascal.
    b. 1 torr.   d. 745 pascal.

    ANS: A   DIF: I   OBJ: 10-2.2

39. Convert the pressure 0.75 atm to mm Hg.
    a. 101.325 mm Hg   c. 570 mm Hg
    b. 430 mm Hg   d. 760 mm Hg

    ANS: C   DIF: III   OBJ: 10-2.3

40. Convert the pressure 0.840 atm to mm Hg.
    a. 365 mm Hg   c. 638 mm Hg
    b. 437 mm Hg   d. 780 mm Hg

    ANS: C   DIF: III   OBJ: 10-2.3

41. Convert the pressure 0.600 atm to mm Hg.
    a. 325 mm Hg   c. 572 mm Hg
    b. 456 mm Hg   d. 708 mm Hg

    ANS: B   DIF: III   OBJ: 10-2.3

42. Convert the pressure 2.50 atm to kPa.
    a. 1 kPa   c. 760 kPa
    b. 253 kPa   d. 1000 kPa

    ANS: B   DIF: III   OBJ: 10-2.3

43. Convert the pressure 1.30 atm to kPa.
    a. 2 kPa
    b. 115 kPa
    c. 132 kPa
    d. 245 kPa

    ANS: C   DIF: III   OBJ: 10-2.3

44. Standard temperature is exactly
    a. 100°C.
    b. 273°C.
    c. 0°C.
    d. 0 K.

    ANS: C   DIF: I   OBJ: 10-2.4

45. Standard pressure is the pressure exerted by a column of mercury exactly
    a. 273 mm high.
    b. 760 mm high.
    c. 760 cm high.
    d. 1.00 m high.

    ANS: B   DIF: I   OBJ: 10-2.4

46. Standard pressure is exactly
    a. 1 atm.
    b. 760 atm.
    c. 101.325 atm.
    d. 101 atm.

    ANS: A   DIF: I   OBJ: 10-2.4

47. If the height of mercury in a barometer at 0°C is less than 760 mm Hg, then
    a. the atmospheric pressure is less than standard atmospheric pressure.
    b. the atmospheric pressure is greater than standard atmospheric pressure.
    c. the atmospheric pressure is equal to standard atmospheric pressure.
    d. the atmospheric pressure cannot be determined.

    ANS: A   DIF: II   OBJ: 10-2.4

48. If the pressure and temperature of a gas are held constant and some gas is added to the container or some is allowed to escape, a change in which of the following can be observed?
    a. kinetic energy
    b. volume
    c. elasticity
    d. fluidity

    ANS: B   DIF: I   OBJ: 10-3.1

49. If the temperature of a fixed quantity and volume of gas changes, what also changes?
    a. pressure
    b. density
    c. mass
    d. formula

    ANS: A   DIF: I   OBJ: 10-3.1

50. To observe the effects of changing pressure on the volume of a gas, factors that must be kept constant are the gas's temperature and
    a. density.
    b. quantity.
    c. elasticity.
    d. All of the above

    ANS: B   DIF: I   OBJ: 10-3.1

51. To study the relationship between the temperature and volume of a gas, which factor must be held constant?

a. elasticity
b. fluidity
c. kinetic energy
d. pressure

ANS: D   DIF: I   OBJ: 10-3.1

52. If the temperature of a fixed quantity of gas decreases and the pressure remains unchanged,

a. its volume increases.
b. its volume is unchanged.
c. its volume decreases.
d. its density decreases.

ANS: C   DIF: I   OBJ: 10-3.1

53. Suppose the temperature of the air in a balloon is increased. If the pressure remains constant, what quantity must change?

a. volume
b. number of molecules
c. compressibility
d. adhesion

ANS: A   DIF: I   OBJ: 10-3.1

54. Two gases have the same temperature but different pressures. The kinetic-molecular theory does NOT predict that

a. molecules in both gases have the same average kinetic energies.
b. molecules in the low-pressure gas travel farther before they collide with other molecules.
c. both gases have the same densities.
d. all collisions of the molecules are elastic.

ANS: C   DIF: II   OBJ: 10-3.1

55. Why does the air pressure inside the tires of a car increase when the car is driven?

a. Some of the air has leaked out.
b. The air particles collide with the tire after the car is in motion.
c. The air particles inside the tire increase their speed because their temperature rises.
d. The atmosphere compresses the tire.

ANS: C   DIF: II   OBJ: 10-3.1

56. If the temperature of a container of gas remains constant, how could the pressure of the gas increase?

a. The mass of the gas molecules increases.
b. The diffusion of the gas molecules increases.
c. The size of the container increases.
d. The number of gas molecules in the container increases.

ANS: D   DIF: I   OBJ: 10-3.1

57. The gas pressure inside a container decreases when

a. the number of gas molecules is increased.
b. the number of gas molecules is decreased.
c. the temperature is increased.
d. the number of molecules is increased and the temperature is increased.

ANS: B   DIF: I   OBJ: 10-3.1

58. If the temperature remains constant, V and P represent the original volume and pressure, and V' and P' represent the new volume and pressure, what is the mathematical expression for Boyle's law?
   a. $P'V = V'P$
   b. $VV' = PP'$
   c. $V'P' = VP$
   d. $V = \dfrac{VP'}{P}$

   ANS: D   DIF: I   OBJ: 10-3.2

59. Pressure and volume changes at a constant temperature can be calculated using
   a. Boyle's law.
   b. Charles's law.
   c. Kelvin's law.
   d. Dalton's law.

   ANS: A   DIF: I   OBJ: 10-3.2

60. The volume of a gas is 400.0 mL when the pressure is 1.00 atm. At the same temperature, what is the pressure at which the volume of the gas is 2.0 L?
   a. 0.5 atm
   b. 5.0 atm
   c. 0.20 atm
   d. 800 atm

   ANS: C   DIF: III   OBJ: 10-3.2

61. The pressure of a sample of helium is 2.0 atm in a 200-mL container. If the container is compressed to 10 mL without changing the temperature, what is the new pressure?
   a. 200 atm
   b. 0.10 atm
   c. 100 atm
   d. 40. atm

   ANS: D   DIF: III   OBJ: 10-3.2

62. A sample of oxygen occupies 560. mL when the pressure is 800.00 mm Hg. At constant temperature, what volume does the gas occupy when the pressure decreases to 700.0 mm Hg?
   a. 80.0 mL
   b. 490. mL
   c. 600. mL
   d. 640. mL

   ANS: D   DIF: III   OBJ: 10-3.2

63. A sample of argon gas at standard pressure occupies 1000. mL. At constant temperature, what volume does the gas occupy if the pressure increases to 800. mm Hg?
   a. 500. mL
   b. 760. mL
   c. 950. mL
   d. 1053. mL

   ANS: C   DIF: III   OBJ: 10-3.2

64. A sample of gas collected at 750. mm Hg occupies 250. mL. At constant temperature, what pressure does the gas exert if the volume increases to 300. mL?
   a. 50. mm Hg
   b. 550. mm Hg
   c. 625. mm Hg
   d. 900. mm Hg

   ANS: C   DIF: III   OBJ: 10-3.2

65. At 710. mm Hg, a sample of nitrogen gas occupies 625 mL. What volume does the gas occupy if the temperature remains constant and the pressure increases to 760. mm Hg?
   a.   135 mL
   b.   584 mL
   c.   600 mL
   d.   669 mL

   ANS: B   DIF: III   OBJ: 10-3.2

66. A 425 mL sample of gas is collected at 780. mm Hg. If the temperature remains constant and the pressure falls to 680. mm Hg, what is the new volume?
   a.   325 mL
   b.   370 mL
   c.   488 mL
   d.   525 mL

   ANS: C   DIF: III   OBJ: 10-3.2

67. If V is the original volume, V' is the new volume, T is the original Kelvin temperature, and T' is the new Kelvin temperature, how is Charles's law expressed mathematically?
   a. $V' = V\dfrac{T'}{T}$
   b. $V = V'\dfrac{T'}{T}$
   c. $V = V - \dfrac{T'}{T}$
   d. $V' = \dfrac{V}{T'} + T$

   ANS: A   DIF: III   OBJ: 10-3.3

68. The volume of a gas is 5.0 L when the temperature is 5.0°C. If the temperature is increased to 10.0°C without changing the pressure, what is the new volume?
   a.   2.5 L
   b.   4.8 L
   c.   5.1 L
   d.   10.0 L

   ANS: C   DIF: III   OBJ: 10-3.3

69. At 7.0°C, the volume of a gas is 49 mL. At the same pressure, its volume is 74 mL at what temperature?
   a.   3.0°C
   b.   16°C
   c.   120°C
   d.   150°C

   ANS: D   DIF: III   OBJ: 10-3.3

70. A 180.0 mL volume of gas is measured at 87.0°C. If the pressure remains unchanged, what is the volume of the gas at standard temperature?
   a.   0.0 mL
   b.   0.5 mL
   c.   136 mL
   d.   410 mL

   ANS: C   DIF: III   OBJ: 10-3.3

71. The volume of a gas is 93 mL when the temperature is 91°C. If the temperature is reduced to 0°C without changing the pressure, what is the new volume of the gas?
   a.   70 mL
   b.   100 mL
   c.   120 mL
   d.   273 mL

   ANS: A   DIF: III   OBJ: 10-3.3

72. The volume of a gas is 400. mL at 30.0°C. If the temperature is increased to 50.0°C without changing the pressure, what is the new volume of the gas?
    a. 375 mL
    c. 426 mL
    b. 400 mL
    d. 600 mL

    ANS: C   DIF: III   OBJ: 10-3.3

73. If a gas occupies 950.0 mL at standard temperature, what volume does it occupy at 25.00°C if the pressure remains constant?
    a. 870.0 mL
    c. 1000.0 mL
    b. 966.0 mL
    d. 1037 mL

    ANS: D   DIF: III   OBJ: 10-3.3

74. If the temperature of 50.0 L of a gas at 40.0°C falls by 10.0C°, what is the new volume of the gas if the pressure is constant?
    a. 45.0 L
    c. 52.0 L
    b. 48.4 L
    d. 55.0 L

    ANS: B   DIF: III   OBJ: 10-3.3

75. Why would the pressure of a sample of gas at a constant volume fall 75 mm Hg?
    a. The container exploded.
    c. The temperature decreased.
    b. The temperature increased.
    d. Fewer particles were present.

    ANS: C   DIF: III   OBJ: 10-3.4

76. On a cold winter morning when the temperature is –13°C, the air pressure in an automobile tire is 1.5 atm. If the volume does not change, what is the pressure after the tire has warmed to 15°C?
    a. –1.5 atm
    c. 3.0 atm
    b. 1.7 atm
    d. 19.5 atm

    ANS: B   DIF: III   OBJ: 10-3.4

77. The pressure of a sample of gas of constant volume is 2.0 atm at 30.°C. What is the pressure of this sample at 20.°C?
    a. 1.0 atm
    c. 2.1 atm
    b. 1.9 atm
    d. 20 atm

    ANS: B   DIF: III   OBJ: 10-3.4

78. The pressure of a 1000. mL sample of gas at 10.0°C increases from 700. mm Hg to 900. mm Hg. If the volume is unchanged, what is the new temperature?
    a. 0°C
    c. 30°C
    b. 24°C
    d. 91°C

    ANS: D   DIF: III   OBJ: 10-3.4

79. The pressure of a sample of gas at a constant volume is 8.0 atm at 70.°C. What is the pressure at 20.°C?
    a. 0.16 atm
    b. 6.8 atm
    c. 9.4 atm
    d. 58 atm

    ANS: B   DIF: III   OBJ: 10-3.4

80. The temperature of a sample of gas at 4.0 atm and 15.°C increases to 30.°C. If the volume is unchanged, what is the new gas pressure?
    a. 3.8 atm
    b. 4.2 atm
    c. 8 atm
    d. 19 atm

    ANS: B   DIF: III   OBJ: 10-3.4

81. The pressure of a sample of gas increases from 450. mm Hg to 500. mm Hg. If the volume is constant and the temperature of the gas was 0.0°C, what is the new gas temperature?
    a. −30.°C
    b. 30.°C
    c. 50.°C
    d. 273°C

    ANS: B   DIF: III   OBJ: 10-3.4

82. As the temperature of a sample of gas falls from 45.0°C to 30.0°C, its pressure falls to 300. mm Hg. If the volume did not change, what was the original gas pressure?
    a. 285 mm Hg
    b. 315 mm Hg
    c. 400. mm Hg
    d. 615 mm Hg

    ANS: B   DIF: III   OBJ: 10-3.4

83. A sample of gas at 6.0 atm and 5.0°C increases in temperature to 35°C. If the volume is unchanged, what is the new pressure?
    a. 5.4 atm
    b. 6.6 atm
    c. 36 atm
    d. 42 atm

    ANS: B   DIF: III   OBJ: 10-3.4

84. If V, P, and T represent the original volume, pressure, and temperature in the correct units, and V', P', and T' represent the new conditions, what is the combined gas law?
    a. $\dfrac{PV}{T'} = \dfrac{P'V'}{T}$
    b. $\dfrac{PV'}{T} = \dfrac{P'V}{T'}$
    c. $\dfrac{P'V}{T} = \dfrac{PV'}{T'}$
    d. $\dfrac{PV}{T} = \dfrac{P'V'}{T'}$

    ANS: D   DIF: I   OBJ: 10-3.5

85. The volume of a sample of oxygen is 300.0 mL when the pressure is 1 atm and the temperature is 27.0°C. At what temperature is the volume 1.00 L and the pressure 0.500 atm?
    a. 22.0°C
    b. 45.0°C
    c. 0.50 K
    d. 227°C

    ANS: D   DIF: III   OBJ: 10-3.5

86. Suppose that the pressure of 1.00 L of gas is 380. mm Hg when the temperature is 200. K. At what temperature is the volume 2.00 L and the pressure 0.750 atm?
   a. 1.00 K
   b. 600. K
   c. 219°C
   d. 67.0 K

   ANS: B   DIF: III   OBJ: 10-3.5

87. The volume of a gas collected when the temperature is 11.0°C and the pressure is 710 mm Hg measures 14.8 mL. What is the calculated volume of the gas at 20.0°C and 740 mm Hg?
   a. 7.8 mL
   b. 13.7 mL
   c. 14.6 mL
   d. 15 mL

   ANS: C   DIF: III   OBJ: 10-3.5

88. The volume of a sample of hydrogen is 798 mL and it exerts 621 mm Hg pressure at 5.00°C. What volume does it occupy at standard temperature and pressure?
   a. 520. mL
   b. 640. mL
   c. 745 mL
   d. 960 mL

   ANS: B   DIF: III   OBJ: 10-3.5

89. A 30.-L sample of gas exerts 200. mm Hg pressure at 10°C. What volume does the gas have at 300. mm Hg and 25°C?
   a. 9.0 L
   b. 17 L
   c. 21 L
   d. 42 L

   ANS: C   DIF: III   OBJ: 10-3.5

90. At 0.500 atm and 15.0°C a sample of gas occupies 120. L. What volume does it occupy at 0.250 atm and 10.0°C?
   a. 60 L
   b. 111 L
   c. 236 L
   d. 480 L

   ANS: C   DIF: III   OBJ: 10-3.5

91. A 70.0 L sample of gas at 20.0°C and 600. mm Hg expands to 90.0 L at 15.0°C. What is the new gas pressure?
   a. 318 mm Hg
   b. 459 mm Hg
   c. 583 mm Hg
   d. 710 mm Hg

   ANS: B   DIF: III   OBJ: 10-3.5

92. A 75.0 mL sample of gas exerts 200. mm Hg pressure at 30°C. What pressure does it exert at 35.0°C if the volume expands to 80.0 mL?
   a. 90.0 mm Hg
   b. 161 mm Hg
   c. 190 mm Hg
   d. 219 mm Hg

   ANS: C   DIF: III   OBJ: 10-3.5

93. A 150.0 L sample of gas is collected at 1.20 atm and 25°C. What volume does the gas have at 1.50 atm and 20.0°C?
   a.   94 L
   b.   120 L
   c.   143 L
   d.   183 L

   ANS: B   DIF: III   OBJ: 10-3.5

94. Who developed the concept that the total pressure of a mixture of gases is the sum of their partial pressures?
   a.   Charles
   b.   Boyle
   c.   Kelvin
   d.   Dalton

   ANS: D   DIF: III   OBJ: 10-3.6

95. To correct for the partial pressure of water vapor, the vapor pressure of H_2O at the collecting temperature is
   a.   divided by 22.4.
   b.   multiplied by 22.4.
   c.   subtracted from the total gas pressure.
   d.   added to the total gas pressure.

   ANS: C   DIF: III   OBJ: 10-3.6

96. Three samples of gas each exert 740. mm Hg in separate 2 L containers. What pressure do they exert if they are all placed in a single 2 L container?
   a.   247 mm Hg
   b.   740 mm Hg
   c.   1480 mm Hg
   d.   2220 mm Hg

   ANS: D   DIF: III   OBJ: 10-3.6

97. A mixture of four gases exerts a total pressure of 860 mm Hg. Gases A and B each exert 220 mm Hg. Gas C exerts 110 mm Hg. What pressure is exerted by gas D?
   a.   165 mm Hg
   b.   310 mm Hg
   c.   860 mm Hg
   d.   cannot be determined

   ANS: B   DIF: III   OBJ: 10-3.6

98. If five gases in a cylinder each exert 1 atm, what is the total pressure exerted by the gases?
   a.   0.2 atm
   b.   0.5 atm
   c.   1 atm
   d.   5 atm

   ANS: D   DIF: III   OBJ: 10-3.6

Water Vapor Pressure

Temperature (°C)	Pressure (mm Hg)
0	4.6
5	6.5
10	9.2
15	12.8
20	17.5
25	23.8
30	31.8
35	42.2
40	55.3
50	92.5

99. What is the partial pressure of water vapor in oxygen gas collected by water displacement at 10°C and 750 mm Hg?
   a. 9.2 mm Hg
   b. 740.8 mm Hg
   c. 750 mm Hg
   d. 759.2 mm Hg

   ANS: A   DIF: III   OBJ: 10-3.6

100. A sample of gas is collected by water displacement at 600.0 mm Hg and 30°C. What is the partial pressure of the gas?
   a. 568.2 mm Hg
   b. 600.0 mm Hg
   c. 630 mm Hg
   d. 631.8 mm Hg

   ANS: A   DIF: III   OBJ: 10-3.6

101. A sample of nitrogen is collected by water displacement at 730.0 mm Hg and 20°C. What is the partial pressure of the nitrogen?
   a. 17.5 mm Hg
   b. 712.5 mm Hg
   c. 717.2 mm Hg
   d. 747.5 mm Hg

   ANS: B   DIF: III   OBJ: 10-3.6

SHORT ANSWER

1. Why are gases described as fluid?

   ANS:
   Gas particles glide easily past one another because the attractive forces between them are insignificant. Because this behavior is similar to liquids, gases are classified as fluids.

   DIF: II   OBJ: 10-1.3

2. Why must a barometer be open to the air at one end and closed at the other?

ANS:
The air pressure at the open end forces the mercury up the tube. A vacuum exists at the closed end so there is no opposing force.

DIF: II   OBJ: 10-2.2

3. What are standard temperature and pressure? Why was a standard needed?

ANS:
Standard temperature is 0°C, and standard pressure is 1 atm. Scientists have agreed upon standard conditions for temperature and pressure to compare volumes of gases.

DIF: II   OBJ: 10-2.4

ESSAY

1. Explain how a barometer works.

ANS:
A tube filled with mercury is inverted into a pool of mercury. The mercury column falls until the weight of the mercury above the surface of the pool balances the force of the atmosphere on the open surface of the mercury in the pool. As air pressure changes, the height of the mercury column rises or falls.

DIF: II   OBJ: 10-2.2

Chapter 11—Molecular Composition of Gases

MULTIPLE CHOICE

1. Gay-Lussac recognized that at constant temperature and pressure, the volumes of gaseous reactants and products
   a. always equal 1 L.
   b. add up to 22.4 L.
   c. equal R.
   d. can be expressed as ratios of small whole numbers.

   ANS: D   DIF: I   OBJ: 11-1.1

2. When Gay-Lussac's law of combining volumes holds, which of the following can be expressed in ratios of small whole numbers?
   a. pressure before and pressure after reaction
   b. volumes of gaseous reactants and products
   c. Kelvin temperatures
   d. molar masses of products and reactants

   ANS: B   DIF: I   OBJ: 11-1.1

3. The law of combining volumes applies only to gas volumes
   a. measured at constant temperature and pressure.
   b. that equal 1 L.
   c. that equal 22.4 L.
   d. measured at STP.

   ANS: A   DIF: I   OBJ: 11-1.1

4. The reaction of two volumes of hydrogen gas with one volume of oxygen gas to produce two volumes of water vapor is an example of
   a. the ideal gas law.
   b. Graham's law of effusion.
   c. Gay-Lussac's law of combining volumes of gases.
   d. Avogadro's law.

   ANS: C   DIF: I   OBJ: 11-1.1

5. In the equation $H_2(g) + Cl_2(g) \rightarrow 2HCl(g)$, one volume of hydrogen yields how many volumes of hydrogen chloride?
   a. 1   c. 3
   b. 2   d. 4

   ANS: B   DIF: II   OBJ: 11-1.1

6. If 0.5 L of $O_2(g)$ reacts with H_2 to produce 1 L of $H_2O(g)$, what is the volume of $H_2O(g)$ obtained from 1 L of $O_2(g)$?
   a. 0.5 L   c. 2 L
   b. 1.5 L   d. Cannot be determined

   ANS: C   DIF: II   OBJ: 11-1.1

7. In the equation $C + O_2(g) \rightarrow CO_2(g)$, one volume of O_2 yields how many volumes of CO_2?
   a. 1
   b. 2
   c. 3
   d. 4

   ANS: A   DIF: II   OBJ: 11-1.1

8. The principle that under similar pressures and temperatures, equal volumes of gases contain the same number of molecules is attributed to
   a. Berthollet.
   b. Proust.
   c. Avogadro.
   d. Dalton.

   ANS: C   DIF: I   OBJ: 11-1.2

9. Equal volumes of diatomic gases under the same conditions of temperature and pressure contain the same number of
   a. protons.
   b. ions.
   c. molecules.
   d. Dalton's "ultimate particles."

   ANS: C   DIF: I   OBJ: 11-1.2

10. If gas A has a molar mass greater than that of gas B and samples of each gas at identical temperatures and pressures contain equal numbers of molecules, then
    a. the volumes of gas A and gas B are equal.
    b. the volume of gas A is greater than that of gas B.
    c. the volume of gas B is greater than that of gas A.
    d. their volumes are proportional to their molar masses.

    ANS: A   DIF: II   OBJ: 11-1.2

11. At constant temperature and pressure, gas volume is directly proportional to the
    a. molar mass of the gas.
    b. number of moles of gas.
    c. density of the gas at STP.
    d. rate of diffusion.

    ANS: B   DIF: I   OBJ: 11-1.2

12. The expression $V = kn$ is a statement of
    a. the ideal gas law.
    b. the law of combining volumes.
    c. Graham's law of effusion.
    d. Avogadro's principle.

    ANS: D   DIF: I   OBJ: 11-1.2

13. Avogadro's principle led to the realization that the molecules of some substances
    a. could not react.
    b. were not composed of atoms.
    c. were invisible.
    d. were made of more than one atom.

    ANS: D   DIF: I   OBJ: 11-1.2

14. In the expression $V = kn$, n represents
    a. the gas constant.
    b. Avogadro's number.
    c. the number of moles of gas.
    d. a constant.

    ANS: C   DIF: I   OBJ: 11-1.2

15. According to Avogadro's law, 1 L of $H_2(g)$ and 1 L of $O_2(g)$ at the same temperature and pressure
    a.   have the same mass.
    b.   have unequal volumes.
    c.   contain 1 mol of gas each.
    d.   contain equal numbers of molecules.

    ANS: D   DIF: I   OBJ: 11-1.2

16. The standard molar volume of a gas is all of the following except
    a.   the volume occupied by 1 mol of a gas at STP.
    b.   equal for all gases under the same conditions.
    c.   22.4 L at STP.
    d.   dependent upon the size of the molecules.

    ANS: D   DIF: I   OBJ: 11-1.3

17. The standard molar volume of a gas at STP is
    a.   22.4 L.   c.   g-mol wt/22.4 L.
    b.   g/22.4 L.   d.   1 L.

    ANS: A   DIF: I   OBJ: 11-1.3

18. The standard molar volume of a gas at STP is all of the following except
    a.   the volume occupied by 1 mol of the gas.
    b.   22.4 g.
    c.   22.4 L.
    d.   the volume occupied by one molar mass of the gas.

    ANS: B   DIF: I   OBJ: 11-1.3

19. What is the volume occupied by 1 mol of oxygen at STP?
    a.   11.2 L   c.   22.4 L
    b.   16.0 L   d.   32.0 L

    ANS: C   DIF: I   OBJ: 11-1.3

20. What is the volume occupied by 1 mol of water vapor at STP?
    a.   11.2 L   c.   22.4 L
    b.   18.0 L   d.   33.6 L

    ANS: C   DIF: I   OBJ: 11-1.3

21. At STP, the standard molar volume of a gas of known volume can be used to calculate the
    a.   number of moles of gas.   c.   gram-molecular weight.
    b.   rate of diffusion.   d.   gram-molecular volume.

    ANS: A   DIF: I   OBJ: 11-1.3

22. If the molecular formula of a gas is known, the molar volume is used directly in solving
    a.   mass-mass problems.   c.   percentage composition problems.
    b.   the volume of any mass of gas.   d.   gas volume–gas volume problems.

    ANS: B   DIF: I   OBJ: 11-1.3

23. Knowing the mass and volume of a gas at STP allows one to calculate the
    a. identity of the gas.
    b. molar mass of the gas.
    c. condensation point of the gas.
    d. rate of diffusion of the gas.

    ANS: B   DIF: I   OBJ: 11-1.4

24. What is the molar mass of gas at STP?
    a. density of the gas multiplied by the mass of 1 mol
    b. density of the gas divided by the mass of 1 mol
    c. density of the gas multiplied by 22.4 L
    d. density of the gas divided by 22.4 L

    ANS: C   DIF: I   OBJ: 11-1.4

25. A 1.00 L sample of a gas has a mass of 1.92 g at STP. What is the molar mass of the gas?
    a. 1.92 g/mol
    b. 19.2 g/mol
    c. 22.4 g/mol
    d. 43.0 g/mol

    ANS: D   DIF: III   OBJ: 11-1.4

26. A 1.00 L sample of a gas has a mass of 0.716 g at STP. What is the molar mass of the gas?
    a. 0.716 g/mol
    b. 1.60 g/mol
    c. 7.16 g/mol
    d. 16.0 g/mol

    ANS: D   DIF: III   OBJ: 11-1.4

27. A 1.00 L sample of a gas has a mass of 1.25 g at STP. What is the mass of 1 mol of this gas?
    a. a little less then 1.0 g
    b. 1.25 g
    c. 22.4 g
    d. 28.0 g

    ANS: D   DIF: III   OBJ: 11-1.4

28. A 1.00 L sample of a gas has a mass of 1.7 g at STP. What is the molar mass of the gas?
    a. 0.076 g/mol
    b. 13.2 g/mol
    c. 38 g/mol
    d. 170 g/mol

    ANS: C   DIF: III   OBJ: 11-1.4

29. The ideal gas law is equivalent to Charles's law when
    a. the number of moles and the pressure are constant.
    b. the number of moles and the temperature are constant.
    c. the volume equals 22.4 L.
    d. R equals zero.

    ANS: A   DIF: I   OBJ: 11-2.1

30. The ideal gas law is equivalent to Boyle's law when
    a. Avogadro's number is reached.
    b. R equals zero.
    c. the pressure is 1 atm.
    d. the number of moles and the temperature are constant.

    ANS: D   DIF: I   OBJ: 11-2.1

31. The ideal gas law is equivalent to Avogadro's law when
    a. the volume is 22.4 L.
    b. 1 mol of a gas is present.
    c. the pressure and the temperature are constant.
    d. the volume and pressure are constant.

    ANS: C   DIF: I   OBJ: 11-2.1

32. The ideal gas law combines Boyle's law, Charles's law, and
    a. Graham's law.   c. Raoult's law.
    b. Avogadro's law.   d. Dalton's principle.

    ANS: B   DIF: I   OBJ: 11-2.1

33. All of the following equations are statements of the ideal gas law except

    a. $P = nRTV$

    c. $\dfrac{P}{n} = \dfrac{RT}{v}$

    b. $\dfrac{PV}{T} = nR$

    d. $R = \dfrac{PV}{nT}$

    ANS: A   DIF: I   OBJ: 11-2.1

34. When pressure, volume, and temperature are known, the ideal gas law can be used to calculate
    a. the chemical formula.   c. molar amount.
    b. the ideal gas constant.   d. compressibility.

    ANS: C   DIF: I   OBJ: 11-2.1

35. Which is a common unit for the ideal gas constant R?

    a. L·atm

    c. $\dfrac{L \cdot atm}{mol \cdot K}$

    b. mol·K

    d. $\dfrac{atm}{K}$

    ANS: C   DIF: I   OBJ: 11-2.2

36. What is the value of the gas constant?

    a. $0.0821 \ \dfrac{L \cdot atm}{mol \cdot K}$

    c. $0.0281 \ \dfrac{L \cdot atm}{mol \cdot K}$

    b. 0.0281 L·atm

    d. 0.0821 mol·K

    ANS: A   DIF: I   OBJ: 11-2.2

37. The value of R, the ideal gas constant, can be calculated from measured values of a gas's pressure, volume, temperature, and
    a. molar amount.
    b. chemical formula.
    c. rate of diffusion.
    d. density.

    ANS: A   DIF: I   OBJ: 11-2.2

38. In the ideal gas law, what is the value that must be calculated from other measurements?
    a. P
    b. V
    c. T
    d. R

    ANS: D   DIF: I   OBJ: 11-2.2

39. The gas constant, R, has the value of
    a. $0.0821 \dfrac{L \cdot atm}{mol \cdot K}$
    b. $0.0821 \dfrac{L \cdot mol}{atm \cdot K}$
    c. $0.0821 \dfrac{L \cdot K}{atm \cdot mol}$
    d. $0.0821 \dfrac{mol \cdot T}{L \cdot atm}$

    ANS: A   DIF: I   OBJ: 11-2.2

40. Calculate the approximate volume of a 0.600 mol sample of gas at 15.0°C and a pressure of 1.10 atm.
    a. 12.9 L
    b. 22.4 L
    c. 24.6 L
    d. 129 L

    ANS: A   DIF: III   OBJ: 11-2.3

41. Calculate the approximate temperature of a 0.50 mol sample of gas at 750 mm Hg and a volume of 12 L.
    a. −7°C
    b. 11°C
    c. 15°C
    d. 288°C

    ANS: C   DIF: III   OBJ: 11-2.3

42. What is the pressure exerted by 1.2 mol of a gas with a temperature of 20.°C and a volume of 9.5 L?
    a. 0.030 atm
    b. 1.0 atm
    c. 3.0 atm
    d. 30. atm

    ANS: C   DIF: III   OBJ: 11-2.3

43. A sample of gas at 25°C has a volume of 11 L and exerts a pressure of 660 mm Hg. How many moles of gas are in the sample?
    a. 0.39 mol
    b. 3.9 mol
    c. 9.3 mol
    d. 87 mol

    ANS: A   DIF: III   OBJ: 11-2.3

44. What is the approximate volume of gas in a 1.50 mol sample that exerts a pressure of 0.922 atm and has a temperature of 10.0°C?
    a. 13 L   c. 37.8 L
    b. 14.2 L   d. 378 L

    ANS: C   DIF: III   OBJ: 11-2.3

45. What pressure is exerted by 0.750 mol of a gas at a temperature of 0.00°C and a volume of 5.00 L?
    a. 2.1 atm   c. 4.98 atm
    b. 3.4 atm   d. 760. atm

    ANS: B   DIF: III   OBJ: 11-2.3

46. A gas sample with a mass of 0.467 g is collected at 20.°C and 732.5 mm Hg. The volume is 200. mL. What is the molar mass of the gas?
    a. 58 g/mol   c. 290 g/mol
    b. 180 g/mol   d. 730 g/mol

    ANS: A   DIF: III   OBJ: 11-2.4

47. A gas sample with a mass of 0.686 g is collected at 20.°C and 722.5 mm Hg. The volume is 350. mL. What is the molar mass of the gas?
    a. 0.31 g/mol   c. 50. g/mol
    b. 2.2 g/mol   d. 720 g/mol

    ANS: C   DIF: III   OBJ: 11-2.4

48. A gas sample with a mass of 2.50 g is collected at 20.0°C and 732.5 mm Hg. The volume is 1.28 L. What is the molar mass of the gas?
    a. 1.26 g/mol   c. 13.7 g/mol
    b. 2.04 g/mol   d. 48.8 g/mol

    ANS: D   DIF: III   OBJ: 11-2.4

49. A gas sample with a mass of 0.934 g is collected at 20.0°C and 733.5 mm Hg. The volume is 200.0 mL. What is the molar mass of the gas?
    a. 116 g/mol   c. 586 g/mol
    b. 358 g/mol   d. 1464.0 g/mol

    ANS: A   DIF: III   OBJ: 11-2.4

50. A gas sample with a mass of 5.16 g is collected at 28°C and 740 mm Hg. The volume is 1.00 L. What is the molar mass of the gas?
    a. 0.395 g/mol   c. 130 g/mol
    b. 0.97 g/mol   d. 300 g/mol

    ANS: C   DIF: III   OBJ: 11-2.4

51. A gas sample with a mass of 0.250 g is collected at 150.0°C and 720. mm Hg. The volume is 85.0 mL. What is the molar mass of the gas?
    a. 0.0023 g/mol
    b. 1.76 g/mol
    c. 108 g/mol
    d. 238.0 g/mol

    ANS: C   DIF: III   OBJ: 11-2.4

52. A gas sample with a mass of 12.8 g exerts a pressure of 1.2 atm at 15°C and a volume of 3.94 L. What is the molar mass of the gas?
    a. 19 g/mol
    b. 32 g/mol
    c. 64 g/mol
    d. 128 g/mol

    ANS: C   DIF: III   OBJ: 11-2.4

53. If n and T are constant, the ideal gas law reduces to
    a. Charles's law.
    b. Boyle's law.
    c. Avogadro's law.
    d. zero.

    ANS: B   DIF: I   OBJ: 11-2.5

54. If n and P are constant, the ideal gas law reduces to
    a. Charles's law.
    b. Boyle's law.
    c. Avogadro's law.
    d. zero.

    ANS: A   DIF: I   OBJ: 11-2.5

55. If P and T are constant, the ideal gas law reduces to
    a. Charles's law.
    b. Boyle's law.
    c. Avogadro's law.
    d. zero.

    ANS: C   DIF: I   OBJ: 11-2.5

56. When the ideal gas law reduces to $PV = k$, the expression is equivalent to
    a. Avogadro's law.
    b. Avogadro's number.
    c. Charles's law.
    d. Boyle's law.

    ANS: D   DIF: I   OBJ: 11-2.5

57. When the ideal gas law reduces to $V = kT$, the expression is equivalent to
    a. Boyle's law.
    b. Charles's law.
    c. Avogadro's principle.
    d. Gay-Lussac's law.

    ANS: B   DIF: I   OBJ: 11-2.5

58. When the ideal gas law reduces to $V = kn$, the expression is equivalent to
    a. Avogadro's law.
    b. Avogadro's number.
    c. Charles's law.
    d. Boyle's law.

    ANS: A   DIF: I   OBJ: 11-2.5

59. For reactants or products that are gases, the coefficients in the chemical equation indicate
    a. the number of grams of each substance.   c. molar volume.
    b. volume.   d. density.

    ANS: B   DIF: I   OBJ: 11-3.1

60. If the volume of the gaseous reactants and products in the equation $2C_2H_6 + 7O_2 \rightarrow 4CO_2 + 6H_2O$ are measured at 1 atm and 25°C, the substance that could not be included in the volume relationship has what formula?
    a. C_2H_6 c. CO_2
    b. O_2 d. H_2O

    ANS: D   DIF: I   OBJ: 11-3.1

61. In a chemical equation, the coefficients for reactants and products that are gases indicate
    a. volumes at STP.   c. molar mass of each substance.
    b. volume ratios.   d. densities.

    ANS: B   DIF: I   OBJ: 11-3.1

62. What law helps explain the volume ratios in a chemical reaction?
    a. Charles's law   c. Boyle's law
    b. Graham's law   d. Gay-Lussac's law of combining volumes

    ANS: D   DIF: I   OBJ: 11-3.1

63. Volumes of gaseous reactants and products in a chemical reaction can be expressed as ratios of small whole numbers
    a. if all reactants and products are gases.
    b. if standard temperature and pressure are maintained.
    c. if constant temperature and pressure are maintained.
    d. if each mass equals 1 mol.

    ANS: C   DIF: I   OBJ: 11-3.1

64. According to Avogadro's law, in a chemical equation equal volumes of gases contain equal
    a. numbers of molecules.   c. temperatures.
    b. pressures.   d. masses.

    ANS: A   DIF: I   OBJ: 11-3.1

65. The ratios of the volumes of the gaseous reactants and products in a chemical reaction at constant temperature and pressure can be determined from the
    a. formulas.   c. subscripts in the balanced equation.
    b. coefficients in the balanced equation.   d. gas constant.

    ANS: B   DIF: I   OBJ: 11-3.1

66. In the reaction $2C + O_2(g) \rightarrow 2CO(g)$, what is the volume ratio of O_2 to CO?
    a. 1:1   c. 1:2
    b. 2:1   d. 2:2

    ANS: C   DIF: II   OBJ: 11-3.2

67. In the reaction $2H_2(g) + O_2(g) \rightarrow 2H_2O(g)$, what is the volume ratio of H_2 to H_2O?
   a.   1:1
   b.   2:3
   c.   4:3
   d.   4:6

   ANS: A   DIF: II   OBJ: 11-3.2

68. In the reaction $N_2(g) + 2O_2(g) \rightarrow 2NO_2(g)$, what is the volume ratio of N_2 to NO_2?
   a.   1:1
   b.   1:2
   c.   2:1
   d.   2:5

   ANS: B   DIF: II   OBJ: 11-3.2

69. In the reaction $2H_2(g) + O_2(g) \rightarrow 2H_2O(g)$, what is the volume ratio of O_2 to H_2O?
   a.   1:1
   b.   2:1
   c.   1:2
   d.   2:2

   ANS: C   DIF: II   OBJ: 11-3.2

70. In the reaction $H_2(g) + Cl_2(g) \rightarrow 2HCl$, what is the volume ratio of Cl_2 to HCl?
   a.   1:1
   b.   1:2
   c.   2:1
   d.   2:2

   ANS: B   DIF: II   OBJ: 11-3.2

71. In the reaction $H_2(g) + Cl_2(g) \rightarrow 2HCl$, what is the volume ratio of H_2 to HCl?
   a.   1:1
   b.   1:2
   c.   2:1
   d.   2:2

   ANS: B   DIF: II   OBJ: 11-3.2

72. In the reaction $H_2(g) + Cl_2(g) \rightarrow 2HCl$, what is the volume ratio of H_2 to Cl_2?
   a.   1:1
   b.   1:2
   c.   2:1
   d.   3:2

   ANS: A   DIF: II   OBJ: 11-3.2

73. The equation for the production of methane is $C + 2H_2(g) \rightarrow CH_4(g)$. How many liters of hydrogen are needed to produce 20 L of methane?
   a.   2.0 L
   b.   20 L
   c.   22.4 L
   d.   40 L

   ANS: D   DIF: III   OBJ: 11-3.3

74. The equation for the complete combustion of methane is $CH_4(g) + 2O_2(g) \rightarrow 2H_2O(g) + CO_2(g)$. If 50 L of methane at STP are burned, what volume of carbon dioxide will be produced at STP?
   a.   16.6 L
   b.   25 L
   c.   50 L
   d.   100 L

   ANS: C   DIF: III   OBJ: 11-3.3

75. When hydrogen burns, water vapor is produced. The equation is $2H_2(g) + O_2(g) \rightarrow 2H_2O(g)$. If 12 L of oxygen are consumed, what volume of water vapor is produced?

   a.   1 L
   b.   2 L
   c.   12 L
   d.   24 L

   ANS: D   DIF: III   OBJ: 11-3.3

76. For the complete combustion of 100. L of CO, the volume of oxygen required is

   a.   23.8 L
   b.   50.0 L
   c.   238 L
   d.   500. L

   ANS: B   DIF: III   OBJ: 11-3.3

77. What is the number of moles of H_2 produced when 23 g of sodium react with water according to the equation $2Na(s) + 2H_2O(l) \rightarrow 2NaOH(aq) + H_2(g)$?

   a.   0.50 mol
   b.   1 mol
   c.   2 mol
   d.   4 mol

   ANS: A   DIF: III   OBJ: 11-3.3

78. Chlorine is produced by the reaction $2HCl(g) \rightarrow H_2(g) + Cl_2(g)$. How many grams of HCl (36.5 g/mol) must be used to produce 10 L of chlorine at STP?

   a.   15.8 g
   b.   30.2 g
   c.   32.6 g
   d.   36.5 g

   ANS: C   DIF: III   OBJ: 11-3.3

79. Iron oxide, FeO_2, is produced by the reaction $Fe + O_2 \rightarrow FeO_2$ (87.8 g/mol). How many grams of FeO_2 can be produced from 50 L of O_2 at STP?

   a.   19.5 g
   b.   37.8 g
   c.   50 g
   d.   196. g

   ANS: D   DIF: III   OBJ: 11-3.3

80. When carbon burns, carbon dioxide is produced by the reaction $C + O_2 \rightarrow CO_2$. If 14 L of CO_2 are produced at STP, how many grams of carbon (atomic mass 12) were used?

   a.   7.5 g
   b.   11.2 g
   c.   17.5 g
   d.   75 g

   ANS: A   DIF: III   OBJ: 11-3.3

81. What is the process by which molecules of a gas randomly encounter and pass through a small opening in a container?

   a.   diffusion
   b.   osmosis
   c.   distillation
   d.   effusion

   ANS: D   DIF: I   OBJ: 11-4.1

82. According to Graham's law, the rates of diffusion of two gases at the same temperature and pressure are inversely proportional to
   a. their volumes.
   b. the square roots of their molar masses.
   c. their compressibilities.
   d. their rates of effusion.

   ANS: B   DIF: I   OBJ: 11-4.1

83. What determines the average kinetic energy of the molecules of any gas?
   a. temperature
   b. pressure
   c. temperature and pressure
   d. molar mass

   ANS: A   DIF: I   OBJ: 11-4.1

84. The equation   is an expression of
   a. Gay-Lussac's law.
   b. Avogadro's law.
   c. Graham's law.
   d. Boyle's law.

   ANS: C   DIF: I   OBJ: 11-4.1

85. According to Graham's law, two gases at the same temperature and pressure will have different rates of diffusion because they have different
   a. volumes.
   b. molar masses.
   c. kinetic energies.
   d. condensation points.

   ANS: B   DIF: I   OBJ: 11-4.1

86. In Graham's equation, the square roots of the molar masses can be substituted with the square roots of the
   a. gas densities.
   b. molar volumes.
   c. compressibilities.
   d. gas constants.

   ANS: A   DIF: I   OBJ: 11-4.1

87. Suppose that two gases with unequal molar masses were injected into opposite ends of a long tube at the same time and allowed to diffuse toward the center. They should begin to mix
   a. in approximately five minutes.
   b. closer to the end that held the heavier gas.
   c. closer to the end that held the lighter gas.
   d. exactly in the middle.

   ANS: B   DIF: II   OBJ: 11-4.1

Element	Atomic mass
Argon	39.948
Bromine	79.904
Carbon	12.011
Chlorine	35.453
Fluorine	18.998
Helium	4.0026
Hydrogen	1.0079
Nitrogen	14.007
Oxygen	15.999

88. How many times greater is the rate of effusion of molecular fluorine than that of molecular bromine at the same temperature and pressure?
    a. 2
    b. 3
    c. 4
    d. 7

    ANS: A   DIF: III   OBJ: 11-4.2

89. How many times greater is the rate of effusion of oxygen gas than that of carbon dioxide gas at the same temperature and pressure?
    a. 1.2
    b. 2.1
    c. 12
    d. They are equal.

    ANS: A   DIF: III   OBJ: 11-4.2

90. A sample of helium diffuses 4.57 times faster than an unknown gas diffuses. What is the molar mass of the unknown gas?
    a. 12 g/mol
    b. 18.2 g/mol
    c. 38.8 g/mol
    d. 83.6 g/mol

    ANS: D   DIF: III   OBJ: 11-4.3

91. A sample of helium diffuses 6.3 times faster than an unknown gas diffuses. What is the molar mass of the unknown gas?
    a. 1.05 g/mol
    b. 3.8 g/mol
    c. 25.2 g/mol
    d. 160 g/mol

    ANS: D   DIF: III   OBJ: 11-4.3

92. A sample of hydrogen gas diffuses 3.8 times faster than an unknown gas diffuses. What is the molar mass of the unknown gas?
    a. 3.8 g/mol
    b. 7.6 g/mol
    c. 22.4 g/mol
    d. 28 g/mol

    ANS: D   DIF: III   OBJ: 11-4.3

PROBLEM

Element	Atomic mass
Argon	39.948
Bromine	79.904
Carbon	12.011
Chlorine	35.453
Fluorine	18.998
Helium	4.0026
Hydrogen	1.0079
Nitrogen	14.007
Oxygen	15.999

1. Methane, CH_4, and propane, C_3H_8, are two components of the natural gas burned in Bunsen burners. If a methane molecule travels at 675 m/s from a gas jet, how fast does the propane molecule travel?

ANS:
407 m/s

$$v_{propane} = 675 \text{ m/s} \times \frac{\sqrt{16.05 \text{ g/mol}}}{\sqrt{44.11 \text{ g/mol}}} = 407 \text{ m/s}$$

DIF: III   OBJ: 11-4.2

2. A nitrogen molecule in the compressed air pumped into your car tires effuses at a velocity of 535 m/s. How fast does the oxygen in the compressed air effuse?

ANS:
501 m/s

$$v_{oxygen} = 535 \text{ m/s} \times \frac{\sqrt{28.02 \text{ g/mol}}}{\sqrt{32.00 \text{ g/mol}}} = 501 \text{ m/s}$$

DIF: III   OBJ: 11-4.2

3. A helium balloon and a balloon filled with CO_2 are punctured with holes of similar sizes. At what speed does the helium effuse if the CO_2 effuses at a speed of 410. m/s?

ANS:
1360 m/s

$$v_{helium} = 410. \text{ m/s} \times \frac{\sqrt{44.01 \text{ g/mol}}}{\sqrt{4.00 \text{ g/mol}}} = 1360 \text{ m/s}$$

DIF: III   OBJ: 11-4.2

4. A gas sprayed from an aerosol can effuses 2.08 times slower than nitrogen diffuses. What is the molar mass of the unknown gas?

ANS:
121 g/mol

$$m = (28.02 \text{ g/mol})(2.08)^2 = 121 \text{ g/mol}$$

DIF: III   OBJ: 11-4.3

5. A gas that effuses 1.19 times as slowly as does nitrogen is added to light bulbs. What is the molar mass of the unknown gas?

ANS:
39.7 g/mol

$$m = (28.02 \text{ g/mol})(1.19)^2 = 39.7 \text{ g/mol}$$

DIF: III   OBJ: 11-4.3

6. A steam vent releases an unknown gas along with steam. This gas travels 1.56 times as slowly as the steam. What is the molar mass of the unknown gas?

ANS:
43.8 g/mol

$$m = (18.02 \text{ g/mol})(1.56)^2 = 43.8 \text{ g/mol}$$

DIF: III   OBJ: 11-4.3

MULTIPLE CHOICE

1. Compared with the particles in a gas, the particles in a liquid
   a. have more energy.
   b. are larger.
   c. move around less.
   d. are farther apart.

   ANS: C   DIF: I   OBJ: 12-1.1

2. The intermolecular forces between particles in a liquid can involve all of the following except
   a. London dispersion forces.
   b. hydrogen bonding.
   c. dipole-dipole attractions.
   d. gravitational forces.

   ANS: D   DIF: I   OBJ: 12-1.1

3. The particles in both gases and liquids
   a. consist only of atoms.
   b. can change positions with other particles.
   c. vibrate in fixed positions.
   d. are packed closely together.

   ANS: B   DIF: I   OBJ: 12-1.1

4. The compressibility of a liquid is generally
   a. less than that of a gas.
   b. more than that of a gas.
   c. equal to that of a gas.
   d. zero.

   ANS: A   DIF: I   OBJ: 12-1.1

5. The intermolecular forces between particles are
   a. less effective in solids than in liquids.
   b. more effective in gases than in solids.
   c. equally effective in gases and in liquids.
   d. more effective in liquids than in gases.

   ANS: D   DIF: I   OBJ: 12-1.1

6. The lower mobility of particles in a liquid compared with those in a gas results in the liquid being
   a. less disordered.
   b. lower in density.
   c. colder.
   d. higher in energy.

   ANS: A   DIF: I   OBJ: 12-1.1

7. The attractive forces in a liquid are
   a. strong enough to prevent the particles from changing positions.
   b. too weak to hold the particles in fixed positions.
   c. more effective than those in a solid.
   d. too weak to limit the movements of the particles.

   ANS: B   DIF: I   OBJ: 12-1.1

8. The particles in a liquid are usually
   a. closer together and lower in energy than those in a solid.
   b. farther apart and higher in energy than those in a gas.
   c. closer together and lower in energy than those in a gas.
   d. farther apart and lower in energy than those in a solid.

   ANS: C   DIF: I   OBJ: 12-1.1

9. What is vaporization?
   a. the process by which a liquid changes to a gas
   b. the process by which a solid changes to a gas
   c. both a and b
   d. neither a nor b

   ANS: C   DIF: I   OBJ: 12-1.2

10. Which term best describes the process by which particles escape from the surface of a nonboiling
    liquid and enter the gas state?
    a. vaporization   c. surface tension
    b. evaporation   d. aeration

    ANS: B   DIF: I   OBJ: 12-1.2

11. What causes particles in a liquid to escape into a gas state?
    a. high kinetic energy   c. surface tension
    b. a freezing temperature   d. the combining of liquids

    ANS: A   DIF: I   OBJ: 12-1.2

12. What is the physical change of a liquid to a solid by the removal of heat?
    a. solidification   c. freezing
    b. particle arrangement   d. both a and c

    ANS: D   DIF: I   OBJ: 12-1.3

13. A solid forms when the average energy of a substance's particles
    a. increases.   c. decreases then increases.
    b. decreases.   d. creates a disorderly arrangement.

    ANS: B   DIF: I   OBJ: 12-1.3

14. Which of the following statements about freezing is NOT correct?
    a. All liquids freeze.   c. Not all liquids freeze.
    b. Water freezes at 0°C.   d. Some liquids freeze at room temperature.

    ANS: C   DIF: I   OBJ: 12-1.3

15. What can happen when the average energy of a liquid's particles decreases?
    a. vaporization   c. a disorderly arrangement
    b. evaporation   d. freezing

    ANS: D   DIF: I   OBJ: 12-1.3

16. Particles within a solid
    a.   do not move.
    b.   vibrate weakly about fixed positions.
    c.   vibrate energetically.
    d.   exchange positions easily.

    ANS: B   DIF: I   OBJ: 12-2.1

17. Forces holding particles together are strongest in a
    a.   solid.
    b.   liquid.
    c.   gas.
    d.   vapor.

    ANS: A   DIF: I   OBJ: 12-2.1

18. The energy level of the particles in a solid is
    a.   higher than the energy of the particles in a gas.
    b.   high enough to allow the particles to interchange with other particles.
    c.   higher than the energy of the particles in a liquid.
    d.   lower than the energy of the particles in liquids and gases.

    ANS: D   DIF: I   OBJ: 12-2.1

19. Compared with the particles in a liquid, the particles in a solid usually are
    a.   higher in energy.
    b.   closer together.
    c.   more massive.
    d.   more fluid.

    ANS: B   DIF: I   OBJ: 12-2.1

20. The intermolecular forces between particles in a solid are
    a.   weaker than those in a gas.
    b.   too weak to hold the particles in fixed positions.
    c.   stronger than those in a liquid.
    d.   of different types than those in a liquid.

    ANS: C   DIF: I   OBJ: 12-2.1

21. The compressibility of solids is generally
    a.   lower than the compressibility of liquids and gases.
    b.   higher than the compressibility of liquids.
    c.   about equal to the compressibility of liquids.
    d.   zero.

    ANS: A   DIF: I   OBJ: 12-2.1

22. Solids have a definite volume because
    a.   the particles do not have a tendency to change positions.
    b.   the particles are far apart.
    c.   they can be easily compressed.
    d.   the energy of the particles is high.

    ANS: A   DIF: I   OBJ: 12-2.1

23. In general, most substances are
    a. least dense in the liquid state.
    c. less dense as solids than as liquids.
    b. more dense as gases than as solids.
    d. most dense in the solid state.

    ANS: D   DIF: I   OBJ: 12-2.1

24. The rate of diffusion in solids is very low because the
    a. particles are not free to move about.
    c. attractive forces are weak.
    b. surfaces of solids usually contact gases.
    d. melting points are high.

    ANS: A   DIF: I   OBJ: 12-2.1

25. Compared with gases, solids are
    a. more easily compressed.
    c. lower in density.
    b. more likely to diffuse.
    d. incompressible.

    ANS: D   DIF: I   OBJ: 12-2.1

26. Which of the following properties do solids share with liquids?
    a. fluidity
    c. definite volume
    b. definite shape
    d. slow rate of diffusion

    ANS: C   DIF: I   OBJ: 12-2.1

27. What causes the high density of solids?
    a. The particles are more massive than those in liquids.
    b. The intermolecular forces between particles are weak.
    c. The particles are packed closely together.
    d. The energy of the particles is very high.

    ANS: C   DIF: I   OBJ: 12-2.1

28. The difference between crystalline and amorphous solids is determined by
    a. temperature changes.
    b. pressure when the substances are formed.
    c. amount of order in particle arrangement.
    d. strength of molecular forces.

    ANS: C   DIF: I   OBJ: 12-2.2

29. When heated, a pure crystalline solid will
    a. gradually soften before it melts.
    b. melt over a wide temperature range.
    c. exhibit a definite melting temperature.
    d. melt at a temperature slightly above its freezing temperature.

    ANS: C   DIF: I   OBJ: 12-2.2

30. Compared with a crystalline solid, the particles in an amorphous solid
    a. occur in a random pattern.
    b. occur in a definite, three-dimensional arrangement.
    c. consist of molecular sheets.
    d. have a more complex unit cell.

    ANS: A   DIF: I   OBJ: 12-2.2

31. Which of the following is an amorphous solid?
    a. ice   c. graphite
    b. diamond   d. glass

    ANS: D   DIF: I   OBJ: 12-2.2

32. Which of the following is not correct about crystalline solids?
    a. They can maintain a definite shape without a container.
    b. They can exist as single crystals.
    c. Their particles are held in relatively fixed positions.
    d. They are geometrically irregular.

    ANS: D   DIF: I   OBJ: 12-2.2

33. Which of the following is a crystalline solid?
    a. a plastic milk container   c. a glass bottle
    b. a quartz rock   d. a three-dimensional glass cube

    ANS: B   DIF: I   OBJ: 12-2.2

34. Which substance's solid state consists of covalent molecular crystals?
    a. salt   c. sodium
    b. water   d. diamond

    ANS: B   DIF: I   OBJ: 12-2.3

35. What type of crystal consists of positive metal cations surrounded by valence electrons that are
    donated by the metal atoms and belong to the crystal as a whole?
    a. ionic   c. metallic
    b. covalent network   d. covalent molecular

    ANS: C   DIF: I   OBJ: 12-2.3

36. What is the total three-dimensional array of points that describes the arrangement of the particles
    of a crystal?
    a. unit cell   c. diffraction pattern
    b. crystal lattice   d. crystalline system

    ANS: B   DIF: I   OBJ: 12-2.3

37. Which of the following is NOT a property of covalent network crystals?
    a. high conductivity   c. high melting point
    b. hardness   d. brittleness

    ANS: A   DIF: I   OBJ: 12-2.3

38. What is the pattern of points that describe the arrangement of particles in the entire crystal structure?
    a. unit cell
    b. cube
    c. crystal lattice
    d. type of symmetry

    ANS: C   DIF: I   OBJ: 12-2.3

39. Which of the following statements about ionic crystals is NOT correct?
    a. Their structure consists of positive and negative ions arranged in a regular pattern.
    b. The strong binding forces between the positive and negative ions in their structure give them certain properties.
    c. Their ions can be monatomic or polyatomic.
    d. They consist of molecules held together by intermolecular forces.

    ANS: D   DIF: I   OBJ: 12-2.3

40. What is the total three-dimensional arrangement of particles of a crystal?
    a. unit cell
    b. crystal structure
    c. crystal lattice
    d. crystalline symmetry

    ANS: B   DIF: I   OBJ: 12-2.3

41. What is the smallest portion of a crystal lattice that reveals the three-dimensional pattern?
    a. unit cell
    b. crystal structure
    c. coordinate system
    d. crystalline symmetry

    ANS: A   DIF: I   OBJ: 12-2.3

42. When melting and freezing proceed at the same rate, the system is
    a. sublimated.
    b. amorphous.
    c. metallic.
    d. in equilibrium.

    ANS: D   DIF: I   OBJ: 12-3.1

43. A system is in equilibrium when
    a. no physical or chemical changes are occurring.
    b. the physical changes counteract the chemical changes.
    c. opposing physical or chemical changes occur at equal rates.
    d. only physical changes are occurring.

    ANS: C   DIF: I   OBJ: 12-3.1

44. Whenever a liquid changes to a vapor, it
    a. absorbs heat energy from its surroundings.
    b. is in equilibrium with its vapor.
    c. is boiling.
    d. is condensing.

    ANS: A   DIF: I   OBJ: 12-3.1

45. If the concentration of vapor above a liquid remains zero, then
    a. no condensation can occur.
    c. the rate of condensation is high.
    b. the rate of evaporation is high.
    d. no further evaporation can occur.

    ANS: A   DIF: I   OBJ: 12-3.1

46. If the rate of evaporation from the surface of a liquid exceeds the rate of condensation,
    a. the system is in equilibrium.
    b. the liquid is boiling.
    c. heat energy is no longer available.
    d. the concentration of the vapor is increasing.

    ANS: D   DIF: I   OBJ: 12-3.1

47. If the temperature and surface area of a liquid remain constant,
    a. the liquid is not in equilibrium with its vapor.
    b. no further evaporation occurs.
    c. the rate of evaporation remains constant.
    d. the rate of condensation is greater than the rate of evaporation.

    ANS: C   DIF: I   OBJ: 12-3.1

48. Molecules at the surface of a liquid can enter the vapor phase only if
    a. equilibrium has not been reached.
    b. the concentration of the vapor is zero.
    c. their energy is high enough to overcome the attractive forces in the liquid.
    d. condensation is not occurring.

    ANS: C   DIF: I   OBJ: 12-3.1

49. When does the concentration of a vapor decrease?
    a. when the rate of evaporation decreases
    b. when the temperature remains constant
    c. when the liquid phase is warmed
    d. when the rate of condensation exceeds the rate of evaporation

    ANS: D   DIF: I   OBJ: 12-3.1

50. A liquid-vapor system at equilibrium is kept at constant temperature while the volume of the
    system is doubled. When equilibrium is restored,
    a. the concentration of vapor molecules has decreased.
    b. the vapor pressure is the same as the original vapor pressure.
    c. the volume of the liquid has increased noticeably.
    d. the number of liquid molecules has increased.

    ANS: B   DIF: I   OBJ: 12-3.2

51. When heat is applied to a liquid-vapor system at equilibrium, a new equilibrium state will have
    a. a higher percentage of liquid.
    c. equal amounts of liquid and vapor.
    b. a higher percentage of vapor.
    d. all liquid.

    ANS: B   DIF: I   OBJ: 12-3.2

52. If the volume of a liquid-vapor system at equilibrium is reduced,
    a. the temperature falls.
    b. a higher percentage of liquid will form.
    c. a higher percentage of vapor will form.
    d. the equilibrium vapor pressure will drop.

    ANS: B   DIF: I   OBJ: 12-3.2

53. If the temperature of a liquid-vapor system at equilibrium is reduced, the
    a. concentration of the vapor will decrease.
    b. rate of evaporation will increase.
    c. equilibrium is unaffected.
    d. percentage of liquid in the system will decrease.

    ANS: A   DIF: I   OBJ: 12-3.2

54. If the volume of a liquid-vapor system increases, the immediate effect will be a net
    a. increase in condensation.
    b. decrease in evaporation.
    c. increase in the vapor pressure.
    d. increase in evaporation.

    ANS: D   DIF: I   OBJ: 12-3.2

55. If the temperature of a liquid-vapor system at equilibrium increases, the new equilibrium
    condition will
    a. have a lower concentration of vapor.
    b. have an increased vapor pressure.
    c. not have equal rates of condensation and evaporation.
    d. be larger in volume.

    ANS: B   DIF: I   OBJ: 12-3.2

56. If the temperature of a closed liquid-vapor equilibrium system is raised, its vapor pressure
    a. decreases.
    b. increases.
    c. remains the same.
    d. shows no correlation.

    ANS: B   DIF: I   OBJ: 12-3.2

57. The equilibrium vapor pressure of a liquid is
    a. the same for all liquids.
    b. measured only at 0°C.
    c. constant for a particular liquid at all temperatures.
    d. the pressure exerted by a vapor in equilibrium with its liquid at a given temperature.

    ANS: D   DIF: I   OBJ: 12-3.3

58. At a given temperature, different liquids will have different equilibrium vapor pressures because
    a. the energy of the particles is the same for different liquids.
    b. diffusion rates differ for the liquids.
    c. the attractive forces between the particles differ among liquids.
    d. they cannot all be in equilibrium at once.

    ANS: C   DIF: I   OBJ: 12-3.3

59. A volatile liquid
    a. has strong attractive forces between particles.
    b. evaporates readily.
    c. has an odor.
    d. is ionic.

    ANS: B   DIF: I   OBJ: 12-3.3

60. The equilibrium vapor pressure of water is
    a. constant at all temperatures.   c. unrelated to temperature.
    b. specific for any given temperature.   d. inversely proportional to the temperature.

    ANS: B   DIF: I   OBJ: 12-3.3

61. The equilibrium vapor pressure of a liquid increases with increasing temperature because
    a. the rate of condensation decreases.
    b. the average energy of the particles in the liquid increases.
    c. the volume decreases.
    d. the boiling point decreases.

    ANS: B   DIF: I   OBJ: 12-3.3

62. The equilibrium vapor pressure of a molten ionic compound is likely to be
    a. lower than that of ether.   c. higher than that of volatile liquids.
    b. zero except when it is boiling.   d. proportional to the volume.

    ANS: A   DIF: II   OBJ: 12-3.3

63. If a solid changes directly to a vapor, its equilibrium vapor pressure
    a. cannot be measured.   c. is determined by the temperature.
    b. is zero.   d. follows Boyle's law.

    ANS: C   DIF: I   OBJ: 12-3.3

64. If the equilibrium vapor pressure is falling,
    a. the rate of evaporation is increasing.
    b. the liquid has begun to boil.
    c. no further condensation can take place.
    d. the temperature of the liquid is decreasing.

    ANS: D   DIF: I   OBJ: 12-3.3

65. What is the process of a substance changing from a solid to a vapor without passing through the liquid phase?
    a. condensation   c. sublimation
    b. evaporation   d. vaporization

    ANS: C   DIF: I   OBJ: 12-3.4

66. At pressures greater than 760 mm Hg, water will boil at
    a. a temperature higher than 100°C.   c. 100°C.
    b. a temperature lower than 100°C.   d. 4°C.

    ANS: A   DIF: I   OBJ: 12-3.4

67. Why would a camper near the top of Mt. Everest find that water boils at less than 100°C?
    a. There is greater atmospheric pressure than at sea level.
    b. The flames are hotter at that elevation.
    c. There is less atmospheric pressure than at sea level.
    d. The atmosphere has less moisture.

    ANS: C   DIF: II   OBJ: 12-3.4

68. Glycerol boils at a slightly higher temperature than does water. This reveals that glycerol's
    attractive forces are
    a. nonexistent.   c. the same as those of water.
    b. weaker than those of water.   d. stronger than those of water.

    ANS: D   DIF: II   OBJ: 12-3.4

69. Diethyl ether's boiling point is about 35°C at 1 atm. At 1.5 atm, what will be ether's approximate
    boiling point?
    a. −10°C   c. 40°C
    b. 20°C   d. 100°C

    ANS: C   DIF: I   OBJ: 12-3.4

70. During boiling, the temperature of a liquid
    a. remains constant.   c. decreases.
    b. increases.   d. approaches the standard boiling point.

    ANS: A   DIF: I   OBJ: 12-3.4

71. During the process of freezing, a liquid
    a. loses kinetic energy.   c. gains potential energy.
    b. loses potential energy.   d. gains kinetic energy.

    ANS: B   DIF: I   OBJ: 12-3.4

72. As the atmospheric pressure on the surface of a liquid decreases, its boiling point
    a. decreases.   c. remains unchanged.
    b. increases.   d. shows no correlation.

    ANS: A   DIF: I   OBJ: 12-3.4

73. At its triple point, water can
    a. have only three pressure values.
    b. exist in equilibrium in three different phases.
    c. only be present as vapor.
    d. exist only as a solid.

    ANS: B   DIF: I   OBJ: 12-3.5

74. A phase diagram indicates the conditions under which
    a. the various states of a substance exist.
    c. Le Châtelier's principle no longer applies.
    b. amorphous solids become crystalline.
    d. all vapors become flammable.

    ANS: A   DIF: I   OBJ: 12-3.5

75. The triple point of a substance is the temperature and pressure conditions at which
    a. density is greatest.
    b. states of a substance coexist at equilibrium.
    c. equilibrium cannot occur.
    d. kinetic energy is at a minimum.

    ANS: B   DIF: I   OBJ: 12-3.5

76. Above the critical temperature, a substance
    a. does not have a vapor pressure.
    c. cannot exist in the liquid state.
    b. sublimes.
    d. is explosive.

    ANS: C   DIF: I   OBJ: 12-3.5

77. What is the critical pressure?
    a. the pressure at which all substances are solids
    b. the pressure at which the attractive forces in matter break down
    c. the highest pressure under which a solid can exist
    d. the lowest pressure under which a substance can exist as a liquid at the critical temperature

    ANS: D   DIF: I   OBJ: 12-3.5

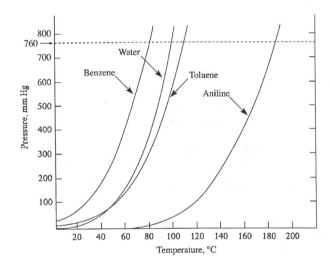

78. According to the figure above, what is the most volatile substance shown?
    a. benzene
    c. toluene
    b. water
    d. aniline

    ANS: A   DIF: III   OBJ: 12-3.5

79. What does the dotted line in the figure above represent?
    a. the number of moles of liquid present
    b. the temperature of all three liquids
    c. standard pressure
    d. boiling temperature

    ANS: C   DIF: III   OBJ: 12-3.5

80. The angle between the two H–O bonds in water is evidence of
    a. sp^3 hybridization.
    b. sp^2 hybridization.
    c. $2s$ orbitals.
    d. $2p$ orbitals.

    ANS: A   DIF: I   OBJ: 12-4.1

81. Water molecules freeze in which definite pattern?
    a. hexagonal
    b. triangular
    c. square
    d. circular

    ANS: A   DIF: I   OBJ: 12-4.1

82. Why are water molecules polar?
    a. They contain two kinds of atoms.
    b. The electrons in the covalent bonds spend more time closer to the oxygen nucleus.
    c. The hydrogen bonds are weak.
    d. They have covalent bonds.

    ANS: B   DIF: I   OBJ: 12-4.1

83. Why is ice less dense than liquid water?
    a. The molecules in liquid water can crowd together more compactly than in ice.
    b. Liquid water's energy level is lower than that of ice.
    c. Liquid water molecules are farther apart than the molecules in ice.
    d. Liquid water has fewer chemical impurities than ice has.

    ANS: A   DIF: I   OBJ: 12-4.1

84. In ice, what holds H_2O molecules together?
    a. ionic bonds
    b. dispersion interaction forces
    c. hydrogen bonds
    d. oxygen-oxygen bonds

    ANS: C   DIF: I   OBJ: 12-4.1

85. If the water molecules were nonpolar, it is likely that
    a. water would be a solid at room temperature.
    b. water would be flammable.
    c. water would be a gas at room temperature.
    d. the hydrogen bonding would be stronger.

    ANS: C   DIF: I   OBJ: 12-4.1

86. The electronegativity difference between oxygen and hydrogen results in the water molecule being
    a. flammable.
    b. polar.
    c. ionically bonded.
    d. linear.

    ANS: B   DIF: I   OBJ: 12-4.1

87. What is the freezing point of water at standard pressure?
    a. −10°C
    b. 0°C
    c. 4°C
    d. 32°C

    ANS: B   DIF: I   OBJ: 12-4.2

88. How does the molar heat of fusion of ice compare with the molar heat of fusion of other solids?
    a. It is about the same.
    b. It is relatively small.
    c. It is relatively large.
    d. It is about the same as that of colorless solids.

    ANS: C   DIF: I   OBJ: 12-4.2

89. Water changes from liquid to solid at 0°C and what pressure?
    a. 1 atm
    b. 706 mm Hg
    c. −1 atm
    d. 1.436 mm Hg

    ANS: A   DIF: I   OBJ: 12-4.2

90. What is the boiling point of water at standard pressure?
    a. 100°C
    b. 112°C
    c. 212°C
    d. 200°C

    ANS: A   DIF: I   OBJ: 12-4.2

91. At about what temperature does water reach its maximum density?
    a. 0°C
    b. 2°C
    c. 4°C
    d. 6°C

    ANS: C   DIF: I   OBJ: 12-4.2

92. What is the mass of 1 mL of water at its temperature of maximum density?
    a. 1 mg
    b. 1.5 mg
    c. 1 g
    d. 1.5 g

    ANS: C   DIF: I   OBJ: 12-4.2

93. When water is warmed from its freezing temperature to its temperature of maximum density, it
    a. contracts.
    b. expands.
    c. maintains a constant volume.
    d. increases in weight.

    ANS: A   DIF: I   OBJ: 12-4.2

94. Why doesn't water in lakes and ponds of temperate climates freeze solid during the winter and kill nearly all the living things it contains?
    a. Water is colorless.
    b. Ice floats.
    c. The molar heat of fusion of ice is relatively low.
    d. Water contracts as it freezes.

    ANS: B   DIF: II   OBJ: 12-4.2

95. What is the characteristic of water that makes steam useful for household heating systems?
    a. high molar heat of vaporization   c. high density
    b. low molar heat of fusion   d. low boiling point

    ANS: A   DIF: I   OBJ: 12-4.2

96. The molar heat of vaporization of water is 40.79 kJ at 100°C. What is the heat of vaporization of 1 g of water?
    a. 40.79 J   c. 500. J
    b. 80.3 J   d. 2.26 kJ

    ANS: D   DIF: III   OBJ: 12-4.3

97. The molar heat of fusion for water is 6.008 kJ/mol. How much energy would be required to melt 94.0 g of ice?
    a. 0.869 kJ   c. 31.3 kJ
    b. 81.7 kJ   d. 282. kJ

    ANS: C   DIF: III   OBJ: 12-4.3

98. The molar heat of fusion for water is 6.008 kJ/mol. What quantity of heat energy is released when 253 g of liquid water freezes?
    a. 759 kJ   c. 2.33 kJ
    b. 0.429 kJ   d. 84.4 kJ

    ANS: D   DIF: III   OBJ: 12-4.3

99. The standard molar heat of vaporization for water is 40.79 kJ/mol. How much energy would be required to vaporize 94.0 g of water?
    a. 0.128 kJ   c. 41.5 kJ
    b. 213 kJ   d. 7.81 kJ

    ANS: B   DIF: III   OBJ: 12-4.3

100. The standard molar heat of vaporization for water is 40.79 kJ/mol. What mass of steam is required to release 500. kJ of heat energy upon condensation?
    a. 221 g   c. 1130 g
    b. 325 g   d. 1660 g

    ANS: A   DIF: III   OBJ: 12-4.3

1. Explain how evaporation occurs.

   ANS:
   The particles in a liquid have different kinetic energies. Particles on the surface of the liquid that have higher-than-average energies can overcome the intermolecular forces that bind them to the liquid. These particles escape the liquid and enter the gas state.

   DIF: II   OBJ: 12-1.2

2. How are vaporization and evaporation related?

   ANS:
   Vaporization is the process by which a liquid or solid changes to a gas. Evaporation is a form of vaporization, and more specifically, the process by which particles escape from the surface of a nonboiling liquid and enter the gas state.

   DIF: II   OBJ: 12-1.2

3. How is a solid formed?

   ANS:
   When a liquid is cooled, the average energy of its particles decreases. When the energy is low enough, attractive forces pull the particles into a more orderly arrangement. This orderly arrangement is a solid.

   DIF: II   OBJ: 12-1.3

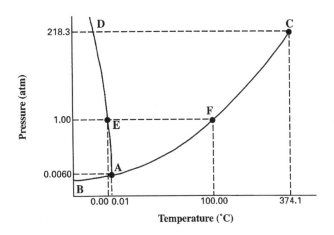

4. What do points E and F represent in the figure above?

   ANS:
   E is the normal freezing point, and F is the normal boiling point.

   DIF: III   OBJ: 12-3.5

5. What does point A represent in the figure above?

ANS:
Point A is the triple point for water, where the solid, liquid, and vapor phases of water exist in equilibrium.

DIF: III   OBJ: 12-3.5

6. Explain what the curves AB, AC, and AD represent in the figure above.

ANS:
Curve AB indicates the conditions under which ice and water vapor coexist at equilibrium. Curve AC indicates the conditions under which liquid water and water vapor coexist at equilibrium. Curve AD indicates conditions under which ice and liquid water coexist at equilbrium.

DIF: III   OBJ: 12-3.5

Modern Chemistry Assessment Item Listing
198

MULTIPLE CHOICE

1. Which of the following has components that are obviously different?
    a. homogeneous mixture
    b. solution
    c. colloid
    d. heterogeneous mixture

    ANS: D   DIF: I   OBJ: 13-1.1

2. A heterogeneous mixture always contains
    a. only one substance.
    b. more than two substances.
    c. two or more substances that are visibly distinguishable.
    d. two or more substances that are not visibly distinguishable.

    ANS: C   DIF: I   OBJ: 13-1.1

3. Which of the following is a pure substance?
    a. water
    b. milk
    c. soil
    d. concrete

    ANS: A   DIF: II   OBJ: 13-1.1

4. Which of the following is a heterogeneous mixture?
    a. water
    b. a sugar-water solution
    c. whole-wheat bread
    d. sugar

    ANS: C   DIF: II   OBJ: 13-1.1

5. Which of the following is a homogeneous mixture?
    a. water
    b. a sugar-water solution
    c. whole-wheat bread
    d. sugar

    ANS: B   DIF: II   OBJ: 13-1.1

6. Which of the following is a thoroughly mixed homogeneous mixture of substances in a single phase?
    a. a solution
    b. a colloid
    c. a compound
    d. a suspension

    ANS: A   DIF: I   OBJ: 13-1.1

7. All of the following are homogeneous mixtures EXCEPT
    a. sodium chloride.
    b. a sugar-water solution.
    c. gasoline.
    d. a salt-water solution.

    ANS: A   DIF: II   OBJ: 13-1.1

8. All of the following are heterogeneous mixtures EXCEPT
   a. whole-wheat bread.
   b. granite.
   c. tap water.
   d. an oil-water mixture.

   ANS: C   DIF: II   OBJ: 13-1.1

9. Which of the following is NOT a solute-solvent combination?
   a. gas-gas combination
   b. gas-liquid combination
   c. gas-solid combination
   d. liquid-solid combination

   ANS: C   DIF: I   OBJ: 13-1.2

10. Water in air is an example of which solute-solvent combination?
    a. gas-liquid
    b. liquid-gas
    c. liquid-liquid
    d. gas-gas

    ANS: B   DIF: II   OBJ: 13-1.2

11. Carbon dioxide in water is an example of which solute-solvent combination?
    a. gas-liquid
    b. liquid-gas
    c. liquid-liquid
    d. cannot be determined

    ANS: A   DIF: II   OBJ: 13-1.2

12. Sugar in water is an example of which solute-solvent combination?
    a. gas-liquid
    b. liquid-liquid
    c. solid-liquid
    d. liquid-solid

    ANS: C   DIF: II   OBJ: 13-1.2

13. Oxygen in nitrogen is an example of which solute-solvent combination?
    a. gas-liquid
    b. liquid-gas
    c. gas-solid
    d. gas-gas

    ANS: D   DIF: II   OBJ: 13-1.2

14. Which mixture contains visible particles that settle out unless the mixture is stirred?
    a. a colloid
    b. a homogeneous mixture
    c. a solution
    d. a suspension

    ANS: D   DIF: I   OBJ: 13-1.3

15. Which mixture contains particles that are in a dispersed phase and do not settle out?
    a. a colloid
    b. a heterogeneous mixture
    c. a solution
    d. a suspension

    ANS: A   DIF: I   OBJ: 13-1.3

16. A metal solution is a(n)
    a. colloid.
    b. alloy.
    c. suspension.
    d. electrolyte.

    ANS: B   DIF: I   OBJ: 13-1.3

17. An aerosol dispenser contains a colloidal dispersion of
    a. two liquids.
    b. two solids.
    c. a solid and a liquid.
    d. a gas and a liquid.

    ANS: D   DIF: I   OBJ: 13-1.3

18. The colloidal particles in a colloid form the
    a. dispersing medium.
    b. dispersed phase.
    c. solvent.
    d. solute.

    ANS: B   DIF: I   OBJ: 13-1.3

19. Colloids
    a. can be separated by filtering.
    b. settle out when allowed to stand.
    c. scatter light.
    d. are heterogeneous.

    ANS: C   DIF: I   OBJ: 13-1.3

20. The Tyndall effect is used to distinguish between
    a. liquids and gases.
    b. solutions and colloids.
    c. colloids and heterogeneous mixtures.
    d. solvents and solutes.

    ANS: B   DIF: I   OBJ: 13-1.3

21. A substance whose water solution conducts a current is a(n)
    a. nonelectrolyte.
    b. electrolyte.
    c. nonpolar substance.
    d. solute.

    ANS: B   DIF: I   OBJ: 13-1.4

22. To conduct electricity, a solution must contain
    a. nonpolar molecules.
    b. polar molecules.
    c. ions.
    d. free electrons.

    ANS: C   DIF: I   OBJ: 13-1.4

23. Which of the following is an electrolyte?
    a. sodium chloride
    b. sugar
    c. water
    d. glass

    ANS: A   DIF: II   OBJ: 13-1.4

24. A substance whose water solution does NOT conduct a current is a(n)
    a. polar substance.
    b. nonelectrolyte.
    c. electrolyte.
    d. ionic substance.

    ANS: B   DIF: I   OBJ: 13-1.4

25. Which of the following is a molecular substance whose water solution conducts electricity?
    a. liquid hydrogen
    b. hydrogen chloride
    c. sugar
    d. iron

    ANS: B   DIF: II   OBJ: 13-1.4

26. Which of the following is a nonelectrolyte?
    a. sodium chloride   c. sugar
    b. hydrogen chloride   d. potassium chloride

    ANS: C   DIF: II   OBJ: 13-1.4

27. Molecules whose water solutions conduct current
    a. require carbon to decompose in water.   c. do not dissolve in water.
    b. ionize in water.   d. can crystallize.

    ANS: B   DIF: I   OBJ: 13-1.4

28. Which of the following does NOT increase the rate of dissolving a solid in water?
    a. raising the temperature   c. using larger pieces of solid
    b. stirring   d. crushing the solid

    ANS: C   DIF: I   OBJ: 13-2.1

29. Increasing the surface area between solute and solvent
    a. increases the rate of dissolution.
    b. decreases the rate of dissolution.
    c. has no effect on the rate of dissolution.
    d. can increase, decrease, or have no effect on the rate of dissolution.

    ANS: A   DIF: I   OBJ: 13-2.1

30. Which of the following decreases the average speed of solvent molecules?
    a. increasing the temperature   c. adding more solvent
    b. stirring the solution   d. decreasing the temperature

    ANS: D   DIF: I   OBJ: 13-2.1

31. Stirring increases the rate of dissolution because it
    a. raises the temperature.
    b. lowers the temperature.
    c. brings fresh solvent into contact with the solute.
    d. decreases surface area of the solute.

    ANS: C   DIF: I   OBJ: 13-2.1

32. Which of the following will dissolve most rapidly?
    a. sugar cubes in cold water   c. powdered sugar in cold water
    b. sugar cubes in hot water   d. powdered sugar in hot water

    ANS: D   DIF: I   OBJ: 13-2.1

33. Which of the following will dissolve most slowly?
    a. large salt crystals in unstirred water   c. small salt crystals in unstirred water
    b. large salt crystals in stirred water   d. small salt crystals in stirred water

    ANS: A   DIF: I   OBJ: 13-2.1

34. Raising the collision rate between solute and solvent
    a. increases the rate of dissolution.
    b. decreases the rate of dissolution.
    c. has no effect on the rate of dissolution.
    d. can increase, decrease, or have no effect on the rate of dissolution.

    ANS: A   DIF: I   OBJ: 13-2.1

35. Raising solvent temperature causes solvent-solute collisions to become
    a. less frequent and more energetic.   c. less frequent and less energetic.
    b. more frequent and more energetic.   d. more frequent and less energetic.

    ANS: B   DIF: I   OBJ: 13-2.1

36. If a solution is not agitated while it is being made, dissolved solute tends to
    a. mix uniformly.   c. build up in the solvent near the solute.
    b. build up in the solvent far from the   d. raise the temperature of the solvent.
        solute.

    ANS: C   DIF: I   OBJ: 13-2.1

37. Which of the following is at equilibrium when undissolved solute is visible?
    a. a saturated solution   c. a supersaturated solution
    b. an unsaturated solution   d. all of the above

    ANS: A   DIF: I   OBJ: 13-2.2

38. If the amount of solute present in a solution at a given temperature is less than the maximum
    amount that can dissolve at that temperature, the solution is said to be
    a. saturated.   c. supersaturated.
    b. unsaturated.   d. concentrated.

    ANS: B   DIF: I   OBJ: 13-2.2

39. If the amount of dissolved solute in a solution at a given temperature is greater than the amount
    that can permanently remain in solution at that temperature, the solution is said to be
    a. saturated.   c. supersaturated.
    b. unsaturated.   d. diluted.

    ANS: C   DIF: I   OBJ: 13-2.2

40. In a solution at equilibrium,
    a. no dissolution occurs.
    b. the rate of dissolution is less than the rate of crystallization.
    c. the rate of dissolution is greater than the rate of crystallization.
    d. the rate of dissolution and the rate of crystallization are equal.

    ANS: D   DIF: I   OBJ: 13-2.2

41. The solubility of a substance at a given temperature can be expressed as
    a. grams of solute.
    b. grams of solvent.
    c. amount of solute per amount of solvent.
    d. grams of water per 100 g of solute.

    ANS: C   DIF: I   OBJ: 13-2.2

42. The solubility of a solute depends on
    a. the nature of the solute and the temperature of the solvent.
    b. the nature of the solute only.
    c. the temperature of the solvent only.
    d. neither the nature of the solute nor the temperature of the solvent.

    ANS: A   DIF: I   OBJ: 13-2.2

43. Which of the following is likely to produce crystals if disturbed?
    a. an unsaturated solution
    b. a supersaturated solution
    c. a saturated solution
    d. a concentrated solution

    ANS: B   DIF: I   OBJ: 13-2.2

44. The rate of dissolution is
    a. directly related to solubility.
    b. inversely related to solubility.
    c. related to the square of the solubility.
    d. not related to solubility.

    ANS: D   DIF: I   OBJ: 13-2.2

45. In the expression like dissolves like, the word like refers to similarity in molecular
    a. mass.
    b. size.
    c. energy.
    d. polarity.

    ANS: D   DIF: I   OBJ: 13-2.3

46. The rule like dissolves like is used to predict
    a. solubility.
    b. equilibrium.
    c. reactivity.
    d. phase.

    ANS: A   DIF: I   OBJ: 13-2.3

47. Which of the following is an example of a polar solvent?
    a. carbon tetrachloride
    b. benzene
    c. water
    d. gasoline

    ANS: C   DIF: II   OBJ: 13-2.3

48. Which of the following is an example of a nonpolar solvent?
    a. water
    b. carbon tetrachloride
    c. both a and b
    d. neither a nor b

    ANS: B   DIF: II   OBJ: 13-2.3

49. A substance that is NOT soluble in a polar solvent is
    a. nonpolar.   c. polar.
    b. ionic.   d. hydrogen bonded.

    ANS: A   DIF: I   OBJ: 13-2.3

50. Which of the following is soluble in water?
    a. potassium nitrate   c. benzene
    b. silver   d. carbon tetrachloride

    ANS: A   DIF: II   OBJ: 13-2.3

51. Sugar is soluble in water because sugar molecules are
    a. massive.   c. nonpolar.
    b. large.   d. polar.

    ANS: D   DIF: I   OBJ: 13-2.3

52. Two immiscible substances are
    a. water and ammonia.   c. carbon tetrachloride and benzene.
    b. water and ethanol.   d. benzene and water.

    ANS: D   DIF: I   OBJ: 13-2.3

53. Which of the following is a solvent for both polar and nonpolar solutes?
    a. water   c. ethanol
    b. carbon tetrachloride   d. benzene

    ANS: C   DIF: I   OBJ: 13-2.3

54. Endothermic dissolution processes
    a. absorb heat and have positive heats of solution.
    b. release heat and have positive heats of solution.
    c. absorb heat and have negative heats of solution.
    d. release heat and have negative heats of solution.

    ANS: A   DIF: I   OBJ: 13-2.4

55. In solvation, which particles interact?
    a. solute only   c. solute and solvent
    b. solvent only   d. none of the above

    ANS: C   DIF: I   OBJ: 13-2.4

56. Which of the following releases energy?
    a. overcoming solute-solute attraction   c. overcoming solvent-solvent attraction
    b. forming solute-solvent attraction   d. all of the above

    ANS: B   DIF: I   OBJ: 13-2.4

57. When the energy released by forming solvent-solute attractions is greater than the energy absorbed by overcoming solute-solute and solvent-solvent attractions, the dissolving process
    a. has a negative heat of solution.
    b. has a positive heat of solution.
    c. occurs rapidly.
    d. does not occur.

    ANS: A   DIF: I   OBJ: 13-2.4

58. Increasing temperature favors dissolution when
    a. the heat of solution is negative.
    b. the heat of solution is positive.
    c. dissolution occurs rapidly.
    d. the dissolution process is exothermic.

    ANS: B   DIF: I   OBJ: 13-2.4

59. The dissolution of gases in liquids is generally
    a. endothermic.
    b. exothermic.
    c. rapid.
    d. impossible.

    ANS: B   DIF: I   OBJ: 13-2.4

60. Heat of solution is generally expressed in
    a. kilocalories.
    b. moles of solute per kilogram.
    c. kilojoules per mole of solute at a specified temperature.
    d. moles of solute in a specified amount of solvent per kilojoule.

    ANS: C   DIF: I   OBJ: 13-2.4

61. Which of the following values for heat of solution at 25°C represents the greatest release of energy?
    a. −3.59 kJ/mol
    b. −0.01 kJ/mol
    c. +1.33 kJ/mol
    d. +12.40 kJ/mol

    ANS: A   DIF: II   OBJ: 13-2.4

62. The formation of solid-liquid solutions
    a. always releases heat.
    b. always absorbs heat.
    c. can either absorb or release heat.
    d. neither absorbs nor releases heat.

    ANS: C   DIF: I   OBJ: 13-2.4

63. Pressure has the greatest effect on the solubility of
    a. solids in liquids.
    b. liquids in liquids.
    c. gases in gases.
    d. gases in liquids.

    ANS: D   DIF: I   OBJ: 13-2.5

64. The solubility of gases in liquids
    a. always increases with increasing pressure.
    b. sometimes increases with increasing pressure.
    c. always decreases with increasing pressure.
    d. does not depend on pressure.

    ANS: A   DIF: I   OBJ: 13-2.5

65. Henry's law relates
    a. pressure to temperature.
    b. pressure to gas-liquid solubility.
    c. temperature to gas-liquid solubility.
    d. pressure to liquid-solid solubility.

    ANS: B   DIF: I   OBJ: 13-2.5

66. For a mixture of gases, the solubility of each gas in water varies
    a. directly with the partial pressure of the gas.
    b. inversely with the partial pressure of the gas.
    c. directly with the total pressure of the mixture.
    d. inversely with the total pressure of the mixture.

    ANS: A   DIF: I   OBJ: 13-2.5

67. Effervescence is the
    a. dissolution of gas in liquid.
    b. escape of gas from a gas-liquid solution.
    c. escape of gas from a container of gas.
    d. escape of solid from a solid-liquid solution.

    ANS: B   DIF: I   OBJ: 13-2.5

68. As temperature increases, solubility of gases in liquids
    a. increases.
    b. decreases.
    c. can increase or decrease.
    d. is not affected.

    ANS: B   DIF: I   OBJ: 13-2.5

69. As temperature increases, solubility of solids in liquids
    a. always increases.
    b. always decreases.
    c. usually increases.
    d. usually decreases.

    ANS: C   DIF: I   OBJ: 13-2.5

70. Cesium sulfate is a typical solid whose solubility decreases
    a. only if the pressure decreases.
    b. with increasing temperature.
    c. only if the pressure increases.
    d. with decreasing temperature.

    ANS: B   DIF: II   OBJ: 13-2.5

71. Which of the following expresses concentration?
    a. molality
    b. molarity
    c. percent concentration by mass
    d. all of the above

    ANS: D   DIF: I   OBJ: 13-3.1

72. Which of the following is expressed in grams of solute instead of moles of solute?
    a. molality
    b. molarity
    c. neither a nor b
    d. a and b

    ANS: C   DIF: I   OBJ: 13-3.1

73. What is the molarity of a solution that contains 0.202 mol KCl in 7.98 L solution?
    a.   0.0132 M   c.   0.459 M
    b.   0.0253 M   d.   1.363 M

    ANS: B   DIF: III   OBJ: 13-3.1

74. What is the molality of a solution that contains 5.10 mol KNO_3 in 4.47 kg water?
    a.   0.315 m c.   1.02 m
    b.   0.779 m d.   1.14 m

    ANS: D   DIF: III   OBJ: 13-3.1

75. What is the molarity of a solution that contains 125 g NaCl in 4.00 L solution?
    a.   0.535 M   c.   8.56 M
    b.   2.14 M   d.   31.3 M

    ANS: A   DIF: III   OBJ: 13-3.1

76. What is the molality of a solution that contains 31.0 g HCl in 5.00 kg water?
    a.   0.062 m c.   0.170 m
    b.   0.425 m d.   15.5 m

    ANS: C   DIF: III   OBJ: 13-3.1

77. How many moles of HCl are present in 0.70 L of a 0.33 M HCl solution?
    a.   0.23 mol   c.   0.38 mol
    b.   0.28 mol   d.   0.47 mol

    ANS: A   DIF: III   OBJ: 13-3.2

78. The concentration of a water solution of NaCl is 2.48 m, and it contains 806 g of water. How much NaCl is in the solution?
    a.   2.00 g   c.   117 g
    b.   89.3 g   d.   224 g

    ANS: C   DIF: III   OBJ: 13-3.2

79. An NaOH solution contains 1.90 mol of NaOH, and its concentration is 0.555 M. What is its volume?
    a.   0.623 L   c.   1.05 L
    b.   0.911 L   d.   3.42 L

    ANS: D   DIF: III   OBJ: 13-3.3

80. What mass of water must be used to make a 1.35 m solution that contains 8.20 mol of NaOH?
    a.   6.07 kg   c.   11.1 kg
    b.   7.44 kg   d.   14.5 kg

    ANS: A   DIF: III   OBJ: 13-3.3

81. How many milliliters water are needed to make a 0.171 M solution that contains 1.00 g of NaCl?
   a.   100 mL   c.   171 mL
   b.   1000 mL   d.   17.1 mL

   ANS: A   DIF: III   OBJ: 13-3.3

PROBLEM

1. A solution contains 85.0 g of $NaNO_3$, and has a volume of 750. mL. Find the molarity of the solution.

   ANS:
   1.33 M $NaNO_3$

   $$\frac{85.0 \text{ g NaNO}_3}{750.\text{ mL}} \times \frac{1 \text{ mol NaNO}_3}{85.0 \text{ g NaNO}_3} \times \frac{1000 \text{ mL}}{1 \text{ L}} = 1.33 \text{ M NaNO}_3$$

   DIF: III   OBJ: 13-3.1

2. What is the molarity of a solution of sucrose, $C_{12}H_{22}O_{11}$, that contains 125 g of sucrose in 3.50 L of solution?

   ANS:
   0.104 M sucrose

   $$\frac{125 \text{ g sucrose}}{3.50 \text{ L}} \times \frac{1 \text{ mol sucrose}}{342.3 \text{ g sucrose}} = 0.104 \text{ M sucrose}$$

   DIF: III   OBJ: 13-3.1

3. How many grams of NaOH are required to prepare 200. mL of a 0.450 M solution?

   ANS:
   3.60 g NaOH

   $$200.\text{ mL} \times \frac{1 \text{ L}}{1000 \text{ mL}} \times \frac{0.450 \text{ mol NaOH}}{1 \text{ L}} \times \frac{40.00 \text{ g NaOH}}{1 \text{ mol NaOH}} = 3.60 \text{ g NaOH}$$

   DIF: III   OBJ: 13-3.2

4. How many grams of $NaC_2H_3O_2$ are needed to prepare 350. mL of a 2.75 M solution?

ANS:
79.0 g $NaC_2H_3O_2$

$$350. \text{ mL} \times \frac{1 \text{ L}}{1000 \text{ mL}} \times \frac{2.75 \text{ mol } NaC_2H_3O_2}{1 \text{ L}} \times \frac{82.04 \text{ g } NaC_2H_3O_2}{1 \text{ mol } NaC_2H_3O_2} = 79.0 \text{ g } NaC_2H_3O_2$$

DIF: III   OBJ: 13-3.2

5. How many grams of Na_2SO_4 are needed to prepare 750. mL of a 0.375 M solution?

ANS:
40.0 g Na_2SO_4

$$750. \text{ mL} \times \frac{1 \text{ L}}{1000 \text{ mL}} \times \frac{0.375 \text{ mol } Na_2SO_4}{1 \text{ L}} \times \frac{142.05 \text{ g } Na_2SO_4}{1 \text{ mol } Na_2SO_4} = 40.0 \text{ g } Na_2SO_4$$

DIF: III   OBJ: 13-3.2

6. Iron(III) chloride can be produced by reacting Fe_2O_3 with a hydrochloric acid solution. How many milliliters of a 6.00 M HCl solution are needed to react with excess Fe_2O_3 to produce 16.5 g of $FeCl_3$?

ANS:
50.9 mL HCl solution

Balanced equation for reaction:
$Fe_2O_3(s) + 6HCl(aq) \rightarrow 2FeCl_3(aq) + 3H_2O(l)$

$$16.5 \text{ g } FeCl_3 \times \frac{1 \text{ mol } FeCl_3}{162.20 \text{ g } FeCl_3} \times \frac{6 \text{ mol HCl}}{2 \text{ mol } FeCl_3} \times \frac{1 \text{ L}}{6.00 \text{ mol HCl}} \times \frac{1000 \text{ mL}}{1 \text{ L}} = 50.9 \text{ mL HCl}$$

DIF: III   OBJ: 13-3.3

7. Muriatic acid is often used to remove rust. A solution of muriatic acid, HCl, reacts with Fe_2O_3 deposits on industrial equipment. How many liters of 5.50 M HCl would be needed to react completely with 439 g Fe_2O_3?

ANS:
2.99 L HCl solution

balanced equation for reaction:
$Fe_2O_3(s) + 6HCl(aq) \rightarrow 2FeCl_3(aq) + 3H_2O(l)$

$$439 \text{ g Fe}_2\text{O}_3 \times \frac{1 \text{ mol Fe}_2\text{O}_3}{160 \text{ g Fe}_2\text{O}_3} \times \frac{6 \text{ mol HCl}}{1 \text{ mol Fe}_2\text{O}_3} \times \frac{1 \text{ L}}{5.50 \text{ mol HCl}} = 2.99 \text{ L HCl}$$

DIF: III   OBJ: 13-3.3

MULTIPLE CHOICE

1. An equation for the dissociation of an ionic solid shows
    a. the solid on the left side and aqueous ions on the right side.
    b. an aqueous solid on the left side and ions on the right side.
    c. aqueous ions on the left side and the solid on the right side.
    d. ions on the left side and an aqueous solid on the right side.

    ANS: A   DIF: I   OBJ: 14-1.1

2. How many ions are produced by each formula unit of solid in a dissociation?
    a. zero   c. two
    b. one   d. two or more

    ANS: D   DIF: I   OBJ: 14-1.1

3. How many moles of ions are produced by the dissociation of 1 mol of $MgCl_2$?
    a. 0   c. 2 mol
    b. 1 mol   d. 3 mol

    ANS: D   DIF: II   OBJ: 14-1.1

4. How many moles of ions are produced by the dissociation of 1 mol of $Al_2(CO_3)_3$?
    a. 2 mol   c. 5 mol
    b. 4 mol   d. 11 mol

    ANS: C   DIF: II   OBJ: 14-1.1

5. How many moles of ions are produced by the dissociation of 1 mol of NH_4Br?
    a. 0   c. 2 mol
    b. 1 mol   d. 5 mol

    ANS: C   DIF: II   OBJ: 14-1.1

6. Which of the following is the right side of the equation for dissolving $Al(NO_3)_3$?
    a. $Al^+ + (NO_3)_3^-$ c. $Al^{3+} + NO_3^{3-}$
    b. $Al^{3+}(aq) + 3NO_3^-(aq)$ d. $Al^{3-}(aq) + 3NO_3^+(aq)$

    ANS: B   DIF: II   OBJ: 14-1.1

7. Which of the following is the right side of the equation for dissolving $K_2S(s)$?
    a. $2K^+(aq) + S^{2-}(aq)$ c. $K^{2+}(aq) + S^{2-}(aq)$
    b. $K_2^{2+}(aq) + S^{2-}(aq)$ d. $2K(aq) + S(aq)$

    ANS: A   DIF: II   OBJ: 14-1.1

8. Which compound dissociates to produce the ions $SO_4^{2-}(aq)$ and $NH_4^+(aq)$?
   a. $NH_4SO_4(s)$
   b. $(NH_4)_2SO_4(s)$
   c. $NH_4(SO_4)_2(s)$
   d. $(NH_4)_3(SO_4)_2(s)$

   ANS: B   DIF: II   OBJ: 14-1.1

9. Which compound dissociates to produce the ions $Ca^{2+}(aq)$ and $NO_3^-(aq)$?
   a. $CaNO_3(s)$
   b. Ca_2NO_3
   c. $Ca(NO_3)_2(s)$
   d. $Ca_2(NO_3)_3$

   ANS: C   DIF: II   OBJ: 14-1.1

10. When solutions of two ionic compounds are combined and a solid forms, the process is called
    a. hydration.
    b. precipitation.
    c. solvation.
    d. dissociation.

    ANS: B   DIF: I   OBJ: 14-1.2

11. Precipitation is an example of what type of reaction?
    a. composition
    b. decomposition
    c. single-replacement
    d. double-replacement

    ANS: D   DIF: I   OBJ: 14-1.2

General Solubility Guidelines

1. Most sodium, potassium, and ammonium compounds are soluble in water.
2. Most nitrates, acetates, and chlorates are soluble.
3. Most chlorides are soluble, except those of silver, mercury(I), and lead. Lead(II) chloride is soluble in hot water.
4. Most sulfates are soluble, except those of barium, strontium, and lead.
5. Most carbonates, phosphates, and silicates are insoluble, except those of sodium, potassium, and ammonium.
6. Most sulfides are insoluble, except those of calcium, strontium, sodium, potassium, and ammonium.

12. Which of the following pairs of solutions produce a precipitate when combined?
    a. $Cu(NO_3)_2$ and $NaCl$
    b. $Fe(NO_3)_3$ and $MgCl_2$
    c. $Cu(NO_3)_2$ and K_2CO_3
    d. $CaCl_2$ and $NaNO_3$

    ANS: C   DIF: II   OBJ: 14-1.2

13. Which of the following pairs of solutions produce a precipitate when combined?
    a. KOH and NH_4Cl
    b. $Fe(NO)_3$ and KCl
    c. Na_2SO_4 and KCl
    d. NH_4Cl and $AgNO_3$

    ANS: D   DIF: II   OBJ: 14-1.2

14. When solutions of $BaCO_3$ and $Fe_2(SO_4)_3$ are combined, what precipitate(s) forms?
    a. $BaSO_4$
    b. $Fe_2(CO_3)_3$
    c. $BaSO_4$ and $Fe_2(CO_3)_3$
    d. none of the above

    ANS: C   DIF: II   OBJ: 14-1.2

15. When solutions of NH_4OH and K_2SO_4 are combined, what precipitate(s) forms?
    a. $(NH_4)_2SO_4$
    b. KOH
    c. $(NH_4)_2SO_4$ and KOH
    d. none of the above

    ANS: D   DIF: II   OBJ: 14-1.2

16. Which of the following is NOT a net ionic equation?
    a. $Ag^+(aq) + Cl^-(aq) \rightarrow AgCl(s)$
    b. $Fe^{2+}(aq) + 2Cl^-(aq) + 2OH^-(aq) \rightarrow Fe(OH)_2(s) + 2Cl^-(aq)$
    c. $3Ca^{2+}(aq) + 2P^{3-}(aq) \rightarrow Ca_3P_2(s)$
    d. $Cu^{2+}(aq) + S^{2-}(aq) \rightarrow CuS(s)$

    ANS: B   DIF: II   OBJ: 14-1.2

17. What is the spectator ion in the equation $Cu^{2+}(aq) + Zn^{2+}(aq) + 2S^{2-}(aq) \rightarrow CuS(s) + ZnS(s)$?
    a. $Cu^{2+}(aq)$
    b. $Zn^{2+}(aq)$
    c. $S^{2-}(aq)$
    d. none of the above

    ANS: D   DIF: II   OBJ: 14-1.2

18. What is the net ionic equation for the precipitation reaction between silver nitrate solution and sodium sulfide solution?
    a. $2Ag^+(aq) + 2NO^{3-}(aq) + 2Na^+(aq) + S^{2-}(aq) \rightarrow Ag_2S(s) + 2Na^+(aq) + 2NO_3^-(aq)$
    b. $2Ag^+(aq) + S^{2-}(aq) \rightarrow Ag_2S(s)$
    c. $Na^+(aq) + NO_3^-(aq) \rightarrow NaNO_3(s)$
    d. $2Ag^+(aq) + 2NO_3^-(aq) + 2Na^+(aq) + S^{2-}(aq) \rightarrow Ag_2S(s) + 2NaNO_3(s)$

    ANS: B   DIF: II   OBJ: 14-1.2

19. What is the net ionic equation for the precipitation reaction between copper(II) chloride and sodium hydroxide?
    a. $Cu^{2+}(aq) + 2OH^-(aq) \rightarrow Cu(OH)_2(s)$
    b. $Na^+(aq) + Cl^-(aq) \rightarrow NaCl(s)$
    c. $Cu^{2+}(aq) + 2OH^-(aq) + 2Cl^-(aq) \rightarrow Cu(OH)_2(s) + 2Cl^-(aq)$
    d. $Cu^{2+}(aq) + 2Cl^-(aq) + 2Na^+(aq) + 2OH^-(aq) \rightarrow Cu(OH)_2(s) + 2NaCl(s)$

    ANS: A   DIF: II   OBJ: 14-1.2

20. Which ions do NOT appear in the net ionic equation for the precipitation reaction involving solutions of $Zn(NO_3)_2$ and Na_3PO_4?
    a. $Zn^{2+}(aq)$ and $NO_3^-(aq)$
    b. $Na^+(aq)$ and $Zn^{2+}(aq)$
    c. $Zn^{2+}(aq)$ and $PO_4^{3-}(aq)$
    d. $Na^+(aq)$ and $NO_3^-(aq)$

    ANS: D   DIF: II   OBJ: 14-1.2

21. Molecular substances contain
    a. ionic bonds.
    b. polar-covalent bonds.
    c. nonpolar-covalent bonds.
    d. either polar-covalent or nonpolar-covalent bonds.

    ANS: D   DIF: I   OBJ: 14-1.3

22. The attraction of water molecules overcomes the strength of covalent bonds in
    a. all molecular substances.
    b. molecular substances that are electrolytes.
    c. molecular substances that are not electrolytes.
    d. none of the above

    ANS: B   DIF: I   OBJ: 14-1.3

23. The separation of ions that occurs when an ionic compound dissolves is called
    a. ionization.   c. precipitaion.
    b. dissociation.   d. oxidation.

    ANS: A   DIF: I   OBJ: 14-1.3

24. Ions are formed from solute molecules by the action of the solvent in a process called
    a. ionization.   c. precipitation.
    b. dissociation.   d. oxidation.

    ANS: B   DIF: I   OBJ: 14-1.3

25. When a molecular substance ionizes in water,
    a. charged particles are formed and the more electronegative atom becomes a negative ion.
    b. charged particles are formed and the less electronegative atom becomes a negative ion.
    c. no charged particles are formed.
    d. ions that were already present are released.

    ANS: A   DIF: I   OBJ: 14-1.3

26. Which of the following is a molecular compound that ionizes in water?
    a. NaCl   c. HCl
    b. Cl_2 d. C_6H_6

    ANS: C   DIF: II   OBJ: 14-1.3

27. Which of the following is an ionic compound that dissociates in water?
    a. NaCl   c. HCl
    b. Cl_2 d. C_6H_6

    ANS: A   DIF: II   OBJ: 14-1.3

28. The hydrogen ion
    a. has a charge of 2+.   c. has a negative charge.
    b. is a proton.   d. is a bare electron.

    ANS: B   DIF: I   OBJ: 14-1.4

29. The formula for the hydronium ion is
    a. H^+.
    b. H_2O^+.
    c. H_3O^-.
    d. H_3O^+.

    ANS: D   DIF: I   OBJ: 14-1.4

30. Dissolving HCl in water produces
    a. H_3O^+ and Cl^-.
    b. H^+ and Cl^-.
    c. $H_3OCl(aq)$.
    d. H_3O^- and Cl^+.

    ANS: A   DIF: I   OBJ: 14-1.4

31. The hydronium ion contains
    a. three equivalent covalent bonds.
    b. three covalent bonds of unequal strengths.
    c. one ionic bond and two covalent bonds.
    d. one covalent bond and two ionic bonds.

    ANS: A   DIF: I   OBJ: 14-1.4

32. In water solutions of HCl, the free H^+ ion is present
    a. in very large quantities.
    b. in moderate quantities.
    c. in small quantities.
    d. is not present.

    ANS: D   DIF: I   OBJ: 14-1.4

33. Formation of a hydronium ion in HCl solution
    a. is favorable and does not release energy.
    b. is unfavorable and releases energy.
    c. is favorable and releases energy.
    d. is unfavorable and does not release energy.

    ANS: C   DIF: I   OBJ: 14-1.4

34. Which of the following is a hydrated proton?
    a. a hydrogen ion
    b. a hydronium ion
    c. a water molecule
    d. a hydrogen chloride molecule

    ANS: B   DIF: I   OBJ: 14-1.4

35. Which of the following is NOT a hydrated proton?
    a. H_3O^+
    b. $H_7O_3^+$
    c. $H_4O_2^+$
    d. $H_9O_4^+$

    ANS: C   DIF: II   OBJ: 14-1.4

36. Which solute is present in aqueous solution as hydrated ions rather than as molecules?
    a. a weak electrolyte
    b. a strong electrolyte
    c. a nonelectrolyte
    d. a covalent electrolyte

    ANS: B   DIF: I   OBJ: 14-1.5

37. A compound of low solubility
    a. is always a strong electrolyte.
    b. is always a weak electrolyte.
    c. may be a strong or a weak electrolyte.
    d. is always a nonelectrolyte.

    ANS: C   DIF: I   OBJ: 14-1.5

38. Which solute is present in aqueous solution mostly as molecules rather than as ions?
    a. a weak electrolyte
    b. a strong electrolyte
    c. a nonelectrolyte
    d. a covalent electrolyte

    ANS: A   DIF: I   OBJ: 14-1.5

39. Which of the following is a strong electrolyte?
    a. $HC_2H_3O_2$
    b. HBr
    c. HF
    d. NH_3

    ANS: B   DIF: II   OBJ: 14-1.5

40. Which of the following is a weak electrolyte?
    a. hydrogen chloride
    b. sugar
    c. sodium chloride
    d. acetic acid

    ANS: D   DIF: II   OBJ: 14-1.5

41. A 15 M acetic acid solution is
    a. concentrated and a strong electrolyte.
    b. dilute and a strong electrolyte.
    c. concentrated and a weak electrolyte.
    d. dilute and a weak electrolyte.

    ANS: C   DIF: II   OBJ: 14-1.5

42. Colligative properties depend on
    a. the identity of the solute particles.
    b. the concentration of the solute particles.
    c. the physical properties of the solute particles.
    d. the boiling point and freezing point of the solution.

    ANS: B   DIF: I   OBJ: 14-2.1

43. Why is vapor-pressure lowering a colligative property?
    a. It depends on the concentration of a nonelectrolyte solute and does not depend on solute identity.
    b. It depends on the concentration of an electrolyte solute and does not depend on solute identity.
    c. It depends on the concentration of a nonelectrolyte solute and on solute identity.
    d. It depends on the concentration of an electrolyte solute and on solute identity.

    ANS: A   DIF: I   OBJ: 14-2.1

44. Why is freezing-point depression a colligative property?
    a. It is inversely proportional to the molal concentration of a solution.
    b. It is directly proportional to the molal concentration of a solution.
    c. It does not depend on a molal freezing-point constant for each solvent.
    d. It depends on the properties of an electrolyte in a solvent.

    ANS: B   DIF: I   OBJ: 14-2.1

45. Why is boiling-point elevation a colligative property?
    a. It is inversely proportional to the molal concentration of a solution.
    b. It is directly proportional to the molal concentration of a solution.
    c. It does not depend on a molal boiling-point constant for each solvent.
    d. It is independent of changes in vapor pressure.

    ANS: B   DIF: I   OBJ: 14-2.1

46. Why is osmotic pressure a colligative property?
    a. It depends on the rate of osmosis.
    b. It depends on the type of solute particles in two solutions of unequal concentrations.
    c. It depends on the concentration of solute particles in a solution.
    d. It depends on the type of solute particles in a solution.

    ANS: C   DIF: I   OBJ: 14-2.1

47. The freezing point of an aqueous solution that contains a nonelectrolyte is –8.0°C. What is the molal concentration of the solution?
    a. $1.86\ m$ c. $4.5\ m$
    b. $4.3\ m$ d. $14.8\ m$

    ANS: B   DIF: III   OBJ: 14-2.2

48. What is the boiling-point elevation of a solution made from 10.0 g of a nonelectrolyte solute and 300.0 g of water? The molar mass of the solute is 50.0 g and the molal boiling-point constant for water is 0.51°C.
    a. 0.01°C   c. 0.32°C
    b. 0.2°C   d. 0.34°C

    ANS: D   DIF: III   OBJ: 14-2.2

49. A water solution containing an unknown quantity of a nonelectrolyte solute has a freezing point of –0.21°C. What is the molal concentration of the solution if $K_f = -1.86$°C/m?
    a. $0.11\ m$ c. $1.11\ m$
    b. $0.12\ m$ d. $1.12\ m$

    ANS: A   DIF: III   OBJ: 14-2.2

50. A water solution containing an unknown quantity of a nonelectrolyte solute has a freezing point of –0.665°C. What is the molal concentration of the solution if $K_f = -1.86$°C/m?
    a. $0.010\ m$ c. $0.358\ m$
    b. $0.355\ m$ d. $2.66\ m$

    ANS: C   DIF: III   OBJ: 14-2.2

51. What is the boiling-point elevation of a solution made from 20.0 g of a nonelectrolyte solute and 300.0 g of water? The molar mass of the solute is 50.0 g and $K_b = 0.51°C/m$.
   a. 0.13°C
   b. 0.38°C
   c. 0.42°C
   d. 0.68°C

   ANS: D   DIF: III   OBJ: 14-2.2

52. What is the approximate freezing-point depression of a 0.020 m aqueous NaBr solution?
   a. −0.0093°C
   b. −0.019°C
   c. −0.037°C
   d. −0.074°C

   ANS: D   DIF: III   OBJ: 14-2.3

53. What is the approximate freezing-point depression of a 0.010 m aqueous $CaCl_2$ solution?
   a. −0.019°C
   b. −0.037°C
   c. −0.056°C
   d. −0.074°C

   ANS: C   DIF: III   OBJ: 14-2.3

54. What is the approximate freezing-point depression of a 0.05 m aqueous $FeCl_3$ solution?
   a. −0.019°C
   b. −0.093°C
   c. −0.28°C
   d. −0.37°C

   ANS: D   DIF: III   OBJ: 14-2.3

55. What is the approximate freezing-point depression of a 0.050 m aqueous Na_2SO_4 solution?
   a. −0.11°C
   b. −0.28°C
   c. −0.22°C
   d. −0.39°C

   ANS: B   DIF: III   OBJ: 14-2.3

56. What is the actual freezing-point depression of a 0.020 m aqueous NaCl solution?
   a. slightly less than −0.0744°C
   b. exactly −0.0744°C
   c. slightly more than −0.0744°C
   d. exactly 0°C

   ANS: A   DIF: III   OBJ: 14-2.3

57. Actual freezing-point depressions of electrolyte solutions are slightly lower than the calculated values because of
   a. ion repulsion.
   b. more complete ionization than is expected.
   c. ion attraction.
   d. higher-than-expected effective concentration.

   ANS: C   DIF: I   OBJ: 14-2.3

58. The Debye-Hückel theory accounts for
    a. attraction between ions in solutions.
    b. repulsion between ions in solution.
    c. attraction between ions in crystals.
    d. the greater freezing-point depressions in electrolyte solutions than in nonelectrolyte solutions.

    ANS: A   DIF: I   OBJ: 14-2.3

59. Compared with the freezing-point depression for a solution of an electrolyte that dissociates into a 3+ and a 3– ion, the freezing-point depression for an equally concentrated solution of an electrolyte that dissociates into a 1+ and a 1– ion is likely to be
    a. the same.   c. much lower.
    b. slightly lower.   d. greater.

    ANS: D   DIF: I   OBJ: 14-2.3

60. Nonvolatile solutes
    a. depress freezing point and elevate boiling point.
    b. elevate freezing point and depress boiling point.
    c. depress both freezing point and boiling point.
    d. elevate both freezing point and boiling point.

    ANS: A   DIF: I   OBJ: 14-2.1

61. Compared with a 0.01 m sugar solution, a 0.01 m KCl solution has
    a. the same freezing-point depression.
    b. about twice the freezing-point depression.
    c. the same freezing-point elevation.
    d. about six times the freezing-point elevation.

    ANS: B   DIF: II   OBJ: 14-2.4

62. Compared with a 0.01 m sugar solution, a 0.01 m $MgCl_2$ solution has
    a. the same freezing-point depression.
    b. about twice the freezing-point depression.
    c. about three times the freezing-point depression.
    d. about four times the freezing-point depression.

    ANS: C   DIF: II   OBJ: 14-2.4

63. Compared with a 0.01 m sugar solution, a 0.01 m KCl solution has
    a. the same boiling-point elevation.   c. the same boiling-point depression.
    b. roughly twice the boiling-point elevation. d. about half the boiling-point depression.

    ANS: B   DIF: II   OBJ: 14-2.4

64. Electrolytes affect colligative properties differently than do nonelectrolytes because electrolytes
    a.   are volatile.
    b.   have lower boiling points.
    c.   produce fewer moles of solute particles per mole of solute.
    d.   produce more moles of solute particles per mole of solute.

    ANS: D   DIF: I   OBJ: 14-2.4

65.

    What is the approximate freezing-point depression of nonvolatile aqueous electrolyte solutions?
    a. $-1.86C°/m \times$ molality of electrolyte
    b. $-1.86C°/m \times$ total molality of solute particles
    c. $-1.86C°$
    d. $0.52C°/m \times$ total molality of solute particles

    ANS: B   DIF: I   OBJ: 14-2.4

66. Compared with the freezing-point depression of a $0.01 \, m$ sugar solution, the freezing-point depression of $0.01 \, m$ HCl solution is
    a.   exactly the same.   c.   exactly twice as great.
    b.   slightly lower.   d.   almost twice as great.

    ANS: D   DIF: II   OBJ: 14-2.4

67. As electrolyte concentration decreases, freezing-point depression
    a.   approaches the value calculated by assuming complete ionization.
    b.   gets farther from the value calculated by assuming complete ionization.
    c.   remains exactly equal to the value calculated by assuming complete ionization.
    d.   approaches 0°C.

    ANS: A   DIF: I   OBJ: 14-2.4

MULTIPLE CHOICE

1. Acids taste
   a. sweet.
   b. sour.
   c. bitter.
   d. salty.

   ANS: B   DIF: I   OBJ: 15-1.1

2. Acetic acid is found in significant quantities in
   a. lemons.
   b. vinegar.
   c. sour milk.
   d. apples.

   ANS: B   DIF: I   OBJ: 15-1.1

3. Acids generally release H_2 gas when they react with
   a. nonmetals.
   b. semimetals.
   c. active metals.
   d. inactive metals.

   ANS: C   DIF: I   OBJ: 15-1.1

4. Acids make litmus paper turn
   a. red.
   b. yellow.
   c. blue.
   d. black.

   ANS: A   DIF: I   OBJ: 15-1.1

5. Acids react with
   a. bases to produce salts and water.
   b. salts to produce bases and water.
   c. water to produce bases and salts.
   d. neither bases, salts, nor water.

   ANS: A   DIF: I   OBJ: 15-1.1

6. Aqueous solutions of acids
   a. always have Faraday properties.
   b. conduct electricity.
   c. have very high boiling points.
   d. cannot be prepared.

   ANS: B   DIF: I   OBJ: 15-1.1

7. Bases taste
   a. soapy.
   b. sour.
   c. sweet.
   d. bitter.

   ANS: D   DIF: I   OBJ: 15-1.1

8. Bases feel
   a. rough.
   b. moist.
   c. slippery.
   d. dry.

   ANS: C   DIF: I   OBJ: 15-1.1

9. Bases make litmus paper turn
   a. blue.
   b. red.
   c. yellow.
   d. black.

   ANS: A   DIF: I   OBJ: 15-1.1

10. Bases react with
    a. acids to produce salts and water.
    b. salts to produce acids and water.
    c. water to produce acids and salts.
    d. neither acids, salts, nor water.

    ANS: A   DIF: I   OBJ: 15-1.1

11. Aqueous solutions of bases
    a. always have Faraday properties.
    b. conduct electricity.
    c. have very high boiling points.
    d. cannot be prepared.

    ANS: B   DIF: I   OBJ: 15-1.1

12. A binary acid contains
    a. two hydrogen atoms.
    b. hydrogen and one other element.
    c. hydrogen and two other elements.
    d. hydrogen and three other elements.

    ANS: B   DIF: I   OBJ: 15-1.2

13. Which of the following is a binary acid?
    a. H_2SO_4
    b. CH_3COOH
    c. HBr
    d. NaOH

    ANS: C   DIF: II   OBJ: 15-1.2

14. The name of a binary acid
    a. has no prefix.
    b. begins with the prefix *bi-*.
    c. ends with the suffix *-ous*.
    d. begins with the prefix *hydro-*.

    ANS: D   DIF: I   OBJ: 15-1.2

15. An oxyacid contains
    a. oxygen and hydrogen only.
    b. oxygen, hydrogen, and one other element.
    c. oxygen, hydrogen, and two other elements.
    d. oxygen and an element other than hydrogen.

    ANS: B   DIF: I   OBJ: 15-1.2

16. Which of the following is NOT an oxyacid?
    a. H_2O_2
    b. H_2SO_4
    c. $HClO_4$
    d. $HClO_2$

    ANS: A   DIF: II   OBJ: 15-1.2

17. Which of the following is perchloric acid?
    a. $HClO$
    b. $HClO_2$
    c. $HClO_3$
    d. $HClO_4$

    ANS: D   DIF: II   OBJ: 15-1.2

18. Which of the following is chlorous acid?
    a. $HClO$
    b. $HClO_2$
    c. $HClO_3$
    d. $HClO_4$

    ANS: B   DIF: II   OBJ: 15-1.2

19. Which of the following is chloric acid?
    a. $HClO$
    b. $HClO_2$
    c. $HClO_3$
    d. $HClO_4$

    ANS: C   DIF: II   OBJ: 15-1.2

20. Compared with acids that have the suffix -ic, acids that have the suffix -ous contain
    a. more hydrogen.
    b. more oxygen.
    c. less oxygen.
    d. the same amount of oxygen.

    ANS: C   DIF: II   OBJ: 15-1.2

21. An acid having the suffix -ic produces an anion having the
    a. suffix -ate.
    b. suffix -ite.
    c. prefix hydro-.
    d. suffix -ous.

    ANS: A   DIF: II   OBJ: 15-1.2

22. What acid is manufactured in largest quantity?
    a. hydrochloric acid
    b. phosphoric acid
    c. nitric acid
    d. sulfuric acid

    ANS: D   DIF: I   OBJ: 15-1.3

23. The acids produced in largest quantity are primarily used to manufacture
    a. agricultural products.
    b. laboratory chemicals.
    c. metal products.
    d. paints and dyes.

    ANS: A   DIF: I   OBJ: 15-1.3

24. Which acid is used in batteries?
    a. hydrochloric acid
    b. phosphoric acid
    c. nitric acid
    d. sulfuric acid

    ANS: D   DIF: I   OBJ: 15-1.3

25. What acid is used to make fertilizers and detergents and is a flavoring agent in beverages?
    a. hydrochloric acid
    b. phosphoric acid
    c. nitric acid
    d. sulfuric acid

    ANS: B   DIF: I   OBJ: 15-1.3

26. What acid is used mainly in the manufacture of explosives, rubber, plastics, dyes, and drugs?
    a. hydrochloric acid
    b. phosphoric acid
    c. nitric acid
    d. sulfuric acid

    ANS: C   DIF: I   OBJ: 15-1.3

27. What acid is produced in the stomach?
    a. hydrochloric acid
    b. phosphoric acid
    c. nitric acid
    d. sulfuric acid

    ANS: A   DIF: I   OBJ: 15-1.3

28. What acid is used to pickle metals, process food, and activate oil wells?
    a. hydrochloric acid
    b. phosphoric acid
    c. nitric acid
    d. sulfuric acid

    ANS: A   DIF: I   OBJ: 15-1.3

29. What acid is found in vinegar?
    a. acetic acid
    b. nitric acid
    c. phosphoric acid
    d. hydrochloric acid

    ANS: A   DIF: I   OBJ: 15-1.3

30. Which acid turns yellowish on standing?
    a. acetic acid
    b. nitric acid
    c. phosphoric acid
    d. hydrochloric acid

    ANS: B   DIF: I   OBJ: 15-1.3

31. The traditional definition of acids is based on the observations of
    a. Brønsted and Lowry.
    b. Lewis.
    c. Arrhenius.
    d. Faraday.

    ANS: C   DIF: I   OBJ: 15-1.4

32. According to the traditional definition, an acid contains
    a. hydrogen and does not ionize.
    b. hydrogen and ionizes to form hydrogen ions.
    c. oxygen and ionizes to form hydroxide ions.
    d. oxygen and ionizes to form oxygen ions.

    ANS: B   DIF: I   OBJ: 15-1.4

33. What is the basic assumption in the Arrhenius theory?
    a. Because acids and bases conduct electric current, they must not produce ions in solution.
    b. Because acids and bases conduct electric current, they must produce ions in solution.
    c. Only acids conduct electric current in solution.
    d. Only bases conduct electric current in solution.

    ANS: B   DIF: I   OBJ: 15-1.4

34. What is an Arrhenius acid?
    a. a chemical compound that increases the concentration of hydrogen ions in aqueous solution
    b. a chemical compound that increases the concentration of hydroxide ions in aqueous solution
    c. a chemical compound that decreases the concentration of hydrogen ions in aqueous solution
    d. a chemical compound that decreases the concentration of hydroxide ions in aqueous solution

    ANS: A   DIF: I   OBJ: 15-1.4

35. What is an Arrhenius base?
    a. a chemical compound that increases the concentration of hydrogen ions in aqueous solution
    b. a chemical compound that increases the concentration of hydroxide ions in aqueous solution
    c. a chemical compound that decreases the concentration of hydrogen ions in aqueous solution
    d. a chemical compound that decreases the concentration of hydroxide ions in aqueous solution

    ANS: B   DIF: I   OBJ: 15-1.4

36. Which statement about Arrhenius acids is FALSE?
    a. Their water solutions are called aqueous acids.
    b. They are molecular compounds with ionizable hydrogen atoms.
    c. Their pure aqueous solutions are electrolytes.
    d. They increase the concentration of hydroxide ions in aqueous solution.

    ANS: D   DIF: I   OBJ: 15-1.4

37. Which statement about Arrhenius bases is FALSE?
    a. Some are ionic hydroxides.
    b. They dissociate in solution to release hydroxide ions into the solution.
    c. They increase the concentration of hydrogen ions in aqueous solution.
    d. Some react with water and remove a hydrogen ion, leaving hydroxide ions.

    ANS: C   DIF: I   OBJ: 15-1.4

38. A substance that ionizes nearly completely in aqueous solutions and produces H_3O^+ is a
    a. weak base.   c. weak acid.
    b. strong base.   d. strong acid.

    ANS: D   DIF: I   OBJ: 15-1.5

39. Strong acids are
    a. strong electrolytes.   c. nonelectrolytes.
    b. weak electrolytes.   d. nonionized.

    ANS: A   DIF: I   OBJ: 15-1.5

40. Which of the following is NOT a strong acid?
    a. HNO_3 c. H_2SO_4
    b. CH_3COOH d. HCl

    ANS: B   DIF: II   OBJ: 15-1.5

41. Which of the following is a strong acid?
    a. HSO_4^- c. CH_3COOH
    b. H_2SO_4 d. H_3PO_4

    ANS: B   DIF: II   OBJ: 15-1.5

42. Which of the following is a triprotic acid?
    a. H_2SO_4 c. HCl
    b. CH_3COOH d. H_3PO_4

    ANS: D   DIF: II   OBJ: 15-1.5

43. Which of the following is a diprotic acid?
    a. H_2SO_4 c. HCl
    b. CH_3COOH d. H_3PO_4

    ANS: A   DIF: II   OBJ: 15-1.5

44. The dilute aqueous solution of a weak base contains
    a. hydronium ions.   c. acid molecules.
    b. anions.   d. all of the above

    ANS: D   DIF: I   OBJ: 15-1.5

45. Strong bases are
    a. strong electrolytes.   c. nonelectrolytes.
    b. weak electrolytes.   d. also strong acids.

    ANS: A   DIF: I   OBJ: 15-1.5

46. Hydroxides of Group 1 metals
    a. are all strong bases.   c. are all acids.
    b. are all weak bases.   d. might be either strong or weak bases.

    ANS: A   DIF: I   OBJ: 15-1.5

47. In water, hydroxides of Group 2 metals
    a. are all strong bases.   c. are all acids.
    b. are all weak bases.   d. might be either strong or weak bases.

    ANS: A   DIF: I   OBJ: 15-1.5

48. Which of the following is a strong base?
    a. NH_3 c. NaOH
    b. aniline   d. acetate ion

    ANS: C   DIF: I   OBJ: 15-1.5

49. Which of the following is a weak base?
    a. KOH
    b. $Ca(OH)_2$
    c. NH_3
    d. HCl

    ANS: C   DIF: II   OBJ: 15-1.5

50. Which of the following is a weak base?
    a. acetate ion
    b. hydroxide ion
    c. hydronium ion
    d. hydrogen ion

    ANS: A   DIF: I   OBJ: 15-1.5

51. Many organic compounds, such as aniline, that contain nitrogen are
    a. strong bases.
    b. weak bases.
    c. strong acids.
    d. weak acids.

    ANS: B   DIF: I   OBJ: 15-1.5

52. Whose definition of acids and bases emphasizes the role of protons?
    a. Brønsted and Lowry
    b. Lewis
    c. Arrhenius
    d. Faraday

    ANS: A   DIF: I   OBJ: 15-2.1

53. A Brønsted-Lowry acid is
    a. an electron-pair acceptor.
    b. an electron-pair donor.
    c. a proton acceptor.
    d. a proton donor.

    ANS: D   DIF: I   OBJ: 15-2.1

54. In the equation $HCl(g) + H_2O(l) \rightarrow H_3O^+(aq) + Cl^-(aq)$, which species is a Brønsted-Lowry acid
    a. HCl
    b. H_2O
    c. Cl^-
    d. none of the above

    ANS: A   DIF: II   OBJ: 15-2.1

55. A Brønsted-Lowry base is a(n)
    a. producer of OH^- ions.
    b. proton acceptor.
    c. electron-pair donor.
    d. electron-pair acceptor.

    ANS: B   DIF: I   OBJ: 15-2.1

56. In the reaction $NH_3 + H_2O \rightleftharpoons NH_4^+ + OH^-$, H_2O is a
    a. Brønsted-Lowry acid.
    b. Lewis base.
    c. Brønsted-Lowry base.
    d. traditional acid.

    ANS: A   DIF: II   OBJ: 15-2.1

57. The reaction $HCl(aq) + NH_3(aq) \rightarrow NH_4^+(aq) + Cl^-(aq)$ is a
    a. traditional acid-base reaction.
    b. Brønsted-Lowry acid-base reaction.
    c. single-replacement reaction.
    d. precipitation reaction.

    ANS: B   DIF: II   OBJ: 15-2.1

58. The reaction $HCl + KOH \rightarrow KCl + H_2O$ is a
    a. single-replacement reaction.
    b. synthesis reaction.
    c. Brønsted-Lowry acid-base reaction.
    d. Lewis acid-base reaction.

    ANS: C   DIF: II   OBJ: 15-2.1

59. A Lewis acid is
    a. an electron-pair acceptor.
    b. an electron-pair donor.
    c. a proton acceptor.
    d. a proton donor.

    ANS: A   DIF: I   OBJ: 15-2.2

60. Which acid definition is the broadest?
    a. traditional
    b. Lewis
    c. Brønsted-Lowry
    d. Faraday

    ANS: B   DIF: I   OBJ: 15-2.2

61. A Lewis base is a(n)
    a. producer of OH^- ions.
    b. proton acceptor.
    c. electron-pair donor.
    d. electron-pair acceptor.

    ANS: C   DIF: I   OBJ: 15-2.2

62. An electron-pair donor is a
    a. traditional base.
    b. Brønsted-Lowry acid.
    c. Brønsted-Lowry base.
    d. Lewis base.

    ANS: D   DIF: I   OBJ: 15-2.2

63. Whenever ammonia forms a covalent bond, it acts as a
    a. Brønsted-Lowry base.
    b. Lewis acid.
    c. Lewis base.
    d. traditional acid.

    ANS: C   DIF: I   OBJ: 15-2.2

64. An electron-pair acceptor is a
    a. Brønsted-Lowry base.
    b. Lewis acid.
    c. Lewis base.
    d. traditional acid.

    ANS: B   DIF: I   OBJ: 15-2.2

65. Which is a Lewis acid but is not a Brønsted-Lowry acid?
    a. HCl
    b. NH_3
    c. BF_3
    d. none of the above

    ANS: C   DIF: II   OBJ: 15-2.3

66. In the reaction $Ni^{2+} + nH_2O \rightleftharpoons [Ni(H_2O)_n]^{2+}$, H_2O is a
    a. Brønsted-Lowry acid.
    b. Brønsted-Lowry base.
    c. Lewis acid.
    d. Lewis base.

    ANS: D   DIF: II   OBJ: 15-2.3

67. The reaction $Ag^+(aq) + 2NH_3(aq) \rightarrow [Ag(NH_3)_2]^+(aq)$ is a
    a. traditional acid-base reaction.
    c. Brønsted-Lowry acid-base reaction.
    b. Lewis acid-base reaction.
    d. None of the above

    ANS: B   DIF: II   OBJ: 15-2.3

68. In the reaction $Ag^+(aq) + 2NH_3(aq) \rightarrow [Ag(NH_3)_2]^+(aq)$, Ag^+ is a
    a. Brønsted-Lowry acid.
    c. Brønsted-Lowry base.
    b. Lewis acid.
    d. Lewis base.

    ANS: B   DIF: II   OBJ: 15-2.3

69. A conjugate base is the species that
    a. remains after a base has given up a proton.
    b. is formed by the addition of a proton to a base.
    c. is formed by the addition of a proton to an acid.
    d. remains after an acid has given up a proton.

    ANS: D   DIF: I   OBJ: 15-3.1

70. A conjugate acid is the species that
    a. remains after a base has given up a proton.
    b. is formed by the addition of a proton to a base.
    c. is formed by the addition of a proton to an acid.
    d. remains after an acid has given up a proton.

    ANS: B   DIF: I   OBJ: 15-3.1

71. A species that remains when an acid has lost a proton is a
    a. conjugate base.
    c. strong base.
    b. conjugate acid.
    d. strong acid.

    ANS: A   DIF: I   OBJ: 15-3.1

72. A species that is formed when a base gains a proton is a
    a. conjugate base.
    c. strong base.
    b. conjugate acid.
    d. strong acid.

    ANS: B   DIF: I   OBJ: 15-3.1

73. How many conjugate acid-base pairs participate in a Brønsted-Lowry acid-base reaction?
    a. none
    c. two
    b. one
    d. four

    ANS: C   DIF: I   OBJ: 15-3.1

74. The members of a conjugate acid-base pair
    a. appear on the same side of the chemical equation.
    b. appear on opposite sides of the chemical equation.
    c. might appear on the same side or on opposite sides of the equation.
    d. are not included in the chemical equation.

    ANS: B   DIF: I   OBJ: 15-3.1

75. In a conjugate acid-base pair, the acid typically has
    a. one more proton than the base.
    b. one fewer proton than the base.
    c. two fewer protons than the base.
    d. the same number of protons as the base.

    ANS: A   DIF: I   OBJ: 15-3.1

76. What theory of acids and bases do conjugate acids and bases belong to?
    a. traditional
    b. Lewis
    c. Brønsted-Lowry
    d. none of the above

    ANS: C   DIF: I   OBJ: 15-3.1

77. The two members of a conjugate acid-base pair differ by a
    a. water molecule.
    b. hydroxide ion.
    c. hydronium ion.
    d. proton.

    ANS: D   DIF: I   OBJ: 15-3.1

78. In the reaction $HF + H_2O \rightleftharpoons H_3O^+ + F^-$, a conjugate acid-base pair is
    a. HF and H_2O.
    b. $F-$ and H_3O^+.
    c. H_3O^+ and H_2O.
    d. HF and H_3O^+.

    ANS: C   DIF: II   OBJ: 15-3.1

79. In the reaction $HF + H_2O \rightleftharpoons H_3O^+ + F^-$, a conjugate acid-base pair is
    a. F^- and H_2O.
    b. HF and F^-.
    c. H_3O^+ and HF.
    d. HF and H_2O.

    ANS: B   DIF: II   OBJ: 15-3.1

80. In the reaction $HClO_3 + NH_3 \rightleftharpoons NH_4^+ + ClO_3^-$, the conjugate acid of NH_3 is
    a. $HClO_3$.
    b. ClO_3^-.
    c. NH_4^+.
    d. not shown.

    ANS: C   DIF: II   OBJ: 15-3.1

81. In the reaction $HClO_3 + NH_3 \rightleftharpoons NH_4^+ + ClO_3^-$, the conjugate base of $HClO_3$ is
    a. ClO_3^-.
    b. NH_3.
    c. NH_4^+.
    d. not shown.

    ANS: A   DIF: II   OBJ: 15-3.1

82. In the reaction $CH_3COOH + H_2O \rightleftharpoons H_3O^+ + CH_3COO^-$, the conjugate acid of CH_3COO^- is
    a. H_2O.
    b. CH_3COOH.
    c. H_3O^+.
    d. not shown.

    ANS: B   DIF: II   OBJ: 15-3.1

83. In the reaction $CH_3COOH + H_2O \rightleftharpoons H_3O^+ + CH_3COO^-$, the conjugate base of H_3O^+ is
    a. H_2O.
    b. CH_3COOH.
    c. CH_3COO^-.
    d. not shown.

    ANS: A   DIF: II   OBJ: 15-3.1

84. The conjugate of a strong base is a
   a. strong acid.
   b. weak acid.
   c. strong base.
   d. weak base.

   ANS: B   DIF: I   OBJ: 15-3.1

85. The conjugate of a strong acid is a
   a. strong acid.
   b. weak acid.
   c. strong base.
   d. weak base.

   ANS: D   DIF: I   OBJ: 15-3.1

86. The conjugate of a weak base is a
   a. strong acid.
   b. weak acid.
   c. strong base.
   d. weak base.

   ANS: A   DIF: I   OBJ: 15-3.1

87. The conjugate of a weak acid is a
   a. strong acid.
   b. weak acid.
   c. strong base.
   d. weak base.

   ANS: C   DIF: I   OBJ: 15-3.1

88. A base is weak if its tendency to
   a. attract a proton is great.
   b. attract a proton is slight.
   c. donate a proton is great.
   d. donate a proton is slight.

   ANS: B   DIF: I   OBJ: 15-3.1

89. If a substance has a great tendency to give up protons, its conjugate has a
   a. great tendency to give up protons.
   b. great tendency to accept protons.
   c. slight tendency to give up protons.
   d. slight tendency to accept protons.

   ANS: D   DIF: I   OBJ: 15-3.1

90. In the equation $HI + H_2O \rightarrow H_3O^+ + I^-$, HI is a strong acid and I^- is a
   a. strong acid.
   b. strong base.
   c. weak acid.
   d. weak base.

   ANS: D   DIF: II   OBJ: 15-3.1

91. In the equation $CH_3COOH + H_2O \rightleftharpoons H_3O^+ + CH_3COO^-$, H_2O is a weak base and H_3O^+ is a
   a. strong acid.
   b. strong base.
   c. weak acid.
   d. weak base.

   ANS: A   DIF: II   OBJ: 15-3.1

92. In the equation $HClO_4 + NH_3 \rightarrow NH_4^+ + ClO_4^-$, ClO_4^- is a weak base and $HClO_4$ is a
   a. strong acid.
   b. strong base.
   c. weak acid.
   d. weak base.

   ANS: A   DIF: II   OBJ: 15-3.1

93. An amphoteric species is one that reacts as a(n)
    a. acid only.
    b. base only.
    c. acid or base.
    d. None of the above

    ANS: C   DIF: I   OBJ: 15-3.1

94. A species that can react as either an acid or a base is a(n)
    a. Lewis acid.
    b. amphoteric substance.
    c. oxyacid.
    d. organic substance.

    ANS: B   DIF: I   OBJ: 15-3.1

95. Which of the following is amphoteric?
    a. H_2SO_4
    b. SO_4^{2-}
    c. H^+
    d. HSO_4^-

    ANS: D   DIF: II   OBJ: 15-3.1

96. Which of the following is amphoteric?
    a. H_3PO_4
    b. H^+
    c. HPO_4^{2-}
    d. PO_4^{3-}

    ANS: C   DIF: II   OBJ: 15-3.1

97. Which of the following is amphoteric?
    a. H_2O
    b. H_3O^+
    c. H^+
    d. O^{2-}

    ANS: A   DIF: II   OBJ: 15-3.1

98. In the reaction $H_2SO_4 + H_2O \rightarrow H_3O^+ + HSO_4^-$, the ion HSO_4^- acts as a(n)
    a. acid.
    b. base.
    c. spectator species.
    d. salt.

    ANS: B   DIF: II   OBJ: 15-3.1

99. In the reaction $HSO_4^- + H_2O \rightleftharpoons H_3O^+ + SO_4^{2-}$, the ion HSO_4^- acts as a(n)
    a. acid.
    b. base.
    c. spectator species.
    d. salt.

    ANS: A   DIF: II   OBJ: 15-3.1

100. In the reaction $H_3PO_4 + H_2O \rightleftharpoons H_3O^+ + H_2PO_4^-$, the ion $H_2PO_4^-$ acts as a(n)
    a. acid.
    b. base.
    c. spectator species.
    d. salt.

    ANS: B   DIF: II   OBJ: 15-3.1

101. In the reaction $H_3PO_4 + H_2O \rightleftharpoons H_3O^+ + H_2PO_4^-$, the molecule H_2O acts as a(n)
    a. acid.
    b. base.
    c. spectator species.
    d. salt.

    ANS: B   DIF: II   OBJ: 15-3.1

102. The substances produced when KOH(*aq*) neutralizes HCl(*aq*) are
    a. HClO(*aq*) and KH(*aq*).
    b. KH_2O^+(*aq*) and Cl^-(*aq*).
    c. H_2O(*l*) and KCl(*aq*).
    d. H_3O^+(*aq*) and KCl(*aq*).

ANS: C   DIF: II   OBJ: 15-3.2

103. What is neutralization?
    a. an acid-base reaction that does not include dissocation of ions
    b. a reaction of hydronium ions and hydroxide ions to form a salt
    c. a reaction of hydronium ions and hydroxide ions to form water molecules
    d. a reaction of hydronium ions and hydroxide ions to form water molecules and a salt

ANS: C   DIF: I   OBJ: 15-3.2

104. A salt is NOT
    a. an ionic compound composed of a metal cation from a base.
    b. an ionic compound composed of an anion from an acid.
    c. a product of neutralization.
    d. a spectator ion.

ANS: D   DIF: I   OBJ: 15-3.2

105. Which of the following is NOT involved in neutralizations?
    a. H_3O^+ ion
    b. OH^- ion
    c. an acid and a base in an aqueous solution
    d. neutral compound

ANS: D   DIF: I   OBJ: 15-3.2

106. Which compound is produced by a neutralization?
    a. H_2O(*l*)
    b. HNO_3(*aq*)
    c. $Ca(OH)_2$(*s*)
    d. H_3PO_4(*aq*)

ANS: A   DIF: I   OBJ: 15-3.2

107. Which of the following gases does NOT dissolve in atmospheric water to produce acidic solutions?
    a. NO
    b. NO_2
    c. O_2
    d. CO_2

ANS: C   DIF: I   OBJ: 15-3.3

108. The reaction of an acid with a carbonate does NOT produce
    a. a salt.
    b. water.
    c. carbon dioxide.
    d. oxygen.

ANS: D   DIF: I   OBJ: 15-3.3

SHORT ANSWER

1. What determines the behavior of an amphoteric compound?

   ANS:
   The behavior of an amphoteric substance is determined by the strength of the acid or base with which it is reacting.

   DIF: I   OBJ: 15-3.1

2. Explain how industrial processes create acid rain.

   ANS:
   Industrial processes produce compounds that dissolve in the atmospheric water in clouds. This creates acidic solutions that fall to the ground in rain or snow.

   DIF: I   OBJ: 15-3.3

3. Use the following equation to explain acid rain: $SO_3(g) + H_2O(l) \rightarrow H_2SO_4(aq)$

   ANS:
   Sulfur trioxide gas is produced in industrial processes and released into the atmosphere. It dissolves in atmospheric water in clouds and produces sulfuric acid, which falls to the earth as acid rain or snow.

   DIF: II   OBJ: 15-3.3

4. Use the following equation to explain how acid rain damages marble structures:

   $CaCO_3(I) + 2H_3O^+(aq) \rightarrow Ca^{2+}(aq) + CO_2(g) + 3H_2O(l)$

   ANS:
   Marble structures are made of calcium carbonate, $CaCO_3$. When acid rain, $2H_3O^+$, falls on marble structures, it reacts with the calcium carbonate and dissolves the marble.

   DIF: II   OBJ: 15-3.3

MULTIPLE CHOICE

1. Pure water contains
    a.   water molecules only.
    b.   hydronium ions only.
    c.   hydroxide ions only.
    d.   water molecules, hydronium ions, and hydroxide ions.

    ANS: D   DIF: I   OBJ: 16-1.1

2. Pure water partially breaks down into charged particles in a process called
    a.   hydration.   c.   self-ionization.
    b.   hydrolysis.   d.   dissociation.

    ANS: C   DIF: I   OBJ: 16-1.1

3. What is the concentration of H_3O^+ in pure water?
    a. 10^{-7} M   c.   55.4 M
    b.   0.7 M   d. 10^7 M

    ANS: A   DIF: I   OBJ: 16-1.1

4. What is the concentration of OH^- in pure water?
    a. 10^{-7} M   c.   55.4 M
    b.   0.7 M   d. 10^7 M

    ANS: A   DIF: I   OBJ: 16-1.1

5. What is the product of H_3O^+ and OH^- concentrations in water?
    a. 10^{-28} c. 10^{-7}
    b. 10^{-14} d.   55.4

    ANS: B   DIF: I   OBJ: 16-1.1

6. Which expression represents the concentration of H_3O^+ in solution?
    a. $10^{-14} - [OH^-]$ c. $10^{-14} \div [OH^-]$
    b. $10^{-14} \times [OH^-]$ d. $[OH^-] \div 10^{-14}$

    ANS: C   DIF: I   OBJ: 16-1.1

7. Which expression represents the concentration of OH^- in solution?
    a. $10^{-14} - [H_3O^+]$ c. $10^{-14} \div [H_3O^+]$
    b. $10^{-14} \times [H_3O^+]$ d. $[OH^-] \div 10^{-14}$

    ANS: C   DIF: I   OBJ: 16-1.1

8. Which expression represents the pH of a solution?
   a. $\log[H_3O^+]$
   b. $-\log[H_3O^+]$
   c. $\log[OH^-]$
   d. $-\log[OH^-]$

   ANS: B   DIF: I   OBJ: 16-1.2

9. If $[H_3O^+]$ of a solution is greater than $[OH^-]$, the solution
   a. is always acidic.
   b. is always basic.
   c. is always neutral.
   d. might be acidic, basic, or neutral.

   ANS: A   DIF: I   OBJ: 16-1.2

10. If $[H_3O^+]$ of a solution is less than $[OH^-]$, the solution
    a. is always acidic.
    b. is always basic.
    c. is always neutral.
    d. might be acidic, basic, or neutral.

    ANS: B   DIF: I   OBJ: 16-1.2

11. The common logarithm of a number, N, is the
    a. inverse of N.
    b. square root of N.
    c. power to which N must be raised to equal 10.
    d. power to which 10 must be raised to equal N.

    ANS: D   DIF: I   OBJ: 16-1.2

12. The antilogarithm of a number, y, is
    a. the inverse of y.
    b. the square root of y.
    c. y raised to the power of 10.
    d. 10 raised to the power of y.

    ANS: D   DIF: I   OBJ: 16-1.2

13. What is the pH of a neutral solution at 25°C?
    a. 0
    b. 1
    c. 7
    d. 14

    ANS: C   DIF: I   OBJ: 16-1.2

14. The pH scale in general use ranges from
    a. 0 to 1.
    b. −1 to 1.
    c. 0 to 7.
    d. 0 to 14.

    ANS: D   DIF: I   OBJ: 16-1.3

15. The pH of an acidic solution is
    a. less than 0.
    b. less than 7.
    c. greater than 7.
    d. greater than 14.

    ANS: B   DIF: I   OBJ: 16-1.3

16. The pH of a basic solution is
    a. less than 0.
    b. less than 7.
    c. greater than 7.
    d. greater than 14.

    ANS: C   DIF: I   OBJ: 16-1.3

17. A water solution whose pH is 4
    a. is always neutral.
    b. is always basic.
    c. is always acidic.
    d. might be neutral, basic, or acidic.

    ANS: C   DIF: I   OBJ: 16-1.3

18. A water solution whose pH is 10
    a. is always neutral.
    b. is always basic.
    c. is always acidic.
    d. might be neutral, basic, or acidic.

    ANS: B   DIF: I   OBJ: 16-1.3

19. A water solution whose pH is 7
    a. is always neutral.
    b. is always basic.
    c. is always acidic.
    d. might be neutral, basic, or acidic.

    ANS: A   DIF: I   OBJ: 16-1.3

20. To calculate the pH of a solution whose [OH$^-$] is known, first calculate
    a. [H$_3$O$^+$].
    b. log[OH$^-$].
    c. antilog[H$_3$O$^+$].
    d. [H$_2$O].

    ANS: A   DIF: I   OBJ: 16-1.4

21. What is the pH of a 10^{-4} M HCl solution?
    a. 4
    b. 6
    c. 8
    d. 10

    ANS: A   DIF: III   OBJ: 16-1.4

22. What is the pH of a 10^{-5} M KOH solution?
    a. 3
    b. 5
    c. 9
    d. 11

    ANS: C   DIF: III   OBJ: 16-1.4

23. If [H$_3$O$^+$] = 1.7 × 10^{-3} M, what is the pH of the solution?
    a. 1.81
    b. 2.13
    c. 2.42
    d. 2.77

    ANS: D   DIF: III   OBJ: 16-1.4

24. If [H$_3$O$^+$] = 8.26 × 10^{-5} M, what is the pH of the solution?
    a. 2.161
    b. 3.912
    c. 4.083
    d. 8.024

    ANS: C   DIF: III   OBJ: 16-1.4

25. What is the pH of a solution whose hydronium ion concentration is 5.03×10^{-1} M?
   a.   0.2984
   b.   0.5133
   c.   1.542
   d.   5.031

   ANS: A   DIF: III   OBJ: 16-1.4

26. What is the pH of a 0.027 M KOH solution?
   a.   6.47
   b.   12.43
   c.   12.92
   d.   14.11

   ANS: B   DIF: III   OBJ: 16-1.4

27. What is the pH of a 0.001 62 M NaOH solution?
   a.   3.841
   b.   5.332
   c.   9.923
   d.   11.210

   ANS: D   DIF: III   OBJ: 16-1.4

28. What is the hydronium ion concentration of a solution whose pH is 4.12?
   a. 4.4×10^{-8} M
   b. 5.1×10^{-6} M
   c. 6.4×10^{-5} M
   d. 7.6×10^{-5} M

   ANS: D   DIF: III   OBJ: 16-1.5

29. What is the hydronium ion concentration of a solution whose pH is 7.30?
   a. 1.4×10^{-11} M
   b. 3.8×10^{-8} M
   c. 5.0×10^{-8} M
   d. 7.1×10^{-6} M

   ANS: C   DIF: III   OBJ: 16-1.5

30. What is the hydroxide ion concentration of a solution whose pH is 12.40?
   a. 2.5×10^{-2} M
   b. 4.4×10^{-2} M
   c. 8.9×10^{-2} M
   d. 1.0×10^{-1} M

   ANS: A   DIF: III   OBJ: 16-1.5

31. Dyes with pH-sensitive colors are used as
   a.   primary standards.
   b.   indicators.
   c.   titrants.
   d.   None of the above

   ANS: B   DIF: I   OBJ: 16-2.1

32. The pH range over which an indicator changes color is its
   a.   equivalence point.
   b.   endpoint.
   c.   transition interval.
   d.   pH interval.

   ANS: C   DIF: I   OBJ: 16-2.1

33. Indicators are classified into three types according to
   a.   their molar mass.
   b.   their polarity.
   c.   their color.
   d.   the pH at which they change color.

   ANS: D   DIF: I   OBJ: 16-2.1

34. What is the transition interval for litmus?
    a. pH 3.2–4.4
    b. pH 5.5–8.0
    c. pH 6.0–7.6
    d. pH 8.2–10.6

    ANS: B   DIF: I   OBJ: 16-2.1

35. What is the transition interval for bromthymol blue?
    a. pH 3.2–4.4
    b. pH 5.5–8.0
    c. pH 6.0–7.6
    d. pH 8.2–10.6

    ANS: C   DIF: I   OBJ: 16-2.1

36. What is the transition interval for phenolphthalein?
    a. pH 3.2–4.4
    b. pH 5.5–8.0
    c. pH 6.0–7.6
    d. pH 8.2–10.6

    ANS: D   DIF: I   OBJ: 16-2.1

37. What is the transition interval for methyl orange?
    a. pH 3.2–4.4
    b. pH 5.5–8.0
    c. pH 6.0–7.6
    d. pH 8.2–10.6

    ANS: A   DIF: I   OBJ: 16-2.1

38. Which indicator is used to study neutralizations of strong acids with strong bases?
    a. phenolphthalein
    b. methyl orange
    c. bromthymol blue
    d. None of the above

    ANS: C   DIF: I   OBJ: 16-2.1

39. Which indicator is used to study neutralizations of weak acids with strong bases?
    a. phenolphthalein
    b. methyl orange
    c. bromthymol blue
    d. None of the above

    ANS: A   DIF: I   OBJ: 16-2.1

40. Which indicator is used to study neutralizations of strong acids with weak bases?
    a. phenolphthalein
    b. methyl orange
    c. bromthymol blue
    d. None of the above

    ANS: B   DIF: I   OBJ: 16-2.1

41. What process measures the amount of a solution of known concentration required to react with a measured amount of a solution of unknown concentration?
    a. autoprotolysis
    b. hydrolysis
    c. neutralization
    d. titration

    ANS: D   DIF: I   OBJ: 16-2.2

42. In an acid-base titration,
    a. base is always added to acid.
    b. acid is always added to base.
    c. base is added to acid or acid is added to base.
    d. None of the above

    ANS: C   DIF: I   OBJ: 16-2.2

43. An acid-base titration involves a
    a. composition reaction.   c. single-replacement reaction.
    b. neutralization reaction.   d. decomposition reaction.

    ANS: B   DIF: I   OBJ: 16-2.2

44. Which quantity is directly measured in a titration?
    a. mass   c. volume
    b. concentration   d. density

    ANS: C   DIF: I   OBJ: 16-2.2

45. An acid-base titration determines the volumes of two solutions that
    a. are chemically equivalent.   c. have equal mass.
    b. have equal molarity.   d. have equal molality.

    ANS: A   DIF: I   OBJ: 16-2.2

46. What unknown quantity can be calculated after performing a titration?
    a. volume   c. mass
    b. concentration   d. density

    ANS: B   DIF: I   OBJ: 16-2.2

47. An acid-base titration is carried out by monitoring
    a. temperature.   c. pressure.
    b. pH.   d. density.

    ANS: B   DIF: I   OBJ: 16-2.2

48. In an acid-base titration, equivalent quantities of hydronium ions and hydroxide ions are present
    a. at the beginning point.   c. at the endpoint.
    b. at the midpoint.   d. throughout the titration.

    ANS: C   DIF: I   OBJ: 16-2.2

49. During an acid-base titration, a very rapid change in pH
    a. occurs when the first addition of the known solution is made.
    b. occurs when the amounts of H_3O^+ and OH^- are nearly equivalent.
    c. occurs at several points during the titration.
    d. does not occur during titration.

    ANS: B   DIF: I   OBJ: 16-2.2

50. What is the molarity of an HCl solution if 50.0 mL is neutralized in a titration by 40.0 mL of 0.400 M NaOH?

    a.   0.200 M   c.   0.320 M
    b.   0.280 M   d.   0.500 M

    ANS: C   DIF: III   OBJ: 16-2.3

51. What is the molarity of an HCl solution if 125 mL is neutralized in a titration by 76.0 mL of 1.22 M KOH?

    a.   0.371 M   c.   0.617 M
    b.   0.455 M   d.   0.742 M

    ANS: D   DIF: III   OBJ: 16-2.3

52. What is the molarity of an NaOH solution if 4.37 mL is titrated by 11.1 mL of 0.0904 M HNO_3?

    a.   0.230 M   c.   0.460 M
    b.   0.355 M   d.   0.620 M

    ANS: A   DIF: III   OBJ: 16-2.3

53. What is the molarity of an H_2SO_4 solution if 49.0 mL is completely titrated by 68.4 mL of an NaOH solution whose concentration is 0.333 M?

    a.   0.116 M   c.   0.465 M
    b.   0.232 M   d.   0.880 M

    ANS: B   DIF: III   OBJ: 16-2.3

54. Calculate the molarity of a $Ba(OH)_2$ solution if 1900 mL is completely titrated by 261 mL of 0.505 M HNO_3.

    a.   0.0173 M   c.   0.0322 M
    b.   0.0254 M   d.   0.0347 M

    ANS: D   DIF: III   OBJ: 16-2.3

55. If 72.1 mL of 0.543 M H_2SO_4 completely titrates 39.0 mL of KOH solution, what is the molarity of the KOH solution?

    a.   0.317 M   c.   1.00 M
    b.   0.502 M   d.   2.01 M

    ANS: D   DIF: III   OBJ: 16-2.3

56. If 114 mL of 0.008 04 M NaOH completely titrates 118 mL of H_3PO_4 solution, what is the molarity of the H_3PO_4 solution?

    a.   0.002 59 M   c.   0.007 77 M
    b.   0.005 18 M   d.   0.0105 M

    ANS: A   DIF: III   OBJ: 16-2.3

57. What is the molarity of a $Ba(OH)_2$ solution if 93.9 mL is completely titrated by 15.3 mL of 0.247 M H_2SO_4?
    a. 0.0101 M
    b. 0.0201 M
    c. 0.0402 M
    d. 0.0805 M

    ANS: C   DIF: III   OBJ: 16-2.3

58. What is the molarity of an H_3PO_4 solution if 358 mL is completely titrated by 876 mL of 0.0102 M $Ba(OH)_2$ solution?
    a. 0.0111 M
    b. 0.0166 M
    c. 0.0250 M
    d. 0.0333 M

    ANS: B   DIF: III   OBJ: 16-2.3

PROBLEM

1. Find $[H_3O^+]$ for a soft drink whose pH is 3.20.

    ANS:
    6.3×10^{-4} M

    DIF: III   OBJ: 16-1.5

2. Find $[H_3O^+]$ for a window-cleaning solution whose pH is 12.

    ANS:
    10^{-12} M

    DIF: III   OBJ: 16-1.5

MULTIPLE CHOICE

1. Which of the following is a measure of the average kinetic energy of the particles in a sample of matter?
    a. chemical kinetics
    b. thermochemistry
    c. reaction rate
    d. temperature

    ANS: D DIF: I OBJ: 17-1.1

2. Which of the following best describes temperature?
    a. heat absorbed or released in a chemical or physical change
    b. a measure of the average kinetic energy of the particles in a sample of matter
    c. heat energy
    d. energy of change

    ANS: B DIF: I OBJ: 17-1.1

3. In what units is temperature measured?
    a. degrees Celsius
    b. kelvins
    c. both degrees Celsius and kelvins
    d. None of the above

    ANS: C DIF: I OBJ: 17-1.1

4. How is a Celsius temperature reading converted to a Kelvin temperature reading?
    a. by adding 273.15
    b. by subtracting 273.15
    c. by dividing by 273.15
    d. by multiplying by 273.15

    ANS: C DIF: I OBJ: 17-1.1

5. The greater the kinetic energy of the particles in a sample of matter,
    a. the higher the temperature is.
    b. the lower the temperature is.
    c. the more energy is absorbed by the sample as heat.
    d. the less energy is released by the sample as heat.

    ANS: A DIF: I OBJ: 17-1.1

6. The energy transferred between samples of matter because of a difference in their temperatures is called
    a. heat.
    b. thermochemistry.
    c. chemical kinetics.
    d. temperature.

    ANS: A DIF: I OBJ: 17-1.2

7. Which of the following best describes heat?
   a. the energy transferred between samples of matter because of a difference in their temperatures
   b. a measure of the average kinetic energy of the particles in a sample of matter
   c. the energy stored in a sample of matter
   d. bond energy

   ANS: A   DIF: I   OBJ: 17-1.2

8. What units are used to measure heat?
   a. joules/mole or kilojoules/mole
   b. kelvins or degrees Celsius
   c. joules or kilojoules
   d. None of the above

   ANS: C   DIF: I   OBJ: 17-1.2

9. What is the heat required to raise the temperature of 1 g of a substance by 1 °C or 1 K?
   a. specific heat
   b. heat energy
   c. heat capacity
   d. heat of formation

   ANS: A   DIF: I   OBJ: 17-1.2

10. What units are used to measure specific heat?
    a. J/g·°C
    b. J/g·K
    c. cal/g·°C
    d. All of the above

    ANS: D   DIF: I   OBJ: 17-1.2

11. Which expression defines specific heat?

    a. heat × mass × temperature change

    c. $\dfrac{\text{heat} \times \text{mass}}{\text{temperature change}}$

    b. $\dfrac{\text{heat}}{\text{mass} \times \text{temperature change}}$

    d. $\dfrac{\text{temperature change}}{\text{heat} \times \text{mass}}$

    ANS: B   DIF: I   OBJ: 17-1.3

12. A 4.0 g sample of iron was heated from 0°C to 20.°C. It absorbed 35.2 J of energy as heat. What is the specific heat of this piece of iron?
    a. 2816 J/g·°C
    b. 2.27 J/g·°C
    c. 2.27 J/g
    d. 0.44 J/g·°C

    ANS: D   DIF: III   OBJ: 17-1.3

13. How much energy does a copper sample absorb as heat if its specific heat is 0.384 J/g·°C, its mass is 8.00 g, and it is heated from 10.0°C to 40.0°C?
    a. 0.0016 J/g·°C
    b. 0.0016 J
    c. 92.2 J
    d. 92.2 J/g·°C

    ANS: C   DIF: III   OBJ: 17-1.3

14. Find the specific heat of a material if a 6-g sample absorbs 50 J when it is heated from 30°C to 50°C.
    a. 6000 J
    b. 6000 J/g·°C
    c. 0.4 J
    d. 0.4 J/g·°C

    ANS: D   DIF: III   OBJ: 17-1.3

15. How much energy is absorbed as heat by 20. g of gold when it is heated from 25°C to 35°C? The specific heat of gold is 0.13 J/g·°C.
    a. 26 J
    b. 26 J/g·°C
    c. 0.0006 J
    d. 0.0006 J/g·°C

    ANS: A   DIF: III   OBJ: 17-1.3

16. A 5.0 g sample of silver is heated from 0°C to 35°C and absorbs 42 J of energy as heat. What is the specific heat of silver?
    a. 0.24 J
    b. 0.24 J/g·°C
    c. 7350 J
    d. 7350 J/g·°C

    ANS: B   DIF: III   OBJ: 17-1.3

17. Enthalpy change is the
    a. pressure change of a system at constant temperature.
    b. entropy change of a system at constant pressure.
    c. temperature change of a system at constant pressure.
    d. amount of energy absorbed or lost by a system as heat during a process at constant pressure.

    ANS: C   DIF: I   OBJ: 17-1.4

18. The quantity of energy released or absorbed as heat during a chemical reaction is called the
    a. enthalpy.
    b. heat of reaction.
    c. entropy.
    d. free energy.

    ANS: B   DIF: I   OBJ: 17-1.4

19. The Greek letter Δ stands for
    a. "heat stored in."
    b. "mass of."
    c. "rate of."
    d. "change in."

    ANS: D   DIF: I   OBJ: 17-1.4

20. What is the difference between the enthalpies of the products and the reactants?
    a. ΔS
    b. ΔG
    c. ΔH
    d. H

    ANS: C   DIF: I   OBJ: 17-1.4

21. What is the energy released or absorbed as heat when one mole of a compound is produced by combination of its elements?
   a. heat of formation
   b. heat of combustion
   c. free energy
   d. entropy

   ANS: A   DIF: I   OBJ: 17-1.4

22. Compounds whose heats of formation are highly negative
   a. do not exist.
   b. are very unstable.
   c. are somewhat stable.
   d. are very stable.

   ANS: D   DIF: I   OBJ: 17-1.4

23. $\Delta H =$ _____ .
   a. $H_{reactants} - H_{products}$
   b. $H_{reactants} + H_{products}$
   c. $H_{products} - H_{reactants}$
   d. $\dfrac{H_{products}}{H_{reactants}}$

   ANS: C   DIF: I   OBJ: 17-1.4

24. Suppose that a chemical equation can be written as the sum of two other chemical equations. If two reactions have ΔH values of −658 kJ and +458 kJ, what is ΔH for the reaction that is their sum?
   a. −1116 kJ
   b. −200 kJ
   c. +200 kJ
   d. +1116 kJ

   ANS: B   DIF: III   OBJ: 17-1.5

25. Suppose reaction A has a ΔH of −200 kJ and reaction B has a ΔH of −100 kJ. If reaction C can be written as the sum of reaction A forward and reaction B reversed, what is ΔH for reaction C?
   a. −300 kJ
   b. −100 kJ
   c. +300 kJ
   d. +100 kJ

   ANS: B   DIF: III   OBJ: 17-1.5

26. The heat of formation of compound X is −612 kJ/mol, and the sole product of its combustion has a heat of formation of −671 kJ/mol. What is the heat of combustion of compound X?
   a. −59 kJ/mol
   b. −40 kJ/mol
   c. +40 kJ/mol
   d. +59 kJ/mol

   ANS: A   DIF: III   OBJ: 17-1.5

Heats of Formation (kJ/mol)			Heats of Combustion (kJ/mol)		
Substance	Formula	ΔH^0_f	Substance	Formula	ΔH^0_c
ammonia(s)	NH_3	-45.9	hydrogen(g)	H_2	-285.8
barium nitrate(s)	$Ba(NO_3)_2$	-992.1	carbon (graphite)(s)	C	-393.5
benzene(l)	C_6H_6	$+49.1$	carbon monoxide (g)	CO	-283.0
calcium chloride(s)	$CaCl_2$	-795.4	methane (g)	CH_4	-890.8
carbon (diamond) (s)	C	$+1.9$	ethane(g)	C_2H_6	-1560.7
carbon (graphite)(s)	C	0.0	propane(g)	C_3H_8	-2219.2
carbon dioxide(g)	CO_2	-393.5	butane(g)	C_4H_{10}	-2877.6
copper(II) sulfate(s)	$CuSO_4$	-771.4	pentane(g)	C_5H_{12}	-3535.6
ethyne (acetylene)(g)	C_2H_2	$+228.2$	hexane(l)	C_6H_{14}	-4163.2
hydrogen chloride(g)	HCl	-92.3	heptane(l)	C_7H_{16}	-4817.0
water(l)	H_2O	-285.8	octane(l)	C_8H_{18}	-5470.5
nitrogen dioxide(g)	NO_2	$+33.2$	ethene (ethylene)(g)	C_2H_4	-1411.2
ozone(g)	O_3	$+142.7$	propene (propylene)(g)	C_3H_6	-2058.0
sodium chloride(s)	$NaCl$	-385.9	ethyne (acetylene)(g)	C_2H_2	-1301.1
sulfur dioxide(g)	SO_2	-296.8	benzene(l)	C_6H_6	-3267.6
zinc sulfate(s)	$ZnSO_4$	-980.1	toluene(l)	C_7H_8	-3910.3

27. What is the heat of combustion of 1 mol of sulfur to form SO_2?
   a. -593.6 kJ/mol   c. 0 kJ/mol
   b. -296.8 kJ/mol   d. $+296.8$ kJ/mol

   ANS: B   DIF: III   OBJ: 17-1.5

28. What is the heat of formation of hexane, C_6H_{14}?
   a. -198.4 kJ/mol   c. $+67.1$ kJ/mol
   b. -67.1 kJ/mol   d. $+198.4$ kJ/mol

   ANS: A   DIF: III   OBJ: 17-1.5

29. What is the heat of formation of ethane, C_2H_6?
   a. -870.7 kJ/mol   c. $+83.7$ kJ/mol
   b. -83.7 kJ/mol   d. $+870.7$ kJ/mol

   ANS: B   DIF: III   OBJ: 17-1.5

30. What is the value of ΔG at 300 K for a reaction in which $\Delta H = -150$ kJ/mol and $\Delta S = +2.00$ kJ/mol·K?
   a. -750 kJ/mol   c. $+750$ kJ/mol
   b. -450 kJ/mol   d. $+450$ kJ/mol

   ANS: A   DIF: III   OBJ: 17-2.3

31. The driving force of a reaction depends mostly on the change in
   a. state or phase.   c. entropy.
   b. enthalpy.   d. product type.

   ANS: B   DIF: I   OBJ: 17-2.1

32. A system that changes spontaneously without an enthalpy change
    a. is impossible.
    b. becomes more ordered.
    c. becomes more disordered.
    d. releases heat.

    ANS: C   DIF: I   OBJ: 17-2.1

33. Spontaneous reactions are driven by
    a. decreasing enthalpy and decreasing entropy.
    b. decreasing enthalpy and increasing entropy.
    c. increasing enthalpy and decreasing entropy.
    d. increasing enthalpy and increasing entropy.

    ANS: B   DIF: I   OBJ: 17-2.1

34. Compared with a single gas, a mixture of gases is
    a. more disordered.
    b. less disordered.
    c. equally disordered.
    d. less favorable.

    ANS: A   DIF: I   OBJ: 17-2.1

35. Why do gases naturally mix with each other when combined?
    a. The enthalpy change is favorable.
    b. The entropy change is favorable.
    c. The enthalpy and entropy changes are both favorable.
    d. The change in partial pressures is favorable.

    ANS: B   DIF: I   OBJ: 17-2.2

36. Which of the following is a measure of the disorder in a system?
    a. entropy
    b. enthalpy
    c. free energy
    d. temperature

    ANS: A   DIF: I   OBJ: 17-2.2

37. If a process increases entropy, the process
    a. is always spontaneous.
    b. is never spontaneous.
    c. is likely to be spontaneous.
    d. is not likely to be spontaneous.

    ANS: C   DIF: I   OBJ: 17-2.2

38. What is the symbol for entropy?
    a. T
    b. H
    c. G
    d. S

    ANS: D   DIF: I   OBJ: 17-2.2

39. What drives spontaneous reactions?
    a. decreasing enthalpy and decreasing entropy
    b. decreasing enthalpy and increasing entropy
    c. increasing enthalpy and decreasing entropy
    d. increasing enthalpy and increasing entropy

    ANS: B   DIF: I   OBJ: 17-2.2

40. What does ΔS stand for?
    a. enthalpy change
    b. free energy change
    c. entropy change
    d. temperature change

    ANS: C   DIF: I   OBJ: 17-2.2

41. Which of the following substances has the highest entropy?
    a. steam
    b. ice water
    c. liquid water
    d. crushed ice

    ANS: A   DIF: II   OBJ: 17-2.2

42. Which of the following systems has the lowest entropy?
    a. salt dissolved in a container of water
    b. sand mixed in a container of water
    c. oil floating in a container of water
    d. a container of frozen water

    ANS: D   DIF: II   OBJ: 17-2.2

43. Entropy decreases when
    a. pressure decreases.
    b. temperature decreases.
    c. the system is agitated.
    d. temperature increases.

    ANS: B   DIF: I   OBJ: 17-2.2

44. Which of the following has the highest entropy when produced in a reaction?
    a. solid
    b. liquid
    c. gas
    d. aqueous solution

    ANS: C   DIF: II   OBJ: 17-2.2

45. The driving force of a reaction is the change in
    a. free energy.
    b. entropy.
    c. enthalpy.
    d. temperature.

    ANS: A   DIF: I   OBJ: 17-2.3

46. What is the symbol for free-energy change?
    a. ΔS
    b. ΔH
    c. ΔG
    d. ΔT

    ANS: C   DIF: I   OBJ: 17-2.3

47. Free-energy change depends on
    a. change of entropy only.
    b. temperature only.
    c. change of enthalpy only.
    d. temperature and changes of entropy and enthalpy.

    ANS: D   DIF: I   OBJ: 17-2.3

48. Which expression defines the change in free energy?
    a. $\Delta H + T\Delta G$
    b. $\Delta H + T\Delta S$
    c. $\Delta H - T\Delta S$
    d. $\Delta S - T\Delta H$

    ANS: C   DIF: I   OBJ: 17-2.3

49. In ΔG calculations, temperature is expressed in
    a. degrees Celsius.
    b. kelvins.
    c. degrees Fahrenheit.
    d. kilojoules.

    ANS: B   DIF: I   OBJ: 17-2.3

50. A reaction is spontaneous if ΔG is
    a. zero.
    b. negative.
    c. positive.
    d. greater than ΔH.

    ANS: B   DIF: I   OBJ: 17-2.3

51. Spontaneity is favored by large positive values of
    a. ΔG.
    b. ΔH.
    c. ΔS.
    d. absolute temperature.

    ANS: C   DIF: I   OBJ: 17-2.3

52. What quantity predicts whether a reaction is spontaneous?
    a. enthalpy
    b. free energy
    c. temperature
    d. entropy

    ANS: B   DIF: I   OBJ: 17-2.4

53. Entropy plays a larger role in determining the free energy of reactions that take place at
    a. high temperatures.
    b. low temperatures.
    c. high pressures.
    d. low pressures.

    ANS: A   DIF: I   OBJ: 17-2.4

54. Which depends only on the initial and final states of a reaction, rather than on the intermediate processes?
    a. ΔH only
    b. ΔS only
    c. both ΔH and ΔS
    d. $T\Delta S$

    ANS: C   DIF: I   OBJ: 17-3.1

55. The equation $H_2(g) + I_2(g) \rightarrow 2HI(g)$ is a(n)
    a. overall reaction.
    b. reaction mechanism.
    c. reaction pathway.
    d. intermediate reaction.

    ANS: A   DIF: I   OBJ: 17-3.1

56. What is the sequence of steps in a reaction called?
    a. heterogeneous reaction
    b. rate law
    c. overall reaction
    d. reaction mechanism

    ANS: D   DIF: I   OBJ: 17-3.1

57. Examining a chemical system before and after reaction reveals the
    a. net chemical change.
    b. reaction mechanism.
    c. intermediates.
    d. activated complex.

    ANS: A   DIF: I   OBJ: 17-3.1

58. Most steps in a reaction mechanism
    a. are complicated.
    b. involve several molecules.
    c. are easily observable.
    d. are simple.

    ANS: D   DIF: I   OBJ: 17-3.1

59. What is the overall equation for the formation of hydrogen iodide from its elements?
    a. $H(g) + I(g) \rightleftharpoons HI(g)$
    b. $H_2(g) + I_2(g) \rightleftharpoons 2HI(g)$
    c. $H(g) + I_2(g) \rightleftharpoons HI(g) + I(g)$
    d. $H_2(g) + I(g) \rightleftharpoons HI(g) + H(g)$

    ANS: B   DIF: II   OBJ: 17-3.1

60. What is the initial step in the formation of hydrogen iodide from its elements?
    a. $H_2(g) + I_2(g) \rightleftharpoons 2HI(g)$
    b. $H(g) + I_2(g) \rightleftharpoons HI(g) + I(g)$
    c. $I_2(g) \rightleftharpoons 2I(g)$
    d. $I(g) + H_2(g) \rightleftharpoons HI(g) + H(g)$

    ANS: C   DIF: II   OBJ: 17-3.1

61. Which of the following is NOT a step in the formation of hydrogen iodide from its elements?
    a. $I_2(g) \rightleftharpoons 2I(g)$
    b. $H_2(g) + I(g) \rightleftharpoons HI(g) + H(g)$
    c. $H(g) + I_2(g) \rightleftharpoons HI(g) + I(g)$
    d. $H_2(g) \rightleftharpoons 2H(g)$

    ANS: D   DIF: II   OBJ: 17-3.1

62. How many steps are in the pathway when hydrogen iodide forms from its elements?
    a. one
    b. two or three
    c. four or five
    d. five or six

    ANS: B   DIF: II   OBJ: 17-3.1

63. To react, gas particles must
    a. be in the same physical state.
    b. have the same energy.
    c. have different energies.
    d. collide.

    ANS: D   DIF: I   OBJ: 17-3.2

64. To be effective, a collision requires
    a. sufficient energy.
    b. a favorable orientation.
    c. sufficient energy and a favorable orientation.
    d. a reaction mechanism.

    ANS: C   DIF: I   OBJ: 17-3.2

65. If colliding molecules have an orientation that favors reaction, they have
    a. the correct angles and distances between atoms.
    b. sufficient energy for each molecule.
    c. speeds that are neither too fast nor too slow.
    d. entropy values in the proper range.

    ANS: A   DIF: I   OBJ: 17-3.2

66. What attempts to explain chemical reactions and physical interactions of molecules?
    a. chemical kinetics   c. thermodynamics
    b. collision theory   d. thermochemistry

    ANS: B   DIF: I   OBJ: 17-3.2

67. Raising the temperature of reactants in a system
    a. increases the average kinetic energy of the molecules.
    b. decreases the average kinetic energy of the molecules.
    c. decreases the rate of collision of molecules.
    d. has no effect on the average kinetic energy of molecules.

    ANS: A   DIF: I   OBJ: 17-3.2

68. Raising the temperature of reactants in a system
    a. increases the average molecular motion.
    b. decreases the average molecular motion.
    c. has no effect on the average molecular motion.
    d. disturbs the system so that the collision theory no longer applies.

    ANS: A   DIF: I   OBJ: 17-3.2

69. Raising the temperature of gas particles
    a. increases both collision energy and favorability of orientation.
    b. increases neither collision energy nor favorability of orientation.
    c. increases collision energy but does not increase favorability of orientation.
    d. increases favorability of orientation but does not increase collision energy.

    ANS: C   DIF: I   OBJ: 17-3.2

70. If a collision between molecules is very gentle, the molecules are
    a. more likely to be favorably oriented.   c. more likely to react.
    b. less likely to be favorably oriented.   d. more likely to rebound without reacting.

    ANS: D   DIF: I   OBJ: 17-3.2

71. The minimum energy required for an effective collision is called
    a. energy of enthalpy.   c. free energy.
    b. activation energy.   d. kinetic energy.

    ANS: B   DIF: I   OBJ: 17-3.2

72. A short-lived structure formed during a collision is a(n)
    a. reagent.   c. activated complex.
    b. catalyst.   d. inhibitor.

    ANS: C   DIF: I   OBJ: 17-3.3

73. How does the energy of the activated complex compare with the energies of reactants and products?
    a. It is lower than the energy of both reactants and products.
    b. It is lower than the energy of reactants but higher than the energy of products.
    c. It is higher than the energy of reactants but lower than the energy of products.
    d. It is higher than the energy of both reactants and products.

    ANS: D   DIF: I   OBJ: 17-3.3

74. Activation energy is
    a. the energy required to form the activated complex.
    b. the net energy required to turn reactants into products.
    c. the heat of reaction.
    d. free energy.

    ANS: A   DIF: I   OBJ: 17-3.3

75. The bonding of the activated complex is characteristic of
    a. reactants only.   c. both reactants and products.
    b. products only.   d. solids only.

    ANS: C   DIF: I   OBJ: 17-3.3

76. In an energy-profile graph, the activated complex is represented at the
    a. left end of the curve.   c. bottom of the curve.
    b. right end of the curve.   d. top of the curve.

    ANS: D   DIF: I   OBJ: 17-3.3

77. An activated complex
    a. always separates into the products.
    b. always re-forms the reactants.
    c. may either separate into the products or re-form the reactants.
    d. always evaporates.

    ANS: C   DIF: I   OBJ: 17-3.3

78. In a diagram of an activated complex, broken lines represent
    a. actual bonds.
    b. partial bonds.
    c. electrons.
    d. bond energies.

    ANS: B   DIF: I   OBJ: 17-3.3

79. What takes place in an activated complex?
    a. Bonds form.
    b. Bonds break.
    c. Some bonds form and other bonds break.
    d. A catalyst is produced.

    ANS: C   DIF: I   OBJ: 17-3.3

80. What is the energy needed to lift reactants from the energy trough?
    a. free energy
    b. activation energy
    c. kinetic energy
    d. energy of reaction

    ANS: B   DIF: I   OBJ: 17-3.4

81. Which statement correctly describes the energy changes that occur when bonds form and when bonds break?
    a. Breaking bonds is endothermic, and forming bonds is exothermic.
    b. Breaking bonds is exothermic, and forming bonds is endothermic.
    c. Both are exothermic.
    d. Both are endothermic.

    ANS: A   DIF: I   OBJ: 17-3.4

82. Which of the following is true in an endothermic reaction?
    a. energy of products < activation energy < energy of reactants
    b. energy of reactants < activation energy < energy of products
    c. energy of products < energy of reactants < activation energy
    d. energy of reactants < energy of products < activation energy

    ANS: D   DIF: I   OBJ: 17-3.4

83. Which of the following is true in an exothermic reaction?
    a. energy of products < activation energy < energy of reactants
    b. energy of reactants < activation energy < energy of products
    c. energy of products < energy of reactants < activation energy
    d. energy of reactants < energy of products < activation energy

    ANS: C   DIF: I   OBJ: 17-3.4

84. What branch of chemistry studies reaction rates?
    a. thermochemistry
    b. thermodynamics
    c. chemical kinetics
    d. calorimetry

    ANS: C   DIF: I   OBJ: 17-4.1

85. What branch of chemistry studies reaction mechanisms?
    a. thermochemistry
    b. thermodynamics
    c. chemical kinetics
    d. calorimetry

    ANS: C   DIF: I   OBJ: 17-4.1

86. The usual conditions for reaction are not always necessary for
    a. synthesis reactions.
    b. decomposition reactions.
    c. single-replacement reactions.
    d. double-replacement reactions.

    ANS: B   DIF: I   OBJ: 17-4.1

87. Chemical kinetics studies
    a. the factors that affect the rate of reaction.
    b. the mathematical expressions for the rate of reaction.
    c. the factors that affect the rate of reaction and the mathematical expressions for the rate of reaction.
    d. the effect of quantum kinetics on chemical reactions.

    ANS: C   DIF: I   OBJ: 17-4.1

88. For most reactions, particles must
    a. collide.
    b. be properly oriented.
    c. be at rest.
    d. collide in the proper orientation.

    ANS: D   DIF: I   OBJ: 17-4.1

89. Reaction rate depends upon
    a. both collision frequency and efficiency.
    b. average kinetic energy.
    c. collision efficiency.
    d. average potential energy.

    ANS: A   DIF: I   OBJ: 17-4.2

90. Which of the following affects reaction rate?
    a. the nature of reactants
    b. surface area of reactants
    c. temperature
    d. All of the above

    ANS: D   DIF: I   OBJ: 17-4.2

91. Which substance naturally combines most rapidly with oxygen?
    a. platinum
    b. sodium
    c. iron
    d. coal

    ANS: B   DIF: II   OBJ: 17-4.2

92. Which of the following burns most slowly?
    a. a large lump of coal
    b. small pieces of coal
    c. powdered coal
    d. All of these burn at the same rate.

    ANS: A   DIF: II   OBJ: 17-4.2

93. If the surface area of reactants is larger,
    a. the reaction rate is generally higher.
    b. the reaction rate is generally lower.
    c. the reaction rate is not affected.
    d. the rate-determining step is eliminated.

    ANS: A   DIF: I   OBJ: 17-4.2

94. In heterogeneous reactions, the reactants
    a. have unequal masses.
    b. are not equally reactive.
    c. have unequal volumes.
    d. are in different phases.

    ANS: D   DIF: I   OBJ: 17-4.2

95. If the concentration of reactants is higher,
    a. the reaction rate is generally higher.
    b. the reaction rate is generally lower.
    c. the reaction rate is not affected.
    d. the rate-determining step is eliminated.

    ANS: A   DIF: I   OBJ: 17-4.2

96. If the temperature of the reactants is lower,
    a. the reaction rate is generally higher.
    b. the reaction rate is generally lower.
    c. the reaction rate is not affected.
    d. the rate-determining step is eliminated.

    ANS: B   DIF: I   OBJ: 17-4.2

97. A sample of a substance burns more rapidly in pure oxygen than in air. Which factor is most responsible for this high rate of reaction?
    a. the properties of the reactants
    b. temperature
    c. concentration of the substance
    d. surface area exposed to air

    ANS: C   DIF: I   OBJ: 17-4.2

98. Changing the temperature affects the rate of reaction because it affects
    a. the energy of the activated complex.
    b. the properties of the reactants.
    c. the heat of reaction.
    d. the frequency of collision and the number of effective collisions.

    ANS: D   DIF: I   OBJ: 17-4.2

99. Which process is used to speed up chemical reactions?
    a. calorimetry
    b. catalysis
    c. activation
    d. inhibition

    ANS: B   DIF: I   OBJ: 17-4.3

100. Which term describes a catalyst in the same phase as the reactants and products?
    a. homogenous
    b. heterogeneous
    c. activated
    d. inhibited

    ANS: A   DIF: I   OBJ: 17-4.3

101. Adsorption of reactants on the surface of a metal catalyst changes the reaction rate by affecting the
  a. concentration of the reactants.
  b. temperature of the system.
  c. properties of the reactants.
  d. surface area of the reactants.

  ANS: A   DIF: I   OBJ: 17-4.3

102. A substance that slows down chemical processes is called a(n)
  a. inhibitor.
  b. reactant.
  c. catalyst.
  d. indicator.

  ANS: A   DIF: I   OBJ: 17-4.3

103. Catalysts generally affect chemical reactions by
  a. increasing the temperature of the system.
  b. increasing the surface area of the reactants.
  c. providing an alternate pathway with a lower activation energy.
  d. providing an alternate pathway with a higher activation energy.

  ANS: C   DIF: I   OBJ: 17-4.3

104. How is a heterogeneous catalyst different from the reactants in a chemical reaction?
  a. The mass of the catalyst is different.
  b. The chemical properties of the catalyst are different.
  c. The energy of the catalyst is different.
  d. The phase of the catalyst is different.

  ANS: D   DIF: I   OBJ: 17-4.3

105. In a net equation, catalysts
  a. are shown with the reactants.
  b. are shown with the products.
  c. are shown with both reactants and products.
  d. are not shown.

  ANS: D   DIF: I   OBJ: 17-4.3

106. A rate law relates
  a. reaction rate and temperature.
  b. reaction rate and concentrations of reactants.
  c. temperature and concentrations of reactants.
  d. energy and concentrations of reactants.

  ANS: B   DIF: I   OBJ: 17-4.4

107. The letter k in a rate law stands for
  a. a proportionality constant.
  b. concentration.
  c. temperature.
  d. reaction rate.

  ANS: A   DIF: I   OBJ: 17-4.4

108. The letter R in a rate law stands for
    a. a proportionality constant.
    b. concentration.
    c. temperature.
    d. reaction rate.

    ANS: D   DIF: I   OBJ: 17-4.4

109. The value of k in a rate law
    a. is the same under all conditions.
    b. varies with concentration.
    c. varies with time.
    d. varies with temperature.

    ANS: D   DIF: I   OBJ: 17-4.4

110. Rate laws are determined by
    a. studying reaction mechanisms.
    b. calculating kinetic energy, frequency of collision, and temperature for a reaction.
    c. applying collision theory to a reaction.
    d. experiment.

    ANS: D   DIF: I   OBJ: 17-4.4

111. If doubling the concentration of a reactant doubles the rate of the reaction, the concentration of the reactant appears in the rate law with a(n)
    a. exponent of 1.
    b. exponent of 2.
    c. exponent of 4.
    d. coefficient of 2.

    ANS: A   DIF: I   OBJ: 17-4.4

112. If doubling the concentration of a reactant quadruples the rate of the reaction, the concentration of the reactant appears in the rate law with a(n)
    a. exponent of 1.
    b. exponent of 2.
    c. exponent of 4.
    d. coefficient of 2.

    ANS: B   DIF: I   OBJ: 17-4.4

113. The rate law for a reaction generally depends most directly on the
    a. net chemical reaction.
    b. first step in the reaction pathway.
    c. rate-determining step.
    d. last step in the reaction pathway.

    ANS: C   DIF: I   OBJ: 17-4.4

114. The rate for a reaction between reactants X, Y, and Z is directly proportional to [X] and to [Y], and proportional to the square of [Z]. What is the rate law for this reaction?
    a. $R = k[X][Y][Z]^2$
    b. $R = k[X][Y][2Z]$
    c. $R = k[X][Y][2Z]^2$
    d. $R = k\dfrac{[X]\,[Y]}{[Z]^2}$

    ANS: A   DIF: I   OBJ: 17-4.4

115. The rate for a reaction between reactants L, M, and N is proportional to the cube of [L] and the square of [M]. What is the rate law for this reaction?
   a. $R = k[3L][2M]$
   b. $R = k[L]^3[M]^2[N]$
   c. $R = k[L]^3[M]^2$
   d. $R = k[3L][2M][N]$

ANS: C   DIF: I   OBJ: 17-4.4

SHORT ANSWER

1. Explain the enthalpy change for a chemical reaction in terms of the enthalpies of its products and reactants.

   ANS:
   The enthalpy change for a reaction equals the total enthalpy of the products minus the total enthalpy of the reactants.

   DIF: II   OBJ: 17-2.1

2. How is a change in free energy related to changes in enthalpy and entropy?

   ANS:
   The change in free energy is the difference between the change in enthalpy and the product of the Kelvin temperature and the entropy change. This relationship can be stated mathematically as $\Delta G = \Delta H - T\Delta S$.

   DIF: II   OBJ: 17-2.4

ESSAY

1. How does temperature affect the free energy of a reaction?

   ANS:
   The formula $\Delta G = \Delta H - T\Delta S$ shows that part of the energy change of a reaction is proportional to the temperature. Therefore, at higher temperatures, the contribution of the entropy change is more important in determining the free energy. At lower temperatures, the enthalpy change has more influence on the free energy.

   DIF: II   OBJ: 17-2.4

2. How does the free-energy equation predict the progress of a chemical reaction?

   ANS:
   The equation $\Delta G = \Delta H - T\Delta S$ shows that the free energy of a reaction depends on the enthalpy change, the entropy change, and the temperature of a reaction. If ΔG is negative, the change is spontaneous. If ΔG is positive, the change will not occur spontaneously. If ΔG is zero, neither the forward nor the reverse reaction is favored, so the result is a mixture of reactants and products.

   DIF: II   OBJ: 17-2.4

3. Explain in terms of particles why some spontaneous exothermic and endothermic reactions have no apparent activation energy.

ANS:
The particles of all substances have an average kinetic energy that depends on the temperature of the substance. Some particles have higher than average kinetic energies, and some have lower than average kinetic energies. The kinetic energy of some particles is large enough to form the activated complex. These particles can react. Then, if the reaction is exothermic, the energy released can activate other particles.

DIF: II   OBJ: 17-3.4

MULTIPLE CHOICE

1. A reaction in which products can react to re-form reactants is
    a.   at equilibrium.
    b.   reversible.
    c.   buffered.
    d.   impossible.

    ANS: B   DIF: I   OBJ: 18-1.1

2. Under suitable conditions, roughly what proportion of all chemical reactions are reversible?
    a.   none
    b.   less than half
    c.   about half
    d.   nearly all

    ANS: D   DIF: I   OBJ: 18-1.1

3. If HgO is heated in a closed container,
    a.   no reaction takes place.
    b.   the HgO decomposes.
    c. Hg_2O_2 forms.
    d.   the HgO decomposes and then recombines.

    ANS: D   DIF: II   OBJ: 18-1.1

4. At equilibrium,
    a.   all reactions have ceased.
    b.   only the forward reaction continues.
    c.   only the reverse reaction continues.
    d.   both the forward and reverse reactions continue.

    ANS: D   DIF: I   OBJ: 18-1.1

5. At equilibrium,
    a.   the forward reaction rate is lower than the reverse reaction rate.
    b.   the forward reaction rate is higher than the reverse reaction rate.
    c.   the forward reaction rate is equal to the reverse reaction rate.
    d.   no reactions take place.

    ANS: C   DIF: I   OBJ: 18-1.1

6. What symbol in a chemical equation indicates equilibrium?
    a. $\rightarrow$
    b. $\leftarrow$
    c. $\rightarrow\leftarrow$
    d.   None of the above

    ANS: D   DIF: I   OBJ: 18-1.1

7. Which two processes are at equilibrium in a saturated sugar solution?
    a.   evaporation and condensation
    b.   dissolving and crystallization
    c.   decomposition and synthesis
    d.   ionization and recombination

    ANS: B   DIF: I   OBJ: 18-1.1

8. At equilibrium, the total amount of the product(s)
    a. is always equal to the total amount of the reactants.
    b. is always greater than the total amount of the reactants.
    c. is always less than the total amount of the reactants.
    d. may be equal to, greater than, or less than the total amount of the reactants.

    ANS: D   DIF: I   OBJ: 18-1.1

9. Which symbol represents the equilibrium constant?
    a. k c. c
    b. K d. R

    ANS: B   DIF: I   OBJ: 18-1.2

10. The value of the equilibrium constant for a reaction
    a. changes with concentration.   c. changes with temperature.
    b. changes with time.   d. is the same under all conditions.

    ANS: C   DIF: I   OBJ: 18-1.2

11. How does the value of K show that a reaction reaches equilibrium very quickly?
    a. K is large.
    b. K is small.
    c. K is zero.
    d. The value of K does not show how quickly a reaction comes to equilibrium.

    ANS: D   DIF: I   OBJ: 18-1.2

12. A very low value of K indicates that
    a. equilibrium is reached slowly.   c. reactants are favored.
    b. products are favored.   d. equilibrium has been reached.

    ANS: C   DIF: I   OBJ: 18-1.2

13. A very high value of K indicates that
    a. equilibrium is reached slowly.   c. reactants are favored.
    b. products are favored.   d. equilibrium has been reached.

    ANS: B   DIF: I   OBJ: 18-1.2

14. A value of K near 1 indicates that at equilibrium probably
    a. only products are present.
    b. only reactants are present.
    c. significant quantities of both products and reactants are present.
    d. the reactions occur at a moderate rate.

    ANS: C   DIF: I   OBJ: 18-1.2

15. The value of K for a system
    a. can be calculated from the molar masses of products and reactants.
    b. can be calculated from the heats of the forward and reverse reactions.
    c. can be calculated from the chemical properties of products and reactants.
    d. must be measured by experiment.

    ANS: D   DIF: I   OBJ: 18-1.2

16. The equilibrium constant depends on changes in
    a. pressure.
    b. concentrations.
    c. temperature.
    d. pressure, concentrations, and temperature.

    ANS: C   DIF: I   OBJ: 18-1.2

17. In the equation $K = \dfrac{[W][X]}{[Y][Z]}$, what represents the concentrations of the reactants?

    a. [Y] and [Z]   c. $\dfrac{[W][X]}{[Y][Z]}$

    b. [W] and [X]   d. $K = \dfrac{[W][X]}{[Y][Z]}$

    ANS: A   DIF: II   OBJ: 18-1.3

18. In the equation $K = \dfrac{[W][X]}{[Y][Z]}$, what represents the concentrations of the products?

    a. [Y] and [Z]   c. $\dfrac{[W][X]}{[Y][Z]}$

    b. [W] and [X]   d. $K = \dfrac{[W][X]}{[Y][Z]}$

    ANS: B   DIF: II   OBJ: 18-1.3

19. How do coefficients from a chemical equilibrium appear when the chemical equilibrium expression is written?
    a. as coefficients   c. as subscripts
    b. as exponents   d. They do not appear.

    ANS: B   DIF: I   OBJ: 18-1.3

20. What is the chemical equilibrium expression for the equation

$2A_2B + 3CD \rightleftharpoons A_4D + C_3B_2$?

a. $\dfrac{6[A_2B][CD]}{[A_4D][C_3B_2]}$

c. $\dfrac{[A_2B]^2[CD]^3}{[A_4D][C_3B_2]}$

b. $\dfrac{[A_4D][C_3B_2]}{6[A_2B][CD]}$

d. $\dfrac{[A_4D][C_3B_2]}{[A_2B]^2[CD]^3}$

ANS: D   DIF: II   OBJ: 18-1.3

21. An equilibrium mixture of SO_2, O_2, and SO_3 gases at 1500 K is determined to consist of 0.344 mol/L SO_2, 0.172 mol/L O_2, and 0.56 mol/L SO_3. What is the equilibrium constant for the system at this temperature? The balanced equation for this reaction is

$2SO_2(g) + O_2(g) \rightleftharpoons 2SO_3(g)$.
a.   0.41   c.   6.7
b.   2.8   d.   0.15

ANS: D   DIF: III   OBJ: 18-1.3

22. If the system $2CO(g) + O_2(g) \rightleftharpoons 2CO_2(g)$ has come to equilibrium and then more $CO(g)$ is added,
a. $[CO_2]$ increases and $[O_2]$ decreases.   c. $[CO_2]$ decreases and $[O_2]$ decreases.
b. $[CO_2]$ increases and $[O_2]$ increases.   d.   both $[CO_2]$ and $[O_2]$ remain the same.

ANS: A   DIF: II   OBJ: 18-2.1

23. If the pressure on the equilibrium system $2CO(g) + O_2(g) \rightleftharpoons 2CO_2(g)$ is increased,
a.   the quantity of $CO(g)$ increases.
b.   the quantity of $CO_2(g)$ decreases.
c.   the quantity of $CO_2(g)$ increases.
d.   the quantities in the system do not change.

ANS: C   DIF: II   OBJ: 18-2.1

24. If the pressure on the equilibrium system $N_2(g) + O_2(g) \rightleftharpoons 2NO(g)$ decreases,
a.   the quantity of $N_2(g)$ decreases.
b.   the quantity of $NO(g)$ increases.
c.   the quantity of $NO(g)$ decreases.
d.   the quantities in the system do not change.

ANS: D   DIF: II   OBJ: 18-2.1

25. If the temperature of the equilibrium system $CH_3OH(g) + 101 \text{ kJ} \rightleftharpoons CO(g) + 2H_2(g)$ increases,
    a. [CH_3OH] increases and [CO] decreases.
    b. [CH_3OH] decreases and [CO] increases.
    c. [CH_3OH] increases and [CO] increases.
    d. the concentrations in the system do not change.

    ANS: B   DIF: II   OBJ: 18-2.1

26. If the temperature of the equilibrium system $CH_3OH(g) + 101 \text{ kJ} \rightleftharpoons CO(g) + 2H_2(g)$ increases, K
    a. increases.   c. increases or decreases.
    b. decreases.   d. does not change.

    ANS: A   DIF: II   OBJ: 18-2.1

27. If more $CO(g)$ is added to the system $2CO(g) + O_2(g) \rightleftharpoons 2CO_2(g)$ at constant temperature, K
    a. increases.   c. increases or decreases.
    b. decreases.   d. does not change.

    ANS: D   DIF: II   OBJ: 18-2.1

28. If the temperature of the equilibrium system $X + Y \rightleftharpoons XY + 25 \text{ kJ}$ decreases,
    a. [X] decreases and [XY] increases.
    b. [X] increases and [XY] decreases.
    c. [X] decreases and [XY] decreases.
    d. the concentrations of reactants and products do not change.

    ANS: A   DIF: II   OBJ: 18-2.1

29. If a reaction system has come to equilibrium, it can be made to run to completion
    a. only if it is not reversible.   c. by applying Le Châtelier's principle.
    b. only if the temperature is low enough.   d. under no circumstances.

    ANS: C   DIF: I   OBJ: 18-2.2

30. Reactions tend to run to completion if
    a. a gaseous product forms and escapes.
    b. a product in the same phase as the reactants forms.
    c. one product is highly ionized.
    d. one product is highly soluble.

    ANS: A   DIF: I   OBJ: 18-2.2

31. Reactions tend to run to completion if a product
    a. has a high melting point.   c. is precipitated as a solid.
    b. is a liquid.   d. is ionic.

    ANS: C   DIF: I   OBJ: 18-2.2

32. If a soluble product forms, a reaction may run to completion
    a. if the product is only slightly ionized.   c. if the product is not gaseous.
    b. if the product is highly soluble.   d. under no circumstances.

    ANS: A   DIF: I   OBJ: 18-2.2

33. Which reaction tends to run to completion?
    a. $N_2(g) + 3H_2(g) \rightarrow 2NH_3(g)$
    b. $Na^+(aq) + Cl^-(aq) \rightarrow NaCl(s)$
    c. $2CO(g) + O_2(g) \rightarrow 2CO_2(g)$
    d. $H_2CO_3(aq) \rightarrow H_2O(l) + CO_2(g)$

    ANS: D   DIF: II   OBJ: 18-2.2

34. Which reaction tends to run to completion?
    a. $K^+(aq) + Cl^-(aq) \rightarrow KCl(s)$ c. $Ag^+(aq) + Cl^-(aq) \rightarrow AgCl(s)$
    b. $2NO_2(g) \rightarrow N_2O_4(g)$ d. $H_2(g) + I_2(g) \rightarrow 2HI(g)$

    ANS: C   DIF: II   OBJ: 18-2.2

35. Which reaction tends to run to completion?
    a. $N_2(g) + O_2(g) \rightarrow 2NO(g)$
    b. $H_3O^+(aq) + OH^-(aq) \rightarrow 2H_2O(l)$
    c. $Na^+(aq) + Br^-(aq) \rightarrow NaBr(s)$
    d. $H_2(g) + CO(g) \rightarrow C(s) + H_2O(g)$

    ANS: D   DIF: II   OBJ: 18-2.2

36. Adding a charged particle common to two solutes decreases solute concentration. This observation demonstrates
    a. the common-ion effect.   c. hydrolysis.
    b. Le Châtelier's principle.   d. buffering.

    ANS: A   DIF: I   OBJ: 18-2.3

37. The common-ion effect promotes
    a. dissolving.   c. boiling.
    b. precipitation.   d. ionization.

    ANS: B   DIF: I   OBJ: 18-2.3

38. The common-ion effect
    a. promotes condensation.   c. reduces ionization.
    b. promotes evaporation.   d. increases solubility.

    ANS: C   DIF: I   OBJ: 18-2.3

39. In which solution does adding hydrogen chloride promote precipitation?
    a. H_2SO_4 c. NaCl
    b. KBr   d. NaH

    ANS: C   DIF: II   OBJ: 18-2.3

40. Adding sodium acetate to an acetic acid, CH_3COOH, solution
    a. precipitates CH_3COOH.   c. increases the ionization of CH_3COOH.
    b. precipitates $NaCH_3COO$.   d. decreases the ionization of CH_3COOH.

    ANS: D   DIF: II   OBJ: 18-2.3

41. Adding sodium acetate to an acetic acid, CH_3COOH, solution
    a. increases pH and lowers $[H^+]$.
    b. increases pH and raises $[H^+]$.
    c. decreases pH and lowers $[H^+]$.
    d. decreases pH and raises $[H^+]$.

    ANS: A   DIF: II   OBJ: 18-2.3

42. The common-ion effect is a consequence of
    a. Boyle's law.
    b. Le Châtelier's principle.
    c. Avogadro's principle.
    d. chemical kinetics.

    ANS: B   DIF: I   OBJ: 18-2.3

43. Adding hydrogen bromide to a solution of KBr precipitates
    a. HBr.
    b. KBr.
    c. KH.
    d. Br_2.

    ANS: B   DIF: II   OBJ: 18-2.3

44. In the equilibrium system $CH_3COOH(aq) + H_2O(l) \rightleftharpoons H_3O^+(aq) + CH_3COO^-(aq)$, which species is present in the highest concentration?
    a. CH_3COOH
    b. H_3O^+
    c. CH_3COO^-
    d. The concentrations of CH_3COOH, H_3O^+ and CH_3COO^- are equal.

    ANS: A   DIF: II   OBJ: 18-3.1

45. In the equilibrium system $CH_3COOH(aq) + H_2O(l) \rightleftharpoons H_3O^+(aq) + CH_3COO^-(aq)$, which reaction proceeds more rapidly?
    a. the forward reaction
    b. the reverse reaction
    c. Neither reaction occurs.
    d. Both reactions occur at equal rates.

    ANS: D   DIF: II   OBJ: 18-3.1

46. What is the equilibrium constant, K, for the ionization of acetic acid, shown in the reaction $CH_3COOH(aq) + H_2O(l) \rightleftharpoons H_3O^+(aq) + CH_3COOH^-(aq)$?

    a. $[H_3O^+] [CH_3COOH^-]$

    b. $\dfrac{[H_3O^+] [CH_3COO^-]}{[CH_3COOH] [H_2O]}$

    c. $\dfrac{[H_3O^+] [CH_3COO^-]}{[CH_3COOH]}$

    d. $\dfrac{[CH_3COOH]}{[H_3O^+] [CH_3COO^-]}$

    ANS: B   DIF: II   OBJ: 18-3.1

47. What is the acid-ionization constant, K_a, for the ionization of acetic acid, shown in the reaction $CH_3COOH(aq) + H_2O(l) \rightleftharpoons H_3O^+(aq) + CH_3COOH^-(aq)$?

a. $[H_3O^+][CH_3COOH^-]$

c. $\dfrac{[H_3O^+][CH_3COO^-]}{[CH_3COOH]}$

b. $\dfrac{[H_3O^+][CH_3COO^-]}{[CH_3COOH][H_2O]}$

d. $\dfrac{[CH_3COOH]}{[H_3O^+][CH_3COO^-]}$

ANS: C   DIF: II   OBJ: 18-3.1

48. What is the relationship between K and K_a?
a. $K = K_a$
c. $K_a = K[H_2O]$

b. $K = K_a[H_3O^+]$
d. $K_a = \dfrac{[H_3O^+]}{K}$

ANS: C   DIF: I   OBJ: 18-3.1

49. How do K_a values for weak and strong acids compare?
a. K_a (weak) $= K_a$ (strong)
c. K_a (weak) $> K_a$ (strong)
b. K_a (weak) $< K_a$ (strong)
d. K_a is not defined for weak acids.

ANS: B   DIF: I   OBJ: 18-3.1

50. What is the equation for the ionization of water?
a. $2H_2O(l) \rightleftharpoons H_3O^+(aq) + OH^-(aq)$
b. $2H_2O(l) \rightleftharpoons H_3O^-(aq) + OH^+(aq)$
c. $2H_2O(l) \rightleftharpoons 2H_2(g) + O_2(g)$
d. $H_2O(l) \rightleftharpoons H^-(aq) + OH^+(aq)$

ANS: A   DIF: I   OBJ: 18-3.2

51. To what degree does water ionize?
a.   completely
c.   slightly
b.   to a large extent
d.   not at all

ANS: C   DIF: I   OBJ: 18-3.2

52. What is the value of the ion-product constant for water?
a.   0
c. 10^{-7}
b. 10^{-14}
d.   55.4

ANS: B   DIF: I   OBJ: 18-3.2

53. What is the symbol for the ion-product constant for water?
   a. K_w
   c. K
   b. K_a
   d. K_{sp}

   ANS: A   DIF: I   OBJ: 18-3.2

54. The pH of a solution is 9. What is its H_3O^+ concentration?
   a. 10^{-9} M
   c. 10^{-5} M
   b. 10^{-7} M
   d. 9 M

   ANS: A   DIF: I   OBJ: 18-3.2

55. The pH of a solution is 10. What is its OH^- concentration?
   a. 10^{-10} M
   c. 10^{-4} M
   b. 10^{-7} M
   d. 10 M

   ANS: C   DIF: I   OBJ: 18-3.2

56. When small amounts of acids or bases are added to a solution of a weak acid and its salt, the pH
   a. always increases considerably.
   b. always decreases considerably.
   c. either increases or decreases considerably.
   d. remains nearly constant.

   ANS: D   DIF: I   OBJ: 18-3.3

57. If an acid is added to a solution of a weak acid and its salt,
   a. more of the nonionized weak acid forms.   c. precipitation occurs.
   b. more of the nonionized acid ionizes.   d. hydronium ion concentration decreases.

   ANS: A   DIF: I   OBJ: 18-3.3

58. If a base is added to a solution of a weak acid and its salt,
   a. more of the nonionized weak acid forms.
   b. more of the nonionized acid ionizes.
   c. precipitation occurs.
   d. the hydronium ion concentration decreases.

   ANS: B   DIF: I   OBJ: 18-3.3

59. If a base is added to a solution of a weak base and its salt,
   a. the hydronium ion concentration increases.
   b. more of the weak base ionizes.
   c. more hydroxide ions form.
   d. more water and nonionized base forms.

   ANS: D   DIF: I   OBJ: 18-3.3

60. If an acid is added to a solution of a weak base and its salt,
    a. more water forms and more weak base ionizes.
    b. the hydronium ion concentration decreases.
    c. more hydroxide ions form.
    d. more nonionized weak base forms.

    ANS: A   DIF: I   OBJ: 18-3.3

61. Which solutions resist changes in pH?
    a. buffered   c. neutral
    b. equilibrium   d. stable

    ANS: A   DIF: I   OBJ: 18-3.3

62. Which reaction describes an anion hydrolysis reaction?
    a. $B^+(aq) + H_2O(l) \rightleftharpoons HB(aq) + OH^+(aq)$ c. $HB(aq) + H_2O(l) \rightleftharpoons H_3O^+(aq) + B^-(aq)$
    b. $B^-(aq) + H_2O(l) \rightleftharpoons HB(aq) + OH^-(aq)$ d. $B^-(aq) + H_2O(l) \rightleftharpoons BOH(aq) + H^-(aq)$

    ANS: B   DIF: II   OBJ: 18-3.4

63. Which of the following is the equilibrium constant for an anion hydrolysis reaction?

    a. $\dfrac{[HB]\,[OH^-]}{[B^-]}$ c. $\dfrac{[HB]}{[B^-]\,[OH^-]}$

    b. $\dfrac{[B^-]}{[HB]\,[OH^-]}$ d. $\dfrac{[B^-]\,[OH^-]}{[HB]}$

    ANS: A   DIF: I   OBJ: 18-3.4

64. What type of reaction occurs in an aqueous solution of the salt of a strong acid and a weak base?
    a. cation hydrolysis   c. both cation and anion hydrolysis
    b. anion hydrolysis   d. buffer hydrolysis

    ANS: A   DIF: I   OBJ: 18-3.4

65. The anion of the salt of a weak acid and a strong base is the
    a. conjugate acid of the strong base.   c. hydronium ion.
    b. conjugate base of the weak acid.   d. hydroxide ion.

    ANS: B   DIF: I   OBJ: 18-3.4

66. What type of reaction occurs in an aqueous solution of the salt of a weak acid and a strong base?
    a. cation hydrolysis   c. both cation and anion hydrolysis
    b. anion hydrolysis   d. buffer hydrolysis

    ANS: B   DIF: I   OBJ: 18-3.4

67. The cation of the salt of a strong acid and a weak base is the
    a. hydronium ion.
    b. hydroxide ion.
    c. conjugate acid of the weak base.
    d. conjugate base of the strong acid.

    ANS: C   DIF: I   OBJ: 18-3.4

68. Basic solutions are generally formed by hydrolysis of anions of salts of
    a. weak acids and weak bases.
    b. weak acids and strong bases.
    c. strong acids and weak bases.
    d. strong acids and strong bases.

    ANS: B   DIF: I   OBJ: 18-3.4

69. Acidic solutions are generally formed by hydrolysis of cations of salts of
    a. weak acids and weak bases.
    b. weak acids and strong bases.
    c. strong acids and weak bases.
    d. strong acids and strong bases.

    ANS: C   DIF: I   OBJ: 18-3.4

70. What is the symbol for the solubility-product constant?
    a. K
    b. K_h
    c. K_a
    d. K_{sp}

    ANS: D   DIF: I   OBJ: 18-4.1

71. A substance with a very low K_{sp} is
    a. insoluble or sparingly soluble.
    b. moderately soluble.
    c. very soluble.
    d. completely miscible in most solvents.

    ANS: A   DIF: I   OBJ: 18-4.1

72. Besides the K_{sp} value of a compound, what other information is needed to calculate solubility in moles per liter?
    a. molar mass
    b. density of the solid
    c. solubility in grams
    d. no other information

    ANS: D   DIF: I   OBJ: 18-4.1

73. If the ion product for two ions whose solutions have just been mixed is greater than the value of K_{sp},
    a. the solution is unsaturated.
    b. precipitation occurs.
    c. equilibrium cannot be achieved.
    d. a decomposition reaction occurs.

    ANS: B   DIF: I   OBJ: 18-4.1

74. Solubility-product principles can be applied very successfully to
    a. only sparingly souble substances.
    b. only slightly soluble substances.
    c. only very soluble substances.
    d. all substances.

    ANS: A   DIF: I   OBJ: 18-4.1

75. What is the solubility-product constant of barium carbonate, $BaCO_3$? Its solubility is 0.0022 g/100 g H_2O.
    a. 1.2×10^{-6}
    b. 1.2×10^{-8}
    c. 2.2×10^{-8}
    d. 4.4×10^{-10}

    ANS: B   DIF: III   OBJ: 18-4.1

76. What is the solubility-product constant of barium sulfate, $BaSO_4$? Its solubility is 2.4×10^{-4} g/100 g H_2O.
    a. 5.6×10^{-6}
    b. 7.8×10^{-8}
    c. 4.0×10^{-9}
    d. 1.1×10^{-10}

    ANS: D   DIF: III   OBJ: 18-4.1

77. What is the solubility-product constant of magnesium hydroxide, $Mg(OH)_2$? Its solubility is 9.0×10^{-4} g/100 g H_2O.
    a. 1.8×10^{-6}
    b. 4.5×10^{-9}
    c. 1.5×10^{-11}
    d. 6.2×10^{-12}

    ANS: C   DIF: III   OBJ: 18-4.1

78. What is the solubility in mol/L of silver iodide, AgI? Its K_{sp} value is 8.3×10^{-17}.
    a. 4.9×10^{-11}
    b. 9.1×10^{-9}
    c. 8.6×10^{-7}
    d. 5.5×10^{-7}

    ANS: B   DIF: III   OBJ: 18-4.2

79. What is the solubility in mol/L of calcium carbonate, $CaCO_3$? Its K_{sp} value is 1.4×10^{-8}.
    a. 1.2×10^{-4}
    b. 2.4×10^{-6}
    c. 6.2×10^{-8}
    d. 8.2×10^{-10}

    ANS: A   DIF: III   OBJ: 18-4.2

80. What is the solubility in mol/L of cobalt(II) sulfide, CoS? Its K_{sp} value is 3.0×10^{-26}.
    a. 5.9×10^{-10}
    b. 1.7×10^{-12}
    c. 1.7×10^{-13}
    d. 1.7×10^{-26}

    ANS: C   DIF: III   OBJ: 18-4.2

81. What is the solubility in mol/L of copper(I) chloride, $CuCl$? Its K_{sp} value is 1.2×10^{-6}.
    a. 1.1×10^{-3}
    b. 1.8×10^{-4}
    c. 2.3×10^{-5}
    d. 2.7×10^{-5}

    ANS: A   DIF: III   OBJ: 18-4.2

82. What is the solubility in mol/L of copper(II) sulfide, CuS? Its K_{sp} value is 6.3×10^{-36}.
    a. 5.5×10^{-13}
    b. 4.1×10^{-14}
    c. 9.9×10^{-16}
    d. 2.5×10^{-18}

    ANS: D   DIF: III   OBJ: 18-4.2

83. What is the solubility in mol/L of copper(I) sulfide, Cu_2S? Its K_{sp} value is 2.5×10^{-48}.
    a. 4.1×10^{-14}
    c. 8.5×10^{-17}
    b. 6.2×10^{-16}
    d. 1.8×10^{-24}

    ANS: C   DIF: III   OBJ: 18-4.2

84. What is the solubility in mol/L of mercury(II) sulfide, HgS? Its K_{sp} value is 1.6×10^{-52}.
    a. 7.3×10^{-22}
    c. 1.3×10^{-26}
    b. 8.9×10^{-25}
    d. 6.7×10^{-30}

    ANS: C   DIF: III   OBJ: 18-4.2

85. What is the solubility in mol/L of silver sulfide, Ag_2S? Its K_{sp} value is 1.6×10^{-49}.
    a. 2.8×10^{-16}
    c. 8.2×10^{-38}
    b. 3.4×10^{-17}
    d. 9.0×10^{-40}

    ANS: B   DIF: III   OBJ: 18-4.2

86. Calculate the ion product for mixing 100.0 mL of 0.0030 M $CaCl_2$ with 100 mL of 0.0020 M Na_2CO_3. K_{sp} for $CaCO_3$ is 1.4×10^{-8}. Does a precipitate form?
    a. 1.8×10^{-9}; no
    c. 1.5×10^{-6}; no
    b. 1.8×10^{-9}; yes
    d. 1.5×10^{-6}; yes

    ANS: D   DIF: III   OBJ: 18-4.3

87. Calculate the ion product for mixing 50. mL of 0.000 70 M $CuNO_3$ with 100. mL of 0.000 10 M $NaCl$. K_{sp} for $CuCl$ is 1.2×10^{-6}. Does a precipitate form?
    a. 1.6×10^{-8}; yes
    c. 1.6×10^{-6}; yes
    b. 1.6×10^{-8}; no
    d. 1.6×10^{-6}; no

    ANS: B   DIF: III   OBJ: 18-4.3

88. Calculate the ion product for mixing 250 mL of $2.1 \times 10{-}14$ M $Cd(NO_3)_2$ with 150 mL of 1.1×10^{-12} M K_2S. K_{sp} for CdS is 8.0×10^{-27}. Does a precipitate form?
    a. 5.4×10^{-27}; no
    c. 4.8×10^{-29}; no
    b. 5.4×10^{-27}; yes
    d. 4.8×10^{-29}; yes

    ANS: A   DIF: III   OBJ: 18-4.3

89. Calculate the ion product for mixing 300 mL of 0.000 30 M $Sr(NO_3)_2$ with 200 mL of 0.000 025 M K_2SO_4. K_{sp} for $SrSO_4$ is 3.2×10^{-7}. Does a precipitate form?
    a. 5.6×10^{-6}; yes
    c. 1.8×10^{-9}; yes
    b. 5.6×10^{-6}; no
    d. 1.8×10^{-9}; no

    ANS: D   DIF: III   OBJ: 18-4.3

90. Calculate the ion product for mixing 100 mL of 0.000 28 M $Pb(NO_3)_2$ with 200 mL of 0.0012 M $NaCl$. K_{sp} for $PbCl_2$ is 1.9×10^{-4}. Does a precipitate form?
    a. 2.3×10^{-4}; no
    c. 6.0×10^{-11}; no
    b. 2.3×10^{-4}; yes
    d. 6.0×10^{-11}; yes

    ANS: C   DIF: III   OBJ: 18-4.3

91. Calculate the ion product for mixing 50 mL of 0.015 M $Ca(NO_3)_2$ with 200 mL of 0.35 M NaOH. K_{sp} for $Ca(OH)_2$ is 5.5×10^{-6}. Does a precipitate form?
   a. 8.5×10^{-7}; no
   b. 8.5×10^{-7}; yes
   c. 2.4×10^{-4}; no
   d. 2.4×10^{-4}; yes

   ANS: D   DIF: III   OBJ: 18-4.3

MULTIPLE CHOICE

1. What is the oxidation number of a monatomic ion?
    a. 0
    b. +1
    c. its charge
    d. its number of electrons

    ANS: C   DIF: I   OBJ: 19-1.1

2. What is the most common oxidation number of combined oxygen?
    a. −2
    b. −1
    c. 0
    d. +1

    ANS: A   DIF: I   OBJ: 19-1.1

3. What is the most common oxidation number of combined hydrogen?
    a. −2
    b. −1
    c. 0
    d. +1

    ANS: D   DIF: I   OBJ: 19-1.1

4. The algebraic sum of the oxidation numbers of the atoms in a compound
    a. is always zero.
    b. is always +1.
    c. is always −1.
    d. can be any whole number.

    ANS: A   DIF: I   OBJ: 19-1.1

5. What are the oxidation numbers in the compound KCl?
    a. $K = 0, Cl = 0$
    b. $K = −1, Cl = +1$
    c. $K = +1, Cl = −1$
    d. $K = +2, Cl = −2$

    ANS: C   DIF: II   OBJ: 19-1.1

6. What are the oxidation numbers in the compound H_2O_2?
    a. $H = +1, O = −2$
    b. $H = −1, O = −2$
    c. $H = +2, O = −2$
    d. $H = +1, O = −1$

    ANS: D   DIF: II   OBJ: 19-1.1

7. What are the oxidation numbers in the compound NO_2?
    a. $N = +2, O = −1$
    b. $N = +2, O = −2$
    c. $N = −2, O = +1$
    d. $N = +4, O = −2$

    ANS: D   DIF: II   OBJ: 19-1.1

8. What are the oxidation numbers in the compound H_3PO_4?
    a. $H = +1, P = 0, O = −2$
    b. $H = +1, P = +5, O = −2$
    c. $H = −1, P = +7, O = −1$
    d. $H = +1, P = +1, O = −1$

    ANS: B   DIF: II   OBJ: 19-1.1

9. What are the oxidation numbers in the ion SO_3^{2-}?
   a. $S = +6, O = -2$
   b. $S = +1, O = -1$
   c. $S = +4, O = -2$
   d. $S = 0, O = -1$

   ANS: C   DIF: II   OBJ: 19-1.1

10. In an oxidation, atoms or ions
    a. increase their oxidation number.
    b. decrease their oxidation number.
    c. do not change their oxidation number.
    d. have a zero oxidation number after the reaction.

    ANS: A   DIF: I   OBJ: 19-1.2

11. In a reduction, atoms or ions
    a. increase their oxidation number.
    b. decrease their oxidation number.
    c. do not change their oxidation number.
    d. have a zero oxidation number after the reaction.

    ANS: B   DIF: I   OBJ: 19-1.2

12. Which of the following are numbers assigned to atoms and ions to keep track of electrons?
    a. charges
    b. coefficients
    c. electrode potentials
    d. oxidation numbers

    ANS: D   DIF: I   OBJ: 19-1.2

13. A species whose oxidation number decreases in a reaction is
    a. oxidized.
    b. reduced.
    c. electrolyzed.
    d. autooxidized.

    ANS: B   DIF: I   OBJ: 19-1.2

14. A species whose oxidation number increases in a reaction is
    a. oxidized.
    b. reduced.
    c. electrolyzed.
    d. autooxidized.

    ANS: A   DIF: I   OBJ: 19-1.2

15. In the reaction $O_2 + 4e^- \rightarrow 2O_2^-$, the species O_2 is
    a. oxidized.
    b. reduced.
    c. electrolyzed.
    d. autooxidized.

    ANS: B   DIF: II   OBJ: 19-1.2

16. In the reaction $Fe \rightarrow Fe^{3+} + 3e^-$, the species Fe is
    a. oxidized.
    b. reduced.
    c. electrolyzed.
    d. autooxidized.

    ANS: A   DIF: II   OBJ: 19-1.2

17. In the reaction $2K + Br_2 \rightarrow 2K^+ + 2Br^-$, which species is reduced?
    a. K only
    b. Br_2 only
    c. both K and Br_2
    d. neither K nor Br_2

    ANS: B   DIF: II   OBJ: 19-1.2

18. In the reaction $F_2 + Mg \rightarrow 2F^- + Mg^{2+}$, which species is oxidized?
    a. F_2 only
    b. Mg only
    c. both Mg and F_2
    d. neither Mg nor F_2

    ANS: B   DIF: II   OBJ: 19-1.2

19. In the reaction $Na^+ + Br^- \rightarrow NaBr$, which species is reduced?
    a. Na^+ only
    b. Br^- only
    c. both Na^+ and Br^-
    d. neither Na^+ nor Br^-

    ANS: D   DIF: II   OBJ: 19-1.2

20. Oxidation and reduction
    a. always occur simultaneously.
    b. always occur at different times.
    c. do not occur in the same reaction.
    d. always occur with oxidation first, then reduction.

    ANS: A   DIF: I   OBJ: 19-1.3

21. If species change their oxidation numbers, the process is a(n)
    a. synthesis.
    b. decomposition.
    c. electrolysis.
    d. oxidation-reduction reaction.

    ANS: D   DIF: I   OBJ: 19-1.3

22. Another name for an oxidation-reduction reaction is a(n)
    a. double-replacement reaction.
    b. redox reaction.
    c. electrochemical reaction.
    d. decomposition reaction.

    ANS: B   DIF: I   OBJ: 19-1.3

23. Which of the following is an oxidation-reduction reaction?
    a. $H_2 \rightarrow 2H$
    b. $2O^- \rightarrow O_2^{2-}$
    c. $H_2 + Cl_2 \rightarrow 2HCl$
    d. $HCl + NaBr \rightarrow HBr + NaCl$

    ANS: C   DIF: II   OBJ: 19-1.3

24. Which of the following is an oxidation-reduction reaction?
    a. $Na \rightarrow Na^+ + e^-$
    b. $2SO_2 + O_2 \rightarrow 2SO_3$
    c. $N_2O_4 \rightarrow 2NO_2$
    d. $2e^- + F_2 \rightarrow 2F^-$

    ANS: B   DIF: II   OBJ: 19-1.3

25. Which of the following is NOT an oxidation-reduction reaction?
    a. $H_2O + SO_2 \rightarrow H_2SO_3$
    b. $N_2 + O_2 \rightarrow 2NO$
    c. $H_2 + Cl_2 \rightarrow 2HCl$
    d. $2NaBr + Cl_2 \rightarrow 2NaCl + Br_2$

    ANS: A   DIF: II   OBJ: 19-1.3

26. Which of the following substances could be produced from SO_3 only by an oxidation-reduction reaction?
    a. H_2SO_4 c. SF_6
    b. H_2SO_3 d. None of the above

    ANS: B   DIF: II   OBJ: 19-1.3

27. In a redox reaction, MnO_4^- is changed to Mn^{2+}. How many electrons must be lost or gained by Mn?
    a. two lost   c. five lost
    b. two gained   d. five gained

    ANS: D   DIF: II   OBJ: 19-1.3

28. In a redox reaction, MnO_4^- is changed to MnO_4^{2-}. How many electrons must be lost or gained by Mn?
    a. none lost or gained   c. one gained
    b. one lost   d. two gained

    ANS: C   DIF: II   OBJ: 19-1.3

29. In a redox reaction, CO is changed to CO_2. How many electrons must be lost or gained by C?
    a. one lost   c. two lost
    b. one gained   d. two gained

    ANS: C   DIF: II   OBJ: 19-1.3

30. In a balanced redox equation, how does the total number of reactant molecules compare with the total number of product molecules?
    a. The two numbers are always equal.
    b. Reactant molecules are always more numerous.
    c. Product molecules are always more numerous.
    d. No relationship exists between the two numbers.

    ANS: D   DIF: I   OBJ: 19-2.1

31. In a balanced redox equation, how does the total number of reactant atoms compare with the total number of product atoms?
    a. The two numbers are always equal.
    b. Reactant atoms are always more numerous.
    c. Product atoms are always more numerous.
    d. No relationship exists between the two numbers.

    ANS: A   DIF: I   OBJ: 19-2.1

32. In a balanced redox equation, how does the total charge of reactants compare with the total charge of products?
    a. The two totals are always equal.
    b. Total reactant charge is always greater.
    c. Total product charge is always greater.
    d. No relationship exists between the two totals.

    ANS: A   DIF: I   OBJ: 19-2.1

33. How does the number of electrons lost in an oxidation compare with the number gained in the simultaneous reduction?
    a. The two numbers are always equal.
    b. The number lost is always greater than the number gained.
    c. The number lost is always less than the number gained.
    d. No relationship exists between the two numbers.

    ANS: A   DIF: I   OBJ: 19-2.1

34. The redox equation $4H^+ + N_2 \rightarrow NH_4^+ + N$ is
    a. correctly balanced.
    b. correctly balanced for number of atoms but not for charge.
    c. correctly balanced for charge but not for number of atoms.
    d. not balanced for number of atoms or charge.

    ANS: B   DIF: II   OBJ: 19-2.1

35. The redox equation $MnO_4^- + 4CO_2 + 4H^+ \rightarrow 4CO^{3+} + Mn^{2+} + 4H_2O$ is
    a. correctly balanced.
    b. correctly balanced for number of atoms but not for charge.
    c. correctly balanced for charge but not for number of atoms.
    d. not balanced for number of atoms or charge.

    ANS: D   DIF: II   OBJ: 19-2.1

36. The redox equation $Cu^{2+} + 2Fe \rightarrow Cu + 2Fe^{2+}$ is
    a. correctly balanced.
    b. correctly balanced for number of atoms but not for charge.
    c. correctly balanced for charge but not for number of atoms.
    d. not balanced for number of atoms or charge.

    ANS: B   DIF: II   OBJ: 19-2.1

37. The redox equation $H_2 + O_2 \rightarrow 2H^+ + O^{2-}$ is
    a. correctly balanced.
    b. correctly balanced for number of atoms but not for charge.
    c. correctly balanced for charge but not for number of atoms.
    d. not balanced for number of atoms or charge.

    ANS: C   DIF: II   OBJ: 19-2.1

38. The redox equation $H_2 + S + 2O_2 \rightarrow 2H^+ + SO_4^{2-}$ is
   a. correctly balanced.
   b. correctly balanced for number of atoms but not for charge.
   c. correctly balanced for charge but not for number of atoms.
   d. not balanced for charge or number of atoms.

   ANS: A   DIF: II   OBJ: 19-2.1

39. After balancing the redox equation $BrO_2 + OH^- + N_2O_4 \rightarrow BrO_3^- + H_2O + NO_2^-$, the coefficients, in order from left to right, are
   a. 1, 4, 3, 1, 4, 6.   c. 2, 6, 2, 2, 4, 12.
   b. 2, 4, 1, 2, 2, 2.   d. 1, 8, 3, 2, 4, 2.

   ANS: B   DIF: III   OBJ: 19-2.2

40. After balancing the redox equation $HBr + NaMnO_4 \rightarrow NaBr + MnBr_2 + Br_2 + H_2O$, the coefficients, in order from left to right, are
   a. 8, 2, 2, 2, 5, 8.   c. 16, 2, 2, 2, 3, 4.
   b. 4, 1, 1, 1, 3, 2.   d. 16, 2, 2, 2, 5, 8.

   ANS: D   DIF: III   OBJ: 19-2.2

41. After balancing the redox equation $FeCl_2 + KMnO_4 + HCl \rightarrow FeCl_3 + MnCl_2 + H_2O + KCl$, the coefficients, in order from left to right, are
   a. 3, 1, 4, 3, 1, 2, 1.   c. 5, 1, 8, 5, 1, 4, 1.
   b. 4, 2, 5, 4, 2, 3, 2.   d. 5, 1, 4, 5, 1, 4, 1.

   ANS: C   DIF: III   OBJ: 19-2.2

42. After balancing the redox equation $Cr_2O_7^{2-} + SO_2 + H^+ \rightarrow Cr^{3+} + HSO_4^- + H_2O$, the coefficients, in order from left to right, are
   a. 1, 3, 5, 2, 3, 1.   c. 2, 5, 10, 4, 5, 2.
   b. 1, 2, 4, 2, 2, 1.   d. 2, 1, 2, 5, 1, 2.

   ANS: A   DIF: III   OBJ: 19-2.2

43. After balancing the redox equation $FeSO_4 + KMnO_4 + H_2SO_4 \rightarrow Fe_2(SO_4)_3 + MnSO_4 + K_2SO_4 + H_2O$, the coefficients, in order from left to right, are
   a. 5, 4, 8, 5, 1, 2, 4.   c. 10, 2, 8, 5, 2, 1, 8.
   b. 2, 5, 3, 2, 5, 1, 4.   d. 5, 1, 4, 3, 1, 1, 4.

   ANS: C   DIF: III   OBJ: 19-2.2

44. After balancing the redox equation in which elemental chlorine reacts with sodium hydroxide to produce sodium hypochlorite (NaClO), water, and sodium chloride, the coefficients of these substances are, respectively,
   a. 1, 1, 1, 1, 1.   c. 2, 4, 2, 1, 2.
   b. 1, 2, 1, 1, 1.   d. 1, 2, 1, 2, 1.

   ANS: B   DIF: III   OBJ: 19-2.2

45. After balancing the redox equation $FeCl_3 + Zn \rightarrow ZnCl_2 + Fe$, the coefficients, in order from left to right, are
   a. 2, 2, 1, 2.
   b. 1, 1, 1, 1.
   c. 4, 3, 3, 4.
   d. 2, 3, 3, 2.

   ANS: D   DIF: III   OBJ: 19-2.2

46. After balancing the redox equation $Bi(OH)_3 + SnO_2^{2-} \rightarrow Bi + SnO_3^{2-} + H_2O$, the coefficients, in order from left to right, are
   a. 4, 5, 4, 6, 6.
   b. 2, 3, 2, 3, 3.
   c. 3, 5, 3, 5, 5.
   d. 2, 1, 2, 1, 3.

   ANS: B   DIF: III   OBJ: 19-2.2

47. During redox reactions, oxidizing agents
   a. increase their oxidation number.
   b. decrease their oxidation number.
   c. keep the same oxidation number.
   d. do not participate.

   ANS: B   DIF: I   OBJ: 19-3.1

48. During redox reactions, reducing agents
   a. increase their oxidation number.
   b. decrease their oxidation number.
   c. keep the same oxidation number.
   d. do not participate.

   ANS: A   DIF: I   OBJ: 19-3.1

49. In redox reactions,
   a. the oxidizing agent is the substance oxidized.
   b. the reducing agent is the substance oxidized.
   c. both oxidizing and reducing agents are oxidized.
   d. the reducing agent is the substance reduced.

   ANS: B   DIF: I   OBJ: 19-3.1

50. In redox reactions,
   a. the oxidizing agent is the substance reduced.
   b. the reducing agent is the substance reduced.
   c. the oxidizing agent is the substance oxidized.
   d. both oxidizing and reducing agents are on the same side of the chemical equation.

   ANS: A   DIF: I   OBJ: 19-3.1

51. Which is the most active reducing agent among the elements?
   a. cesium
   b. iodine
   c. fluorine
   d. lithium

   ANS: D   DIF: II   OBJ: 19-3.1

52. Which is the most active oxidizing agent among the elements?
   a. cesium
   b. iodine
   c. fluorine
   d. lithium

   ANS: C   DIF: II   OBJ: 19-3.1

Relative Strength of Oxidizing and Reducing Agents

	Reducing Agents	Oxidizing Agents	
S	Li	Li^+	W
T	K	K^+	E
R	Ca	Ca^{2+}	A
O	Na	Na^+	K
N	Mg	Mg^{2+}	E
G	Al	Al^{3+}	R
E	Zn	Zn^{2+}	
R	Cr	Cr^{3+}	
	Fe	Fe^{2+}	
	Ni	Ni^{2+}	
	Sn	Sn^{2+}	
	Pb	Pb^{2+}	
	H_2	H_3O^+	
	H_2S	S	
	Cu	Cu^{2+}	
	I^-	I_2	
	MnO_4^{2-}	MnO_4^-	
	Fe^{2+}	Fe^{3+}	
	Hg	Hg_2^{2+}	
	Ag	Ag^+	
	NO_2^-	NO_3^-	S
	Br^-	Br_2	T
W	Mn^{2+}	MnO_2	R
E	SO_2	H_2SO_4 (conc.)	O
A	Cr^{3+}	$Cr_2O_7^{2-}$	N
K	Cl^-	Cl_2	G
E	Mn^{2+}	MnO_4^-	E
R	F^-	F_2	R

53. In the figure above, which ion oxidizes Sn to Sn^{2+} but does NOT oxidize Hg to Hg_2^{2+}?
   a. NO_3^-
   b. Al^{3+}
   c. $Cr_2O_7^{2-}$
   d. Cu^{2+}

   ANS: D   DIF: III   OBJ: 19-3.1

54. In the figure above, which ion is reduced by Zn but reduces Ag^+ to Ag?
   a. Mg^{2+}
   b. Fe^{3+}
   c. Fe^{2+}
   d. NO_3^-

   ANS: C   DIF: III   OBJ: 19-3.1

55. In the figure above, which element displaces Cu^{2+} ions from solution but is displaced by Ni metal when it is in ionic form?

   a. Al
   b. Ag
   c. Fe
   d. Pb

ANS: D   DIF: III   OBJ: 19-3.1

56. What is the oxidation number of a free element?

   a. its group number
   b. its total number of valence electrons
   c. +1
   d. 0

ANS: D   DIF: I   OBJ: 19-1.1

57. In which process does a substance act as both an oxidizing agent and a reducing agent and oxidize itself?

   a. electrolysis
   b. autooxidation
   c. autoreduction
   d. double replacement

ANS: B   DIF: I   OBJ: 19-3.2

58. What is the formula of the peroxide ion?

   a. O^{2-}
   b. O^-
   c. O_2^-
   d. O_2^{2-}

ANS: D   DIF: I   OBJ: 19-3.2

59. Which of the following describes the bond in a peroxide ion?

   a. a double bond
   b. highly stable
   c. somewhat unstable
   d. a triple bond

ANS: C   DIF: I   OBJ: 19-3.2

60. When hydrogen peroxide decomposes, oxygen is

   a. reduced only.
   b. oxidized only.
   c. both oxidized and reduced.
   d. electrolyzed.

ANS: C   DIF: I   OBJ: 19-3.2

61. Which of the following reactions is an autooxidation?

   a. $2H_2O_2 \rightarrow 2H_2O + O_2$
   b. $2H_2O \rightarrow H_3O^+ + OH^-$
   c. $Cu + 2AgNO_3 \rightarrow Cu(NO_3)_2 + 2Ag$
   d. $2Li + 2H_2O \rightarrow 2LiOH + H_2$

ANS: A   DIF: II   OBJ: 19-3.2

62. When hydrogen peroxide decomposes, usually the only products are

   a. hydrogen and oxygen.
   b. hydrogen and water.
   c. oxygen and water.
   d. hydrogen, oxygen, and water.

ANS: C   DIF: II   OBJ: 19-3.2

63. Which of the following reactions is an autooxidation?
    a. $2NO_2 \rightarrow N_2O_4$
    b. $N_2O_3 \rightarrow NO_2 + NO$
    c. $2NO + O_2 \rightarrow 2NO_2$
    d. $N_2 + O_2 \rightarrow 2NO$

    ANS: B   DIF: II   OBJ: 19-3.2

64. Give the oxidation numbers for the oxygen atoms in H_2O_2, H_2O, and O_2, respectively.
    a. $-2, -1, 0$
    b. $-1, -2, 0$
    c. $-2, -2, 0$
    d. $-1, 0, 0$

    ANS: B   DIF: II   OBJ: 19-3.2

65. In which system does a spontaneous redox reaction produce electrical energy?
    a. electrochemical cell
    b. electrolytic cell
    c. electroplating cell
    d. half-cell

    ANS: A   DIF: I   OBJ: 19-4.1

66. If the reactants in a spontaneous energy-releasing redox reaction are in direct contact, the energy is released in the form of
    a. light.
    b. electrical energy.
    c. heat.
    d. mechanical energy.

    ANS: C   DIF: I   OBJ: 19-4.1

67. If the reactants in a spontaneous energy-releasing redox reaction are connected externally by a wire conductor, the energy is released in the form of
    a. light.
    b. electrical energy.
    c. heat.
    d. mechanical energy.

    ANS: B   DIF: I   OBJ: 19-4.1

68. Where does oxidation take place in an electrochemical cell?
    a. the anode
    b. the cathode
    c. the anode or the cathode
    d. the half-cell

    ANS: A   DIF: I   OBJ: 19-4.1

69. Where does reduction take place in an electrochemical cell?
    a. the anode
    b. the cathode
    c. the anode or the cathode
    d. the half-cell

    ANS: B   DIF: I   OBJ: 19-4.1

70. In an electrochemical cell,
    a. both the cathode and anode are positively charged.
    b. both the cathode and anode are negatively charged.
    c. the cathode is negatively charged and the anode is positively charged.
    d. the cathode is positively charged and the anode is negatively charged.

    ANS: C   DIF: I   OBJ: 19-4.1

71. In a zinc-carbon dry cell, oxidation of
    a. zinc occurs at the anode.
    c. zinc occurs at the cathode.
    b. manganese occurs at the anode.
    d. manganese occurs at the cathode.

    ANS: A   DIF: II   OBJ: 19-4.1

72. In a zinc-carbon dry cell,
    a. the zinc electrode is the cathode and the carbon electrode is the anode.
    b. the zinc electrode is the anode and the carbon electrode is the cathode.
    c. both electrodes are auto-oxidizing and serve as both cathodes and anodes.
    d. neither electrode can be considered a cathode or an anode.

    ANS: B   DIF: I   OBJ: 19-4.1

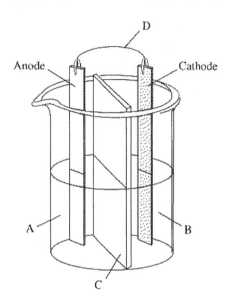

73. In the figure of a voltaic cell shown above, where are electrons gained?
    a. A
    c. C
    b. B
    d. D

    ANS: B   DIF: II   OBJ: 19-4.2

74. An electrochemical cell that produces electricity is a(n)
    a. alkaline cell.
    c. electrolytic cell.
    b. dry cell.
    d. voltaic cell.

    ANS: D   DIF: I   OBJ: 19-4.2

75. If the reactants of a voltaic cell are in contact,
    a. most of the energy produced is an electric current.
    b. most of the energy produced is heat.
    c. oxidation and reduction do not occur.
    d. hydrogen gas is released.

    ANS: B   DIF: I   OBJ: 19-4.2

76. The voltage of a voltaic cell is determined by the E^0 value(s) of the
   a. half-reaction at the anode.   c. half-reaction at the cathode.
   b. half-reactions at the cathode and anode.   d. standard hydrogen electrode.

   ANS: B   DIF: I   OBJ: 19-4.2

77. Electrons in a voltaic cell normally flow
   a. from cathode to anode.
   b. through a porous barrier.
   c. in both directions through the external circuit.
   d. from anode to cathode.

   ANS: D   DIF: I   OBJ: 19-4.2

78. For electricity to flow in a voltaic cell, the two half-cells must be
   a. connected by a wire and a porous barrier.   c. in the same solution.
   b. completely isolated from one another.   d. connected to a dry cell.

   ANS: A   DIF: I   OBJ: 19-4.2

79. In which cell does a current drive a nonspontaneous redox reaction?
   a. electrolytic cell   c. electrochemical cell
   b. dry cell   d. voltaic cell

   ANS: A   DIF: I   OBJ: 19-4.3

80. In an electrolytic cell, reduction occurs
   a. at the cathode.   c. at either the cathode or the anode.
   b. at the anode.   d. between the cathode and anode.

   ANS: A   DIF: I   OBJ: 19-4.3

81. In an electrolytic cell, oxidation occurs
   a. at the cathode.   c. at either the cathode or the anode.
   b. at the anode.   d. between the cathode and the anode.

   ANS: B   DIF: I   OBJ: 19-4.3

82. In an electrolytic cell, the anode
   a. can be either positively or negatively charged.
   b. is not charged.
   c. is positively charged.
   d. is negatively charged.

   ANS: C   DIF: I   OBJ: 19-4.3

83. In an electrolytic cell, the cathode
   a. can be either positively or negatively charged.
   b. is not charged.
   c. is positively charged.
   d. is negatively charged.

   ANS: D   DIF: I   OBJ: 19-4.3

84. In an electrolytic cell, negative ions move toward the
    a. negative electrode, and positive ions move toward the negative electrode.
    b. negative electrode, and positive ions move toward the positive electrode.
    c. positive electrode, and positive ions move toward the positive electrode.
    d. positive electrode, and positive ions move toward the negative electrode.

    ANS: D   DIF: I   OBJ: 19-4.3

85. What is the source of energy for an electrolytic cell?
    a. the reaction occurring in the electrolytic cell
    b. an external direct-current source, such as a battery
    c. ion migration in the electrolyte
    d. electron migration in the electrolyte

    ANS: B   DIF: I   OBJ: 19-4.3

86. Which process deposits metal onto a surface?
    a. electrolysis   c. autooxidation
    b. electroplating   d. oxidation

    ANS: B   DIF: I   OBJ: 19-4.4

87. Electroplating is an application of which reaction?
    a. electrolytic reactions   c. auto-oxidation reactions
    b. electrochemical reactions   d. voltaic reactions

    ANS: A   DIF: I   OBJ: 19-4.4

88. An electroplating cell contains a solution of
    a. a salt of the plating metal.
    b. a salt of the metal of the object that is to be plated.
    c. a substance that does not conduct electricity.
    d. H_2SO_4.

    ANS: A   DIF: I   OBJ: 19-4.4

89. In an electroplating cell, the object to be plated is the
    a. external circuit.   c. anode.
    b. electrolyte.   d. cathode.

    ANS: D   DIF: I   OBJ: 19-4.4

90. In an electroplating cell, the metal used to plate the object is the
    a. external circuit.   c. anode.
    b. electrolyte.   d. cathode.

    ANS: C   DIF: I   OBJ: 19-4.4

91. In an electroplating cell, a solution of the salt of the plating metal is the
    a. external circuit.   c. anode.
    b. electrolyte.   d. cathode.

    ANS: B   DIF: I   OBJ: 19-4.4

92. In a cell used to electroplate silver onto an object, Ag^+ is
    a. oxidized at the anode.
    b. reduced at the anode.
    c. oxidized at the cathode.
    d. reduced at the cathode.

    ANS: D   DIF: I   OBJ: 19-4.4

93. In a cell used to electroplate silver onto an object, Ag is
    a. oxidized at the anode.
    b. reduced at the anode.
    c. oxidized at the cathode.
    d. plated out at the anode.

    ANS: A   DIF: I   OBJ: 19-4.4

94. The oxidation-reduction reactions in a rechargeable cell are the same as in
    a. an electrochemical cell.
    b. an electrolysis cell.
    c. an electrochemical cell and an electrolysis cell.
    d. a half-cell.

    ANS: C   DIF: I   OBJ: 19-4.5

95. When a rechargeable cell produces electrical energy, it acts as a(n)
    a. electrochemical cell.
    b. electrolytic cell.
    c. voltaic cell.
    d. half-cell.

    ANS: C   DIF: I   OBJ: 19-4.5

96. A rechargeable cell produces energy when
    a. it is charging.
    b. it is discharging.
    c. its external circuit is not closed.
    d. the porous barrier is in place.

    ANS: B   DIF: I   OBJ: 19-4.5

97. When electrical energy is provided to a rechargeable cell from an external source, the cell acts as a(n)
    a. electrochemical cell.
    b. electrolytic cell.
    c. voltaic cell.
    d. half-cell.

    ANS: B   DIF: I   OBJ: 19-4.5

98. Electrical energy is provided to a rechargeable cell from an outside source when
    a. it is charging.
    b. it is discharging.
    c. its external circuit is not closed.
    d. the porous barrier is in place.

    ANS: A   DIF: I   OBJ: 19-4.5

99. What is the voltage of the standard automobile battery?
    a. 1.5 V
    b. 6 V
    c. 12 V
    d. 50 V

    ANS: C   DIF: I   OBJ: 19-4.5

100. When an automobile battery is charging,
  a. heat energy is converted to energy of motion.
  b. energy of motion is converted to heat energy.
  c. chemical energy is converted to electrical energy.
  d. electrical energy is converted to chemical energy.

  ANS: D   DIF: I   OBJ: 19-4.5

101. When an automobile battery is discharging,
  a. heat energy is converted to energy of motion.
  b. energy of motion is converted to heat energy.
  c. chemical energy is converted to electrical energy.
  d. electrical energy is converted to chemical energy.

  ANS: C   DIF: I   OBJ: 19-4.5

102. Which substances react in the standard automobile battery?
  a. lead(IV) oxide, lead, and sulfuric acid   c. zinc oxide, zinc, and sulfuric acid
  b. copper(II) oxide, copper, and sulfuric   d. iron(III) oxide, iron, and sulfuric acid
      acid

  ANS: A   DIF: I   OBJ: 19-4.5

Standard Reduction Potentials

Half-cell reaction	Standard electrode potential, E^0 (in volts)	Half-cell reaction	Standard electrode potential, E^0 (in volts)
$F_2 + 2e^- \rightleftharpoons F^-$	+2.87	$Fe^{3+} + 3e^- \rightleftharpoons Fe$	−0.04
$MnO_4^- + 8H^+ + 5e^- \rightleftharpoons Mn^{2+} + 4H_2O$	+1.50	$Pb^{2+} + 2e^- \rightleftharpoons Pb$	−0.13
$Au^{3+} + 3e^- \rightleftharpoons Au$	+1.50	$Sn^{2+} + 2e^- \rightleftharpoons Sn$	−0.14
$Cl_2 + 2e^- \rightleftharpoons 2Cl^-$	+1.36	$Ni^{2+} + 2e^- \rightleftharpoons Ni$	−0.26
$Cr_2O_7^{2-} + 14H^+ + 6e^- \rightleftharpoons 2Cr^{3+} + 7H_2O$	+1.23	$Co^{2+} + 2e^- \rightleftharpoons Co$	−0.28
$MnO_2 + 4H^+ + 2e^- \rightleftharpoons Mn^{2+} + 2H_2O$	+1.22	$Cd^{2+} + 2e^- \rightleftharpoons Cd$	−0.40
$Br_2 + 2e^- \rightleftharpoons 2Br^-$	+1.07	$Fe^{2+} + 2e^- \rightleftharpoons Fe$	−0.45
$Hg^{2+} + 2e^- \rightleftharpoons Hg$	+0.85	$S^{2+} + 2e^- \rightleftharpoons S^{2-}$	−0.48
$Ag^+ + e^- \rightleftharpoons Ag$	+0.80	$Cr^{3+} + 3e^- \rightleftharpoons Cr$	−0.74
$Hg_2^{2+} + 2e^- \rightleftharpoons 2Hg$	+0.80	$Zn^{2+} + 2e^- \rightleftharpoons Zn$	−0.76
$Fe^{3+} + e^- \rightleftharpoons Fe^{2+}$	+0.77	$Al^{3+} + 3e^- \rightleftharpoons Al$	−1.66
$MnO_4^- + e^- \rightleftharpoons MnO_4^{2-}$	+0.56	$Mg^{2+} + 2e^- \rightleftharpoons Mg$	−2.37
$I_2 + 2e^- \rightleftharpoons 2I^-$	+0.54	$Na^+ + e^- \rightleftharpoons Na$	−2.71
$Cu^{2+} + 2e^- \rightleftharpoons Cu$	+0.34	$Ca^{2+} + 2e^- \rightleftharpoons Ca$	−2.87
$Cu^{2+} + e^- \rightleftharpoons Cu^+$	+0.15	$Ba^{2+} + 2e^- \rightleftharpoons Ba$	−2.91
$S + 2H^{2+}(aq) + 2e^- \rightleftharpoons H_2S(aq)$	+0.14	$K^+ + e^- \rightleftharpoons K$	−2.93
$2H^+(aq) + 2e^- \rightleftharpoons H_2$	0.00	$Li^+ + e^- \rightleftharpoons Li$	−3.04

103. Calculate E^0 for the spontaneous reaction when an Ag^+/Ag half-cell is joined to an Hg^{2+}/Hg half-cell. Name the neutral metal produced.
    a.   +1.65 V; Ag
    b.   +1.65 V; Hg
    c.   +0.05 V; Ag
    d.   +0.05 V; Hg

    ANS: D   DIF: III   OBJ: 19-4.6

104. Calculate E^0 for the spontaneous reaction when a Co^{2+}/Co half-cell is joined to a Cu^{2+}/Cu half-cell. Name the neutral metal produced.
    a.   +0.62 V; Cu
    b.   +0.62 V; Co
    c.   +0.06 V; Cu
    d.   +0.06 V; Co

    ANS: A   DIF: III   OBJ: 19-4.6

105. Calculate E^0 for the reaction $3Ni^{2+} + 2Cr \rightarrow 3Ni + 2Cr^{3+}$. Is the reaction spontaneous?
    a.   −1.00 V; yes
    b.   −0.48 V; no
    c.   +0.48 V; yes
    d.   +0.48 V; no

    ANS: C   DIF: III   OBJ: 19-4.6

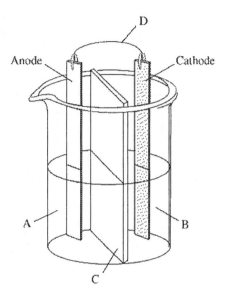

1. In the figure above, what is the purpose of the part of the voltaic cell labeled C?

    ANS:
    The porous barrier allows ions to migrate and equalize charge between the two half-cells but prevents ions and atoms from exchanging electrons directly.

    DIF: II   OBJ: 19-4.2

2. In which direction do electrons flow in a voltaic cell?

    ANS:
    Electrons flow from the anode to the cathode.

    DIF: I   OBJ: 19-4.2

PROBLEM

Standard Reduction Potentials

Half-cell reaction	Standard electrode potential, E^0 (in volts)	Half-cell reaction	Standard electrode potential, E^0 (in volts)
$F_2 + 2e^- \rightleftharpoons F^-$	+2.87	$Fe^{3+} + 3e^- \rightleftharpoons Fe$	−0.04
$MnO_4^- + 8H^+ + 5e^- \rightleftharpoons Mn^{2+} + 4H_2O$	+1.50	$Pb^{2+} + 2e^- \rightleftharpoons Pb$	−0.13
$Au^{3+} + 3e^- \rightleftharpoons Au$	+1.50	$Sn^{2+} + 2e^- \rightleftharpoons Sn$	−0.14
$Cl_2 + 2e^- \rightleftharpoons 2Cl^-$	+1.36	$Ni^{2+} + 2e^- \rightleftharpoons Ni$	−0.26
$Cr_2O_7^{2-} + 14H^+ + 6e^- \rightleftharpoons 2Cr^{3+} + 7H_2O$	+1.23	$Co^{2+} + 2e^- \rightleftharpoons Co$	−0.28
$MnO_2 + 4H^+ + 2e^- \rightleftharpoons Mn^{2+} + 2H_2O$	+1.22	$Cd^{2+} + 2e^- \rightleftharpoons Cd$	−0.40
$Br_2 + 2e^- \rightleftharpoons 2Br^-$	+1.07	$Fe^{2+} + 2e^- \rightleftharpoons Fe$	−0.45
$Hg^{2+} + 2e^- \rightleftharpoons Hg$	+0.85	$S^{2+} + 2e^- \rightleftharpoons S^{2-}$	−0.48
$Ag^+ + e^- \rightleftharpoons Ag$	+0.80	$Cr^{3+} + 3e^- \rightleftharpoons Cr$	−0.74
$Hg_2^{2+} + 2e^- \rightleftharpoons 2Hg$	+0.80	$Zn^{2+} + 2e^- \rightleftharpoons Zn$	−0.76
$Fe^{3+} + e^- \rightleftharpoons Fe^{2+}$	+0.77	$Al^{3+} + 3e^- \rightleftharpoons Al$	−1.66
$MnO_4^- + e^- \rightleftharpoons MnO_4^{2-}$	+0.56	$Mg^{2+} + 2e^- \rightleftharpoons Mg$	−2.37
$I_2 + 2e^- \rightleftharpoons 2I^-$	+0.54	$Na^+ + e^- \rightleftharpoons Na$	−2.71
$Cu^{2+} + 2e^- \rightleftharpoons Cu$	+0.34	$Ca^{2+} + 2e^- \rightleftharpoons Ca$	−2.87
$Cu^{2+} + e^- \rightleftharpoons Cu^+$	+0.15	$Ba^{2+} + 2e^- \rightleftharpoons Ba$	−2.91
$S + 2H^{2+}(aq) + 2e^- \rightleftharpoons H_2S(aq)$	+0.14	$K^+ + e^- \rightleftharpoons K$	−2.93
$2H^+(aq) + 2e^- \rightleftharpoons H_2$	0.00	$Li^+ + e^- \rightleftharpoons Li$	−3.04

1. Use the figure above to calculate the E^0_{cell} for a calcium-chlorine cell.

   ANS:
   4.23 V

   DIF: III   OBJ: 19-4.6

2. Use the figure above to calculate the E^0_{cell} for an aluminum-gold cell.

   ANS:
   3.16 V

   DIF: III   OBJ: 19-4.6

3. Use the figure above to calculate the E^0_{cell} for a chromium-fluorine cell.

   ANS:
   3.61 V

   DIF: III   OBJ: 19-4.6

4. Use the figure above to calculate the E^0_{cell} for a lithium-tin cell.

ANS:
2.91 V

DIF: III   OBJ: 19-4.6

5. Use the figure above to calculate the E^0_{cell} for a cell made from iodine and bromine.

ANS:
0.53 V

DIF: III   OBJ: 19-4.6

ESSAY

1. What observations suggest that a chemical reaction is occurring in a voltaic cell?

ANS:
Answers will vary. Some examples: electrical energy is produced, gas evolves at the electrodes, the color of the solution changes, the masses of the electrodes change.

DIF: II   OBJ: 19-4.2

MULTIPLE CHOICE

1. Which of the following is the electron configuration of carbon in the ground state?
   a. $1s^2 2s^2 2p^2$
   b. $1s^2 2s^1 2p^3$
   c. $1s^2 2s^2 2p^3$
   d. $1s^2 2s^2 2p^6$

   ANS: A   DIF: I   OBJ: 20-1.1

2. Carbon shows a very strong tendency to form
   a. ionic bonds.
   b. covalent bonds.
   c. hydrogen bonds.
   d. highly polar bonds.

   ANS: B   DIF: I   OBJ: 20-1.1

3. How many valence electrons does a carbon atom have?
   a. 3
   b. 4
   c. 5
   d. 6

   ANS: B   DIF: I   OBJ: 20-1.1

4. How many covalent bonds can a carbon atom usually form?
   a. 2
   b. 3
   c. 4
   d. 5

   ANS: C   DIF: I   OBJ: 20-1.1

5. When a carbon atom forms four covalent bonds, the bonds are directed toward the corners of a
   a. triangle.
   b. pyramid.
   c. square.
   d. tetrahedron.

   ANS: D   DIF: I   OBJ: 20-1.1

6. When carbon forms four covalent bonds, what is the orbital hybridization?
   a. sp
   b. sp^2
   c. sp^3
   d. $s^2 p^2$

   ANS: C   DIF: I   OBJ: 20-1.1

7. Carbon atoms readily join with atoms of
   a. metals.
   b. carbon only.
   c. both other elements and carbon.
   d. nonmetals.

   ANS: C   DIF: I   OBJ: 20-1.1

8. What are forms of the same element in the same state that have different properties?
   a. monomers
   b. isotopes
   c. allotropes
   d. isomers

   ANS: C   DIF: I   OBJ: 20-1.2

9. What are two allotropic forms of carbon?
   a.   carbon-12 and carbon-14
   b.   alkanes and alkenes
   c.   solid carbon and liquid carbon
   d.   diamond and graphite

   ANS: D   DIF: I   OBJ: 20-1.2

10. Which of the following is an allotropic form of carbon?
    a.   carbon-12
    b.   alkanes
    c.   solid carbon
    d.   fullerenes

    ANS: D   DIF: I   OBJ: 20-1.2

11. Which statement about the hardness of diamond and graphite is correct?
    a.   Diamond is very soft, and graphite is very hard.
    b.   Diamond is very hard, and graphite is very soft.
    c.   Both diamond and graphite are very hard.
    d.   Both diamond and graphite are very soft.

    ANS: B   DIF: I   OBJ: 20-1.2

12. Which allotrope of carbon feels greasy and crumbles easily?
    a.   fullerenes
    b.   graphite
    c.   carbon-12
    d.   solid carbon

    ANS: B   DIF: I   OBJ: 20-1.2

13. Which allotrope of carbon was the most recently discovered?
    a.   diamond
    b.   carbon-14
    c.   fullerenes
    d.   carbon-12

    ANS: C   DIF: I   OBJ: 20-1.2

14. Which allotrope of carbon is found in the soot that forms when carbon-containing materials are burned with limited oxygen?
    a.   graphite
    b.   carbon-14
    c.   fullerenes
    d.   carbon-12

    ANS: C   DIF: I   OBJ: 20-1.2

15. Diamond is a good conductor of
    a.   heat only.
    b.   electricity only.
    c.   both heat and electricity.
    d.   magnetic forces.

    ANS: A   DIF: I   OBJ: 20-1.3

16. Graphite is a good conductor of
    a.   heat only.
    b.   electricity only.
    c.   both heat and electricity.
    d.   water only.

    ANS: B   DIF: I   OBJ: 20-1.3

17. The atoms in graphite
    a. form a tetrahedral network.
    b. are arranged in squares.
    c. are arranged in layers that form thin hexagonal plates.
    d. form a face-centered cube.

    ANS: C   DIF: I   OBJ: 20-1.3

18. The atoms in a fullerene
    a. form a tetrahedral network.
    b. form near-spherical cage-like structures.
    c. are arranged in layers of thin hexagonal plates.
    d. either form a body-centered cube or exhibit closest hexagonal packing.

    ANS: B   DIF: I   OBJ: 20-1.3

19. Which of the following statements about diamond and graphite is true?
    a. In diamond, each atom is bonded to four other atoms. In graphite, each atom is bonded to three other atoms.
    b. In diamond, each atom is bonded to three other atoms. In graphite, each atom is bonded to four other atoms.
    c. In both diamond and graphite, each atom is bonded to four other atoms.
    d. In both diamond and graphite, each atom is bonded to three other atoms.

    ANS: A   DIF: I   OBJ: 20-1.3

20. The bonds between atoms in a layer of graphite are
    a. all single bonds.
    b. all double bonds.
    c. alternating single and double bonds.
    d. intermediate in character between single and double bonds.

    ANS: D   DIF: I   OBJ: 20-1.3

21. Which of the following statements about fullerenes is correct?
    a. The number of carbon atoms in the molecule is the same for all fullerenes.
    b. The number of carbon atoms in the molecule varies.
    c. The atoms are arranged in layers of thin, hexagonal plates.
    d. Each atom is arranged tetrahedrally around its four nearest neighbors.

    ANS: B   DIF: I   OBJ: 20-1.3

22. What do all organic compounds contain?
    a. hydrogen   c. oxygen
    b. water   d. carbon

    ANS: D   DIF: I   OBJ: 20-2.1

23. Organic compounds are defined as all covalently bonded compounds containing carbon EXCEPT
    a. oxides
    b. carbonates
    c. Both a and b
    d. There are no exceptions.

    ANS: C   DIF: I   OBJ: 20-2.1

24. Organic compounds are so numerous because carbon atoms
    a. are highly electronegative.
    b. can bond with each other in many ways.
    c. are very small.
    d. are stable and very common in nature.

    ANS: B   DIF: I   OBJ: 20-2.1

25. The diversity of organic compounds is so great because carbon
    a. has many allotropes.
    b. has many isotopes.
    c. compounds have many isomers.
    d. compounds are very reactive.

    ANS: C   DIF: I   OBJ: 20-2.1

26. As the number of carbon atoms in a molecular formula increases, the number of possible isomers
    a. increases rapidly.
    b. increases slowly.
    c. remains the same.
    d. decreases.

    ANS: A   DIF: I   OBJ: 20-2.1

27. Which formula is most useful in distinguishing isomers?
    a. molecular formula
    b. structural formula
    c. empirical formula
    d. ionic formula

    ANS: B   DIF: I   OBJ: 20-2.2

28. Which formula shows the numbers and types of atoms in a molecule but not the bonds?
    a. molecular formula
    b. structural formula
    c. empirical formula
    d. ionic formula

    ANS: A   DIF: I   OBJ: 20-2.2

29. Which formula shows the numbers and types of atoms in a molecule and the bonds?
    a. molecular formula
    b. structural formula
    c. empirical formula
    d. ionic formula

    ANS: B   DIF: I   OBJ: 20-2.2

30. Which of the following representations best shows the shape of a molecule?
    a. molecular formula
    b. structural formula
    c. drawing or model
    d. empirical formula

    ANS: C   DIF: I   OBJ: 20-2.2

31. Which molecular representation cannot distinguish isomers of a compound?
    a. molecular formula
    b. structural formula
    c. ionic formula
    d. three-dimensional drawing or model

    ANS: A   DIF: I   OBJ: 20-2.2

32. Isomers are compounds that have
    a. the same molecular formula but different structures.
    b. the same molecular formula and the same structure.
    c. different molecular formulas and different structures.
    d. different molecular formulas but the same structure.

    ANS: A   DIF: I   OBJ: 20-2.3

33. In which of the following are the atoms bonded in a different order?
    a. structural isomers
    b. geometric isomers
    c. three-dimensional models
    d. organic compounds

    ANS: A   DIF: I   OBJ: 20-2.3

34. In which of the following are atoms bonded in the same order but arranged differently in space?
    a. structural isomers
    b. geometric isomers
    c. three-dimensional models
    d. organic compounds

    ANS: B   DIF: I   OBJ: 20-2.3

35. What type of bond is rigid and does not allow atoms to rotate freely?
    a. ionic bond
    b. double bond
    c. single bond
    d. hydrogen bond

    ANS: B   DIF: I   OBJ: 20-2.3

36. Atoms are bonded together in a different order in
    a. structural isomers.
    b. geometric isomers.
    c. organic compounds.
    d. three-dimensional models.

    ANS: A   DIF: I   OBJ: 20-2.3

37. When two identical atoms are on the same side, the molecule is
    a. cis.
    b. trans.
    c. structural.
    d. rigid.

    ANS: A   DIF: I   OBJ: 20-2.3

38. When two identical atoms are on opposite sides, the molecule is
    a. cis.
    b. trans.
    c. structural.
    d. free to rotate.

    ANS: B   DIF: I   OBJ: 20-2.3

39. How many structural isomers does CH_4 have?
    a. one   c. three
    b. two   d. five

    ANS: A   DIF: II   OBJ: 20-2.3

40. How many structural isomers does C_3H_8 have?
    a. one   c. three
    b. two   d. five

    ANS: A   DIF: II   OBJ: 20-2.3

41. How many structural isomers does C_4H_{10} have?
    a. one   c. three
    b. two   d. five

    ANS: B   DIF: II   OBJ: 20-2.3

42. How many structural isomers does C_5H_{12} have?
    a. one   c. three
    b. two   d. five

    ANS: C   DIF: II   OBJ: 20-2.3

43. Each carbon atom in a molecule forms four single covalent bonds with other atoms in a(n)
    a. aromatic hydrocarbon.   c. structural isomer.
    b. saturated hydrocarbon.   d. geometric isomer.

    ANS: B   DIF: I   OBJ: 20-3.1

44. Hydrocarbons are grouped into series according to the
    a. number of carbon atoms.   c. bonds between carbon atoms.
    b. isotope of carbon.   d. mass of the compounds.

    ANS: C   DIF: I   OBJ: 20-3.1

45. Hydrocarbons in which carbon atoms are connected by only single bonds in straight chains or branched chains are called
    a. aromatic hydrocarbons.   c. alkenes.
    b. alkynes.   d. alkanes.

    ANS: D   DIF: I   OBJ: 20-3.1

46. Which of the following is an alkane?
    a. propyne   c. propene
    b. propane   d. propyl bromide

    ANS: B   DIF: II   OBJ: 20-3.1

47. Hydrocarbons in which carbon atoms form only single bonds and are arranged in a ring are called
    a. cycloalkanes.
    b. alkynes.
    c. alkenes.
    d. aromatic hydrocarbons.

    ANS: A   DIF: I   OBJ: 20-3.1

48. In alkanes with three or fewer carbon atoms,
    a. the chains can be straight or branched.
    b. structural isomers exist.
    c. only one molecular structure is possible.
    d. usually one but sometimes more than one molecular structure is possible.

    ANS: C   DIF: I   OBJ: 20-3.1

49. In alkanes with more than three carbon atoms,
    a. the chains can be straight or branched.
    b. structural isomers exist.
    c. only one molecular structure is possible.
    d. both a and b

    ANS: D   DIF: I   OBJ: 20-3.1

50. Adjacent members differ by a constant unit in a(n)
    a. homologous series.
    b. saturated hydrocarbon.
    c. alkane.
    d. cycloalkane.

    ANS: A   DIF: I   OBJ: 20-3.1

A.
$$H-\underset{\underset{H}{|}}{\overset{\overset{H}{|}}{C}}-\underset{\underset{H}{|}}{\overset{\overset{H}{|}}{C}}-\underset{\underset{H}{|}}{\overset{\overset{H}{|}}{C}}-H$$

B.
$$H-\underset{\underset{H}{|}}{\overset{\overset{H}{|}}{C}}-\underset{\underset{\underset{\underset{H}{|}}{H-\overset{|}{C}-H}}{|}}{\overset{\overset{H}{|}}{C}}-\underset{\underset{H}{|}}{\overset{\overset{H}{|}}{C}}-H$$

C.
$$H-\underset{\underset{H}{|}}{\overset{\overset{H}{|}}{C}}-\underset{\underset{H}{|}}{\overset{\overset{H}{|}}{C}}-\underset{\underset{H}{|}}{\overset{\overset{H}{|}}{C}}-\underset{\underset{H}{|}}{\overset{\overset{H}{|}}{C}}-H$$

D.
$$H-\underset{\underset{H}{|}}{\overset{\overset{H}{|}}{C}}-\underset{\underset{\underset{\underset{H}{|}}{H-\overset{|}{C}-H}}{|}}{\overset{\overset{H}{|}}{C}}-\underset{\underset{H}{|}}{\overset{\overset{H}{|}}{C}}-\underset{\underset{H}{|}}{\overset{\overset{H}{|}}{C}}-H$$

51. In the figure above, what is the structural formula for 2-methylpropane?
    a. A
    b. B
    c. C
    d. D

    ANS: B   DIF: II   OBJ: 20-3.2

52. In the figure above, what is the structural formula for butane?
    a. A
    b. B
    c. C
    d. D

ANS: C   DIF: II   OBJ: 20-3.2

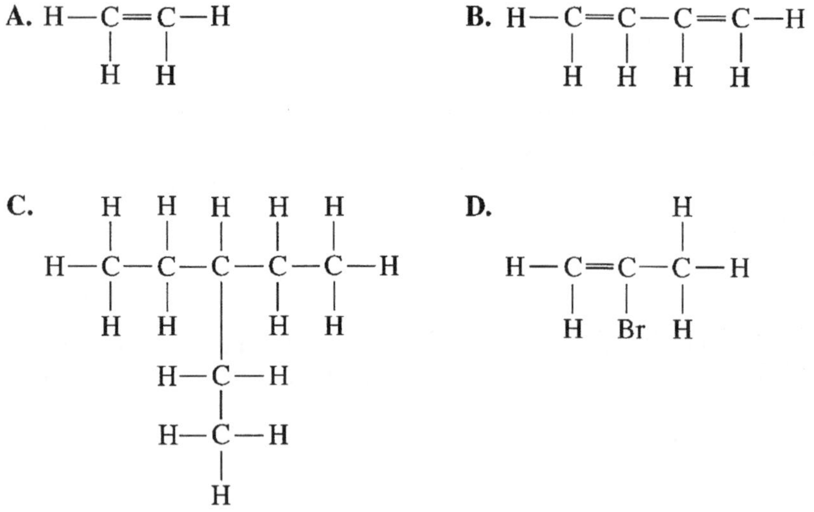

A. H—C=C—H
     | |
     H H

B. H—C=C—C=C—H
     | | | |
     H H H H

C.
      H H H H H
      | | | | |
H—C—C—C—C—C—H
      | | | | |
      H H | H H
          H—C—H
            |
          H—C—H
            |
            H

D.
               H
               |
H—C=C—C—H
     | | |
     H Br H

53. In the figure above, name the compound in diagram C.
    a. diethylpentane
    b. 3-isoheptane
    c. 3-methylpentane
    d. 3-ethylpentane

ANS: D   DIF: II   OBJ: 20-3.2

54. In the figure above, name the compound in diagram D.
    a. 2-bromopropyne
    b. 2-bromopropene
    c. 2-bromopropane
    d. 2,2-bromopropene

ANS: B   DIF: II   OBJ: 20-4.2

55. In the figure above, name the compound in diagram B.
    a. 1,3-butadiene
    b. 1,4-butadiene
    c. 2,3-butadiene
    d. 1,3-butene

ANS: A   DIF: II   OBJ: 20-4.2

56. In the figure above, name the compound in diagram A.
    a. ethane
    b. ethene
    c. ethyne
    d. ethadiene

ANS: B   DIF: II   OBJ: 20-4.2

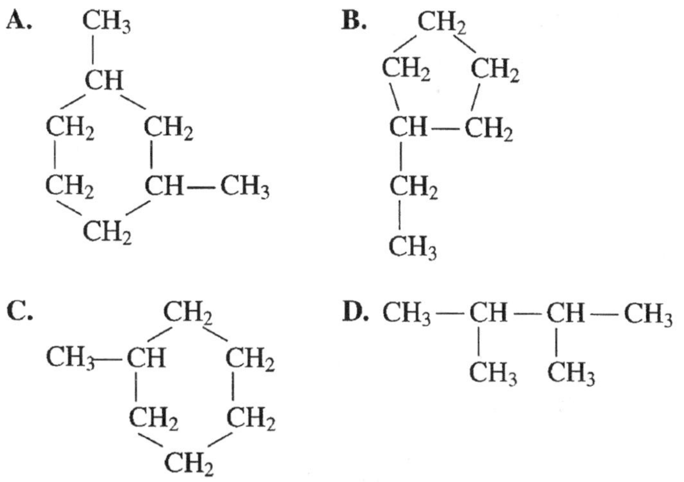

A.

$$CH_3$$
$$|$$
$$CH$$
$$CH_2 \quad CH_2$$
$$|$$
$$CH_2 \quad CH—CH_3$$
$$CH_2$$

B.

$$CH_2$$
$$CH_2 \quad CH_2$$
$$CH—CH_2$$
$$|$$
$$CH_2$$
$$|$$
$$CH_3$$

C.

$$CH_2$$
$$CH_3—CH \quad CH_2$$
$$|$$
$$CH_2 \quad CH_2$$
$$CH_2$$

D. $CH_3—CH—CH—CH_3$
$\qquad \quad | \qquad |$
$\qquad \; CH_3 \quad CH_3$

57. In the figure above, name the compound in diagram B.
    a. 1,1-dimethylcyclopentane
    b. ethylcyclopentane
    c. cycloheptane
    d. propylcyclohexane

    ANS: B   DIF: I   OBJ: 20-3.2

58. In the figure above, which structural formula represents methylcyclohexane?
    a. A
    b. B
    c. C
    d. D

    ANS: C   DIF: I   OBJ: 20-3.2

59. The carbon-hydrogen bonds of alkanes are
    a. polar.
    b. nonpolar.
    c. nonhomologous.
    d. arranged in a ring structure.

    ANS: B   DIF: I   OBJ: 20-3.3

60. Which statement about London dispersion forces is correct?
    a. Their strength increases as the mass of the molecule increases.
    b. Their strength decreases as the mass of the molecule increases.
    c. Their strength increases as the mass of the molecule decreases.
    d. Their strength does not depend on the mass of the molecule.

    ANS: A   DIF: I   OBJ: 20-3.3

61. The attractive forces between molecules in alkanes are
    a. strong intermolecular forces.
    b. weak intermolecular forces.
    c. London dispersion forces.
    d. Both b and c

    ANS: D   DIF: I   OBJ: 20-3.3

62. What properties of alkanes make them useful for paving roads?
    a. Their long carbon-atom chains have low melting points.
    b. Their long carbon-atom chains have high melting points.
    c. Their short carbon-atom chains have low melting points.
    d. Their short carbon-atom chains have high melting points.

    ANS: A   DIF: II   OBJ: 20-3.3

63. Paraffin is a solid because alkanes have
    a. high molecular masses.   c. strong London dispersion forces.
    b. low molecular masses.   d. Both a and c

    ANS: D   DIF: II   OBJ: 20-3.3

64. Which of the following statements about alkanes is correct?
    a. They have higher boiling points and higher condensation temperatures.
    b. They have higher boiling points and lower condensation temperatures.
    c. They have lower boiling points and higher condensation temperatures.
    d. The boiling point is not related to the condensation temperature of alkanes.

    ANS: A   DIF: I   OBJ: 20-3.3

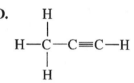

65. In the figure above, what is the structural formula for propene?
    a. A   c. C
    b. B   d. D

    ANS: C   DIF: II   OBJ: 20-4.1

66. In the figure above, what is the structural formula for propyne?
    a. A   c. C
    b. B   d. D

    ANS: D   DIF: II   OBJ: 20-4.1

67. Which hydrocarbons do NOT have double or triple bonds?
    a. alkanes   c. alkynes
    b. alkenes   d. alkadienes

    ANS: A   DIF: I   OBJ: 20-4.1

68. Which hydrocarbons have double covalent bonds?
    a. alkanes
    b. alkenes
    c. alkynes
    d. aromatic hydrocarbons

    ANS: B   DIF: I   OBJ: 20-4.1

69. Which hydrocarbons have triple covalent bonds?
    a. alkanes
    b. alkynes
    c. alkenes
    d. aromatic hydrocarbons

    ANS: B   DIF: I   OBJ: 20-4.1

70. Which hydrocarbons have six-membered carbon rings and delocalized electrons?
    a. alkanes
    b. alkenes
    c. alkynes
    d. aromatic hydrocarbons

    ANS: D   DIF: I   OBJ: 20-4.1

71. Which hydrocarbons are saturated?
    a. alkenes
    b. alkanes
    c. alkynes
    d. aromatic hydrocarbons

    ANS: B   DIF: I   OBJ: 20-4.1

72. In which hydrocarbons do carbon atoms bond to fewer atoms than the maximum possible number?
    a. alkanes
    b. alkenes
    c. alkynes
    d. both b and c

    ANS: D   DIF: I   OBJ: 20-4.1

73. Two carbon atoms are bound by a linear triple-covalent bond in
    a. alkanes.
    b. alkenes.
    c. alkynes.
    d. both b and c

    ANS: C   DIF: I   OBJ: 20-4.1

74. The two resonance structures of benzene have
    a. only single bonds.
    b. only double bonds.
    c. alternating double and triple bonds.
    d. alternating single and double bonds.

    ANS: D   DIF: II   OBJ: 20-4.2

75. In a structural formula, the C_6H_5 group is a(n)
    a. hexyl group.
    b. phenyl group.
    c. benzoyl group.
    d. alkyl group.

    ANS: B   DIF: II   OBJ: 20-4.2

76. Name the alkene in the figure above.
   a. ethene
   b. propene
   c. trans-2-butene
   d. cis-2-butene

   ANS: B   DIF: II   OBJ: 20-4.2

$$H-C\equiv C-H$$

77. Name the alkyne in the figure above.
   a. ethane
   b. ethene
   c. ethyne
   d. None of the above

   ANS: C   DIF: II   OBJ: 20-4.2

78. Which alkene is found in the wax covering of apples?
   a. ethene
   b. α-farnesene
   c. pentene
   d. paraffin

   ANS: B   DIF: II   OBJ: 20-4.3

79. Which alkene is used to synthesize plastics?
   a. ethene
   b. α-farnesene
   c. pentene
   d. paraffin

   ANS: A   DIF: II   OBJ: 20-4.3

80. Which alkyne combusts with pure oxygen to release intense heat in welding torches?
   a. ethyne
   b. 1-pentyne
   c. 3-methyl-1-butyne
   d. None of the above

   ANS: A   DIF: II   OBJ: 20-4.3

81. What are the properties of alkynes?
   a. They are nonpolar.
   b. They have a boiling point similar to those of other hydrocarbons.
   c. Both a and b are correct.
   d. Neither a nor b are correct.

   ANS: C   DIF: I   OBJ: 20-4.3

82. What are the properties of aromatic hydrocarbons?
   a. They are nonpolar.
   b. They have limited solubility. in water
   c. They are chemically very stable.
   d. All of the above

   ANS: D   DIF: I   OBJ: 20-4.3

83. Which of the following aromatic hydrocarbons is found in tar from cigarette smoke and in soot in heavily polluted areas?
   a. benzene
   b. 3,4-benzpyrene
   c. methylbenzene
   d. None of the above

ANS: B   DIF: II   OBJ: 20-4.3

MULTIPLE CHOICE

1. Which of the following is an atom or a group of atoms responsible for the specific properties of an organic compound?
   a. isomer
   b. hydrocarbon
   c. substituted hydrocarbon
   d. functional group

   ANS: D   DIF: I   OBJ: 21-1.1

2. A functional group is
   a. a group of organic compounds with similar structural formulas
   b. a group of organic compounds that undergo similar chemical reactions
   c. a group of atoms that help determine the properties of an organic compound
   d. Both a and b

   ANS: C   DIF: I   OBJ: 21-1.1

3. A functional group determines
   a. the properties of an organic compound.
   b. how a compound is classified.
   c. the molecular mass of a compound.
   d. Both a and b

   ANS: D   DIF: I   OBJ: 21-1.1

4. Why are functional groups important?
   a. The properties of functional groups help to systematically classify compounds.
   b. Reactions of a compound involve the bonds within a functional group.
   c. Both a and b are correct.
   d. Neither a nor b is correct.

   ANS: C   DIF: I   OBJ: 21-1.1

5. Which of the following are often the sites of chemical reactions within functional groups?
   a. bonds
   b. carbon atoms
   c. hydrogen atoms
   d. None of the above.

   ANS: A   DIF: I   OBJ: 21-1.1

6. All organic compounds containing the same functional group
   a. have the same name.
   b. are classified together.
   c. undergo different chemical reactions.
   d. behave differently.

   ANS: B   DIF: I   OBJ: 21-1.1

7. Which of the following is the functional group in alcohols?
   a. –COOH
   b. –OH
   c. –CO
   d. –O–

   ANS: B   DIF: I   OBJ: 21-1.2

8. What is the name of the functional group –OH?
   a. hydroxyl group
   b. base unit
   c. carboxyl group
   d. hydroxide ion

   ANS: A   DIF: I   OBJ: 21-1.2

9. The systematic names of alcohols end in
   a. -al.
   b. -one.
   c. -ol.
   d. -oic.

   ANS: C   DIF: I   OBJ: 21-1.2

10. What are alkanes called when atoms of fluorine, chlorine, bromine, or iodine are substituted for hydrogen atoms?
    a. aldehydes
    b. ketones
    c. alkyl halides
    d. carboxylic acids

    ANS: C   DIF: I   OBJ: 21-1.2

11. How is the functional group in alkyl halides often written?
    a. –OX
    b. –XO
    c. –X
    d. –O–

    ANS: C   DIF: I   OBJ: 21-1.2

12. In which of the following compounds are two hydrocarbon groups bonded to the same oxygen atom?
    a. ketones
    b. aldehydes
    c. carboxylic acids
    d. ethers

    ANS: D   DIF: I   OBJ: 21-1.2

13. What is the general formula for ethers?
    a. R–O–R'
    b. R–COOH
    c. R–COO–R'
    d. R–CHO

    ANS: A   DIF: I   OBJ: 21-1.2

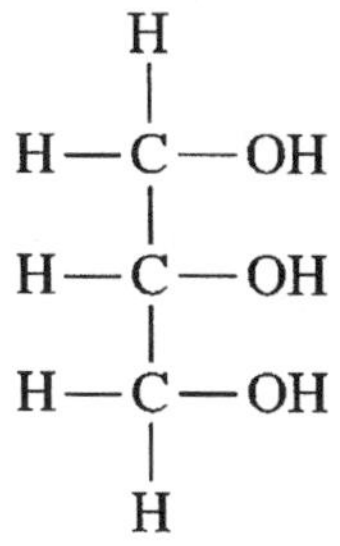

$$\begin{array}{c}
\text{H} \\
| \\
\text{H}-\text{C}-\text{OH} \\
| \\
\text{H}-\text{C}-\text{OH} \\
| \\
\text{H}-\text{C}-\text{OH} \\
| \\
\text{H}
\end{array}$$

14. In the figure above, what class does the organic compound belong to?
    a. alcohols
    b. alkyl halides
    c. ethers
    d. aldehydes

    ANS: A   DIF: II   OBJ: 21-1.2

15. Name the compound in the figure above.
    a. 1,2,3-propanetriol
    b. 1,2-ethanediol
    c. ethanol
    d. 2-butanol

ANS: A   DIF: II   OBJ: 21-1.3

16. In the figure above, what class does the organic compound belong to?
    a. alcohols
    b. alkyl halides
    c. ethers
    d. aldehydes

ANS: B   DIF: II   OBJ: 21-1.2

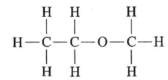

17. In the figure above, what class does the organic compound belong to?
    a. alcohols
    b. alkyl halides
    c. ethers
    d. aldehydes

ANS: C   DIF: II   OBJ: 21-1.2

18. What is the systematic name for the two-carbon alcohol?
    a. ethanol
    b. ethanal
    c. 2-propanol
    d. 2-propanal

ANS: A   DIF: I   OBJ: 21-1.3

19. What is the systematic name for the one-carbon alcohol?
    a. ethanol
    b. ethanal
    c. methanol
    d. methanal

ANS: C   DIF: I   OBJ: 21-1.3

20. What is the systematic name for wood alcohol?
    a. 1-propanol
    b. methanol
    c. glycerol
    d. ethanol

ANS: B   DIF: I   OBJ: 21-1.3

21. What is the systematic name of the compound C_2H_5Br?
    a. ethyl bromide
    b. bromoethyne
    c. ethylene bromide
    d. bromoethene

ANS: A   DIF: I   OBJ: 21-1.3

22. What is the systematic name of the compound CCl_3F?
    a. trichlorofluoromethane
    b. 3-chloro-1-fluoromethane
    c. chloromethylfluoride
    d. chlorofluorocarbon

ANS: B   DIF: I   OBJ: 21-1.3

A.
```
    H   H
    |   |
H — C — C — H
    |   |
    Cl  Cl
```

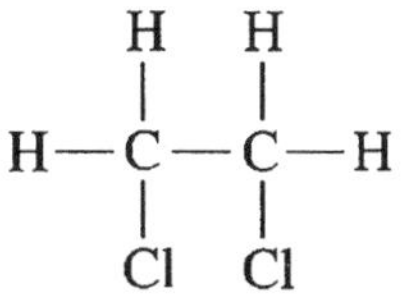

B.
```
            H
            |
H — C = C — C — H
    |   |   |
    Cl  Cl  H
```

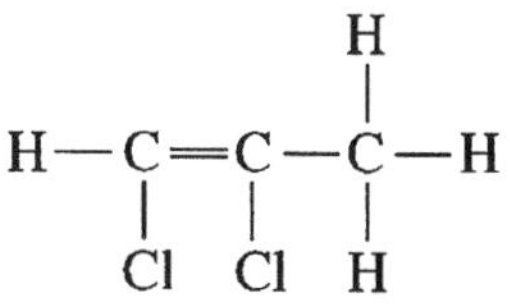

C.
```
    H   H   H
    |   |   |
H — C — C — C — Cl
    |   |   |
    H   H   Cl
```

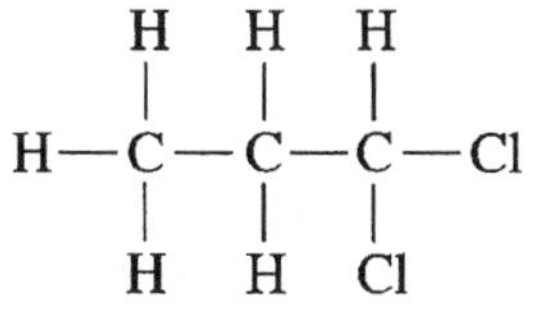

D.
```
    H   H   H
    |   |   |
H — C — C — C — H
    |   |   |
    H   Cl  Cl
```

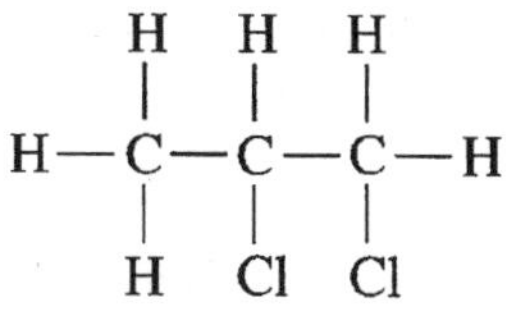

23. Which figure above represents the structural formula of 1,2-dichloropropane?
    a. A   c. C
    b. B   d. D

ANS: D   DIF: II   OBJ: 21-1.3

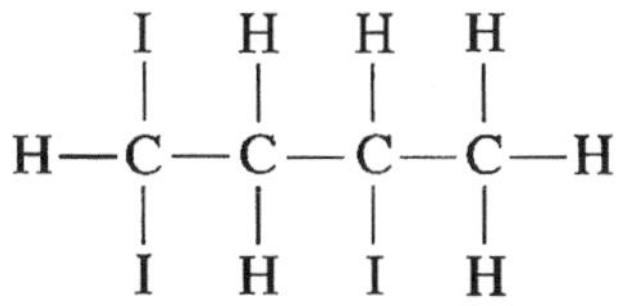

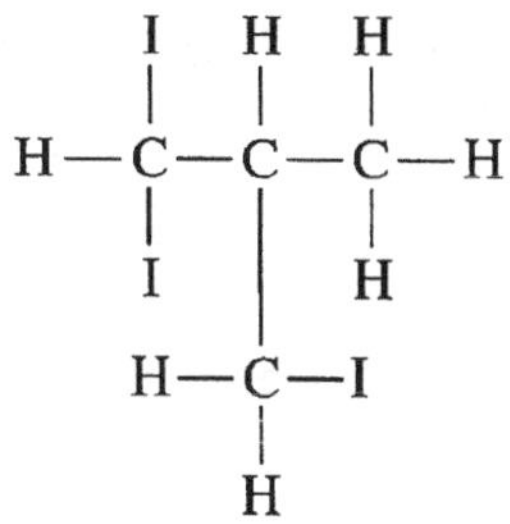

A.
```
    I   H   H   H
    |   |   |   |
H — C — C — C — C — H
    |   |   |   |
    I   H   I   H
```

B.
```
    I   H   H
    |   |   |
H — C — C — C — H
    |   |   |
    I   |   H
        |
    H — C — I
        |
        H
```

C.
```
I       H   H
|       |   |
C = C — C — C — H
|   |   |   |
I   H   I   H
```

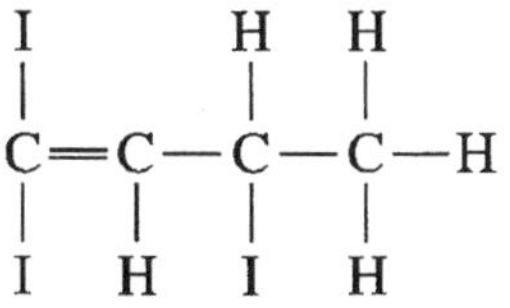

D.
```
    H   H   I   H
    |   |   |   |
H — C — C — C — C — H
    |   |   |   |
    I   H   I   H
```

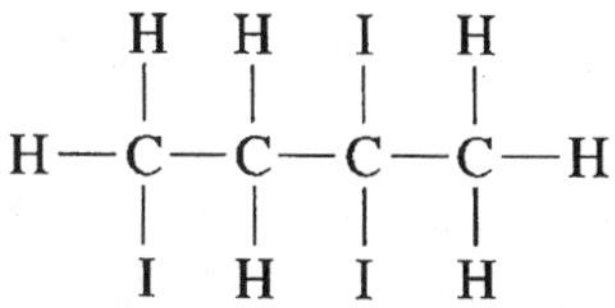

24. Which figure above represents the structural formula of 1,1,3-triiodobutane?
    a. A   c. C
    b. B   d. D

ANS: A   DIF: II   OBJ: 21-1.3

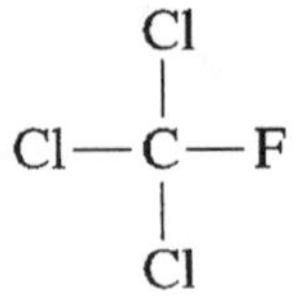

25. Name the compound in the figure above.
    a.   trichlorofluoromethane
    b.   3-chloro-1-fluoromethane
    c.   chloromethylfluoride
    d.   chlorofluorocarbon

ANS: A   DIF: II   OBJ: 21-1.3

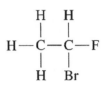

26. Name the compound in the figure above.
    a.   2-bromo-2-fluoroethane
    b.   1-bromo-1-fluoroethane
    c.   bromoethane fluoride
    d.   fluoroethane bromide

ANS: B   DIF: II   OBJ: 21-1.3

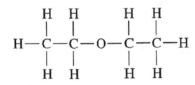

27. Name the compound in the figure above.
    a.   acetone
    b.   diethyl ketone
    c.   ethyl acetate
    d.   diethyl ether

ANS: D   DIF: II   OBJ: 21-1.3

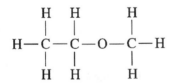

28. Name the compound in the figure above.
    a.   ethyl ether
    b.   ethyl methyl ether
    c.   ethyl formate
    d.   methyl ethyl ketone

ANS: B   DIF: II   OBJ: 21-1.3

A.

$$H-\underset{\underset{H}{|}}{\overset{\overset{H}{|}}{C}}-\underset{\underset{H}{|}}{\overset{\overset{H}{|}}{C}}-O-\underset{\underset{H}{|}}{\overset{\overset{H}{|}}{C}}-\underset{\underset{H}{|}}{\overset{\overset{H}{|}}{C}}-\underset{\underset{H}{|}}{\overset{\overset{H}{|}}{C}}-H$$

B.

$$H-C-O\ H$$

(H at top, O—C ring, C with H below)

C.

$$H-\underset{\underset{H}{|}}{\overset{\overset{H}{|}}{C}}-O-\underset{\underset{H}{|}}{\overset{\overset{H}{|}}{C}}-H$$

D.

$$H-\underset{\underset{H}{|}}{\overset{\overset{H}{|}}{C}}-\underset{\underset{O}{\|}}{C}-\underset{\underset{H}{|}}{\overset{\overset{H}{|}}{C}}-H$$

29. Which of the figures above represents the structural formula of dimethyl ether?
   a. A
   b. B
   c. C
   d. D

ANS: C   DIF: II   OBJ: 21-1.3

A.

$$H-\underset{\underset{H}{|}}{\overset{\overset{H}{|}}{C}}-\underset{\underset{H}{|}}{\overset{\overset{H}{|}}{C}}-\underset{\underset{H}{|}}{\overset{\overset{H}{|}}{C}}-O-\underset{\underset{H}{|}}{\overset{\overset{H}{|}}{C}}-\underset{\underset{H}{|}}{\overset{\overset{H}{|}}{C}}-H$$

B.

$$H-\underset{\underset{H}{|}}{\overset{\overset{H}{|}}{C}}-\underset{\underset{H}{|}}{\overset{\overset{H}{|}}{C}}-\underset{\underset{H}{|}}{\overset{\overset{H}{|}}{C}}-\underset{\underset{O}{\|}}{C}-\underset{\underset{H}{|}}{\overset{\overset{H}{|}}{C}}-\underset{\underset{H}{|}}{\overset{\overset{H}{|}}{C}}-H$$

C.

$$H-\underset{\underset{H}{|}}{\overset{\overset{H}{|}}{C}}-\underset{\underset{H}{|}}{\overset{\overset{H}{|}}{C}}-\overset{\overset{O}{\|}}{C}$$

$$O-\underset{\underset{H}{|}}{\overset{\overset{H}{|}}{C}}-\underset{\underset{H}{|}}{\overset{\overset{H}{|}}{C}}-H$$

D.

$$H-\underset{\underset{H}{|}}{\overset{\overset{H}{|}}{C}}-\underset{\underset{H}{|}}{\overset{\overset{H}{|}}{C}}-\underset{\underset{H}{|}}{\overset{\overset{H}{|}}{C}}-\underset{\underset{O}{|}}{\overset{\overset{H}{|}}{C}}-\underset{\underset{H}{|}}{\overset{\overset{H}{|}}{C}}-\underset{\underset{H}{|}}{\overset{\overset{H}{|}}{C}}-H$$

(with H attached below O)

30. Which of the figures above represents the structural formula of ethyl propyl ether?
   a. A
   b. B
   c. C
   d. D

ANS: A   DIF: II   OBJ: 21-1.3

$$H-\underset{\underset{OH}{|}}{\overset{\overset{H}{|}}{C}}-\underset{\underset{OH}{|}}{\overset{\overset{H}{|}}{C}}-H$$

31. Name the compound in the figure above.
   a. 1,2,3-propanetriol
   b. 1,2-ethanediol
   c. ethanol
   d. 2-butanol

ANS: B   DIF: II   OBJ: 21-1.3

32. Which of the following compounds is lethal in large quantities and used as a solvent and a fuel?
    a. ethanol
    b. ethylene glycol
    c. glycerol
    d. methanol

    ANS: A   DIF: I   OBJ: 21-1.4

33. Which of the following compounds contains exactly three hydroxyl groups?
    a. ethanediol
    b. ethylene glycol
    c. glycerol
    d. methanol

    ANS: C   DIF: I   OBJ: 21-1.4

34. Which of the following compounds is used in cold creams, body lotions, and lipsticks?
    a. ethanol
    b. ethylene glycol
    c. glycerol
    d. methanol

    ANS: C   DIF: I   OBJ: 21-1.4

35. Which of the following compounds is found in alcoholic beverages?
    a. ethanol
    b. ethylene glycol
    c. glycerol
    d. methanol

    ANS: A   DIF: I   OBJ: 21-1.4

36. Which of the following is the common name for 1,2,3-propanetriol?
    a. ethanol
    b. ethylene glycol
    c. glycerol
    d. methanol

    ANS: C   DIF: I   OBJ: 21-1.4

37. Which compound is inactive and stable to about 325°C?
    a. dichlorodifluoromethane
    b. tetrafluoroethene polymer
    c. tetrachloromethane
    d. polyvinylchloride

    ANS: A   DIF: I   OBJ: 21-1.4

38. Which of the following compounds attacks ozone in the upper atmosphere?
    a. dichlorodifluoromethane
    b. tetrafluoroethene polymer
    c. tetrachloromethane
    d. polyvinylchloride

    ANS: A   DIF: I   OBJ: 21-1.4

39. Which of the following compounds is used for nonstick surfaces?
    a. dichlorodifluoromethane
    b. tetrafluoroethene
    c. tetrachloromethane
    d. polyvinylchloride

    ANS: B   DIF: I   OBJ: 21-1.4

40. The solubility of ethers in water is similar to the solubility of
    a. alkanes.
    b. alkyl halides.
    c. alcohols.
    d. inorganic acids.

    ANS: C   DIF: I   OBJ: 21-1.4

41. Compared with the boiling points of alcohols, the boiling points of ethers of similar molar mass
    a. are much lower.   c. are about the same.
    b. are much higher.   d. may be higher or lower.

    ANS: A   DIF: I   OBJ: 21-1.4

42. Compared with the boiling points of alkanes, the boiling points of ethers of similar molar mass
    a. are much lower.   c. are about the same.
    b. are much higher.   d. may be higher or lower.

    ANS: C   DIF: I   OBJ: 21-1.4

43. In many organic reactions, an ether is used as a solvent instead of
    a. an alkane.   c. an alkyl halide.
    b. an alcohol.   d. water.

    ANS: D   DIF: I   OBJ: 21-1.4

44. Ethers are mainly used as
    a. medicines.   c. antiseptics.
    b. fuels.   d. solvents.

    ANS: D   DIF: I   OBJ: 21-1.4

45. The systematic name for the compound commonly called ether is
    a. diethyl ether.   c. ethyl methyl ether.
    b. dimethyl ether.   d. butyl methyl ether.

    ANS: A   DIF: I   OBJ: 21-1.4

46. Which ether is often added to gasoline to increase its octane rating?
    a. diethyl ether   c. ethyl methyl ether
    b. dimethyl ether   d. methyl-tertiary-butyl ether

    ANS: D   DIF: I   OBJ: 21-1.4

47. A carbonyl group consists of
    a. one carbon atom and one oxygen atom.   c. one carbon atom and two oxygen atoms.
    b. two carbon atoms and one oxygen atom. d. two carbon atoms and two oxygen atoms.

    ANS: A   DIF: I   OBJ: 21-2.1

48. Which class of compounds contains a carbonyl group bonded to a hydrogen atom?
    a. alcohols   c. ethers
    b. aldehydes   d. ketones

    ANS: B   DIF: I   OBJ: 21-2.1

49. Which class of compounds contains a carbonyl group bonded to two carbon groups?
    a. alcohols   c. ethers
    b. aldehydes   d. ketones

    ANS: D   DIF: I   OBJ: 21-2.1

50. The systematic names of aldehydes end in
    a.   -one.
    c.   -ol.
    b.   -al.
    d.   -ane.

    ANS: B   DIF: I   OBJ: 21-2.1

51. The systematic names of ketones end in
    a.   -one.
    c.   -ol.
    b.   -al.
    d.   -ane.

    ANS: A   DIF: I   OBJ: 21-2.1

52. Organic acids contain the
    a.   propyl group.
    c.   carboxyl group.
    b.   amine group.
    d.   formyl group.

    ANS: C   DIF: I   OBJ: 21-2.1

53. What is the formula of the carboxyl group?
    a.   –CHO
    c.   –OH
    b.   –COOH
    d.   –CO

    ANS: B   DIF: I   OBJ: 21-2.1

54. What is the general formula of esters?
    a.   R–COO–R'
    c.   R–CO–R'
    b.   R–COOH
    d.   R–CHO

    ANS: A   DIF: I   OBJ: 21-2.1

55. Which group of organic compounds can be considered derivatives of ammonia, NH_3?
    a.   ester
    c.   aldehyde
    b.   ketone
    d.   amine

    ANS: D   DIF: I   OBJ: 21-2.1

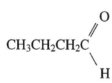

56. Name the organic compound shown in the figure above.
    a.   aldehyde
    c.   carboxylic acid
    b.   ketone
    d.   ester

    ANS: A   DIF: II   OBJ: 21-2.1

57. Name the compound shown in the figure above.
    a.   methanal
    c.   propanal
    b.   ethanal
    d.   butanal

    ANS: D   DIF: II   OBJ: 21-2.2

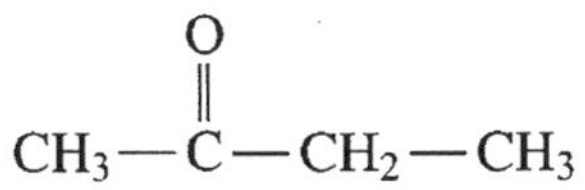

$$CH_3-C-CH_2-CH_3$$

58. Name the organic compound shown in the figure above.
    a. aldehyde   c. ketone
    b. carboxylic acid   d. ester

    ANS: C   DIF: II   OBJ: 21-2.1

59. Name the compound shown in the figure above.
    a. propanone   c. 2-butanone
    b. 3-pentanone   d. None of the above

    ANS: C   DIF: II   OBJ: 21-2.2

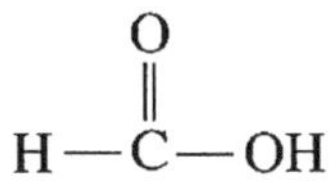

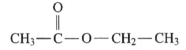
$$H-C-OH$$

60. Name the organic compound shown in the figure above.
    a. aldehyde   c. carboxylic acid
    b. ketone   d. ester

    ANS: C   DIF: II   OBJ: 21-2.1

$$CH_3-C-O-CH_2-CH_3$$

61. Name the organic compound shown in the figure above.
    a. aldehyde   c. amine
    b. ketone   d. ester

    ANS: D   DIF: II   OBJ: 21-2.1

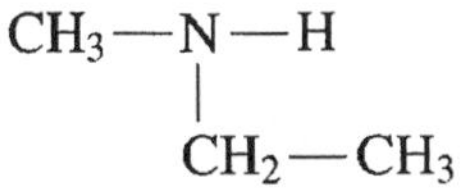

$$CH_3-N-H$$
$$CH_2-CH_3$$

62. Name the organic compound shown in the figure above.
    a. aldehyde   c. amine
    b. carboxylic acid   d. ester

    ANS: C   DIF: II   OBJ: 21-2.1

63. Name the compound shown in the figure above.
    a. methylamine   c. trimethylamine
    b. ethylmethylamine   d. epinephrine

    ANS: B   DIF: II   OBJ: 21-2.2

64. What is the systematic name for the compound formaldehyde?
a. ethanal
b. propanal
c. methanal
d. benzaldehyde

ANS: C   DIF: I   OBJ: 21-2.2

65. What is the common name for propanone?
a. dimethyl ketone
b. 3-pentanone
c. acetone
d. butanone

ANS: C   DIF: I   OBJ: 21-2.2

66. Which is the simplest carboxylic acid?
a. ethanedoic acid
b. butanoic acid
c. methanoic acid
d. carbonic acid

ANS: C   DIF: I   OBJ: 21-2.2

67. Which is the systematic name for acetic acid?
a. ethanoic acid
b. methanoic acid
c. propanoic acid
d. butanoic acid

ANS: A   DIF: I   OBJ: 21-2.2

68. Which ester contains an ethyl group and an acetate group?
a. acetic ethane
b. ethyl ethanoate
c. methyl acetate
d. acetic ethanoate

ANS: B   DIF: I   OBJ: 21-2.2

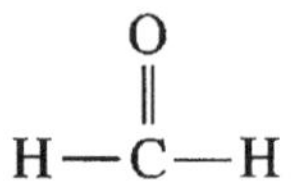

$$\begin{array}{c} O \\ \parallel \\ H-C-H \end{array}$$

69. Name the compound shown in the figure above.
a. methanal
b. formaldehyde
c. propanal
d. both a and b

ANS: D   DIF: II   OBJ: 21-2.2

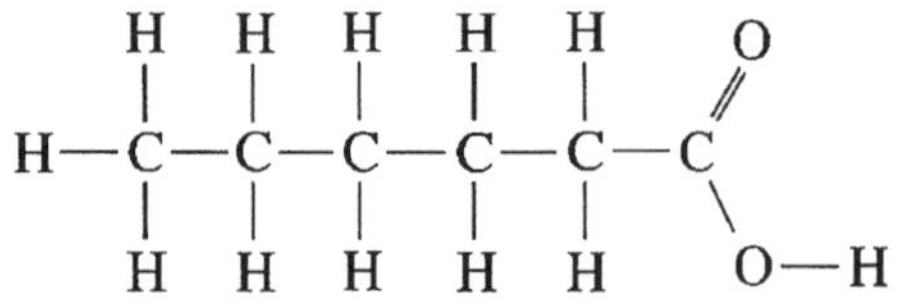

$$\begin{array}{ccccccc} H & H & H & H & H & & O \\ | & | & | & | & | & & \parallel \\ H-C-C-C-C-C-C & & & & & & \\ | & | & | & | & | & & \backslash \\ H & H & H & H & H & & O-H \end{array}$$

70. Name the compound shown in the figure above.
a. heptanoic acid
b. hexanoic acid
c. hexanone
d. hexanal

ANS: B   DIF: II   OBJ: 21-2.2

A.

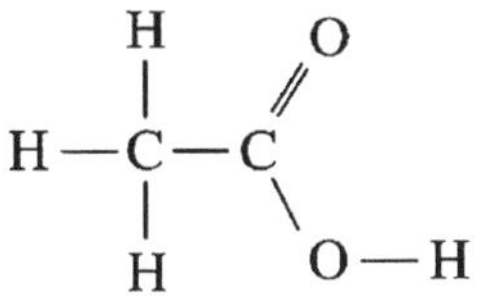

```
      H    O
      |    ‖
  H—C—C
      |    \
      H    O—H
```

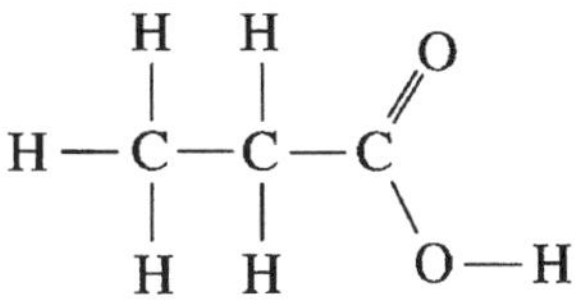

B.

```
      H   H
      |   |    O
  H—C—C—C
      |   |    \
      H   H    O—H
```

C.

```
         O
         ‖
  H—C
         \
         O—H
```

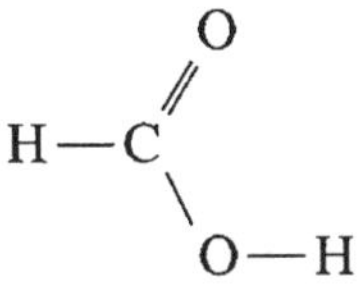

D.

```
      H   H   H
      |   |   |    O
  H—C—C—C—C
      |   |   |    \
      H   H   H    O—H
```

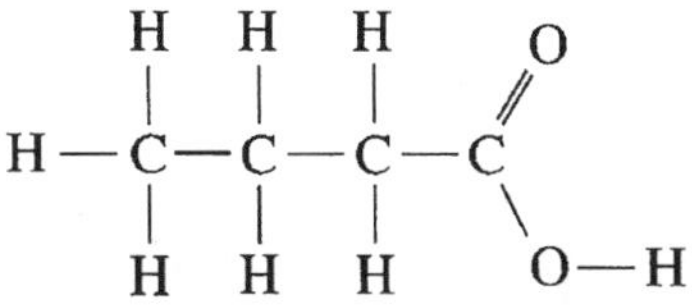

71. Which of the figures above represents the structural formula of butanoic acid?
    a. A   c. C
    b. B   d. D

    ANS: D   DIF: II   OBJ: 21-2.2

72. Which of the figures above represents the structural formula of acetic acid?
    a. A   c. C
    b. B   d. D

    ANS: A   DIF: II   OBJ: 21-2.2

```
      H    O
      |    ‖
  H—C—C         H   H   H   H
      |    \    |   |   |   |
      H    O—C—C—C—C—H
               |   |   |   |
               H   H   H   H
```

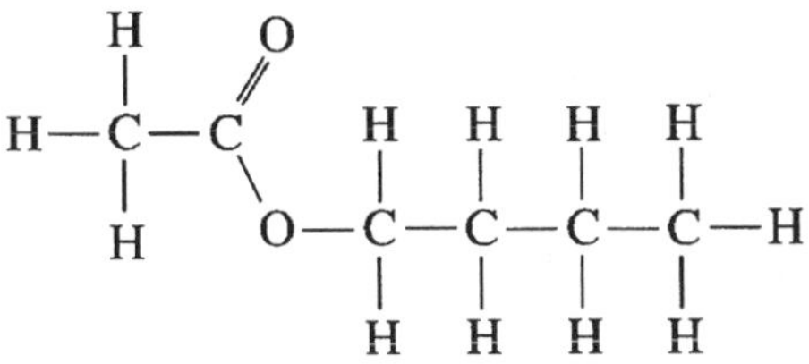

73. Name the compound shown in the figure above.
    a. ethyl butanoate   c. butyl ethanoate
    b. butyl formate   d. methyl butanoate

    ANS: C   DIF: II   OBJ: 21-2.2

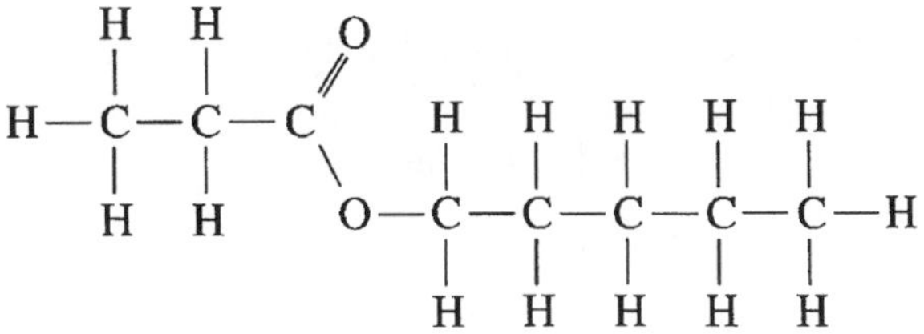

```
      H   H    O
      |   |    ‖
  H—C—C—C             H   H   H   H   H
      |   |    \      |   |   |   |   |
      H   H    O—C—C—C—C—C—H
                     |   |   |   |   |
                     H   H   H   H   H
```

74. Name the compound shown in the figure above.
    a. pentyl propanoate   c. pentyl acetate
    b. propyl pentanoate   d. butyl propanoate

    ANS: A   DIF: II   OBJ: 21-2.2

A.

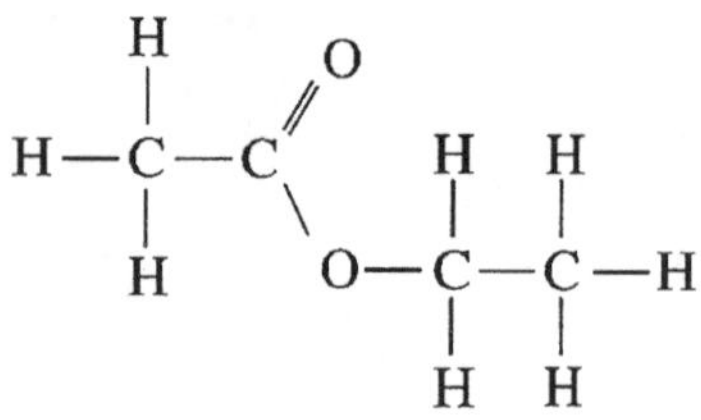

B.

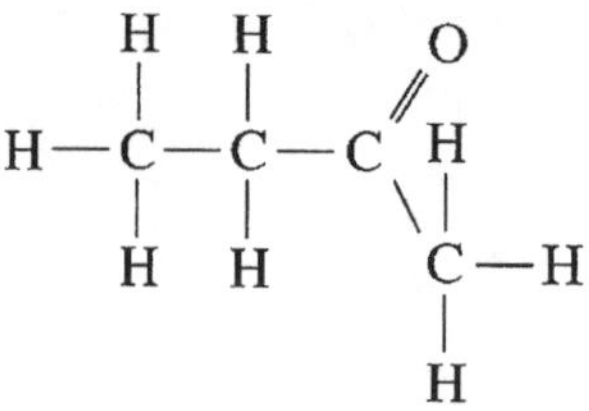

C.

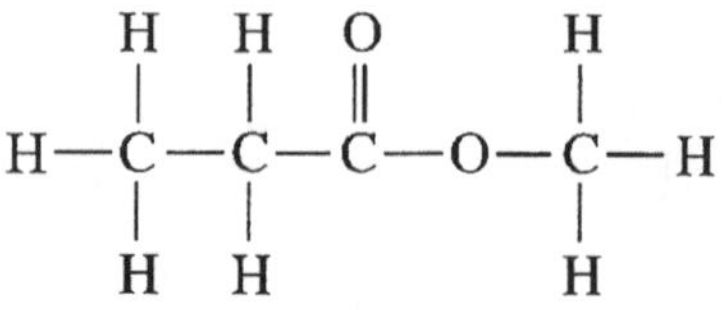

D.

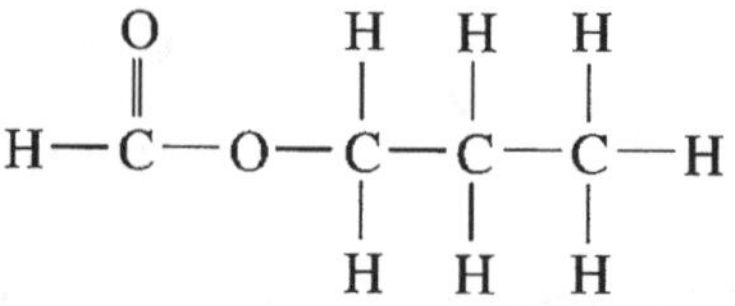

75. Which of the figures above represents the structural formula of propyl methanoate?
    a. A   c. C
    b. B   d. D

    ANS: D   DIF: II   OBJ: 21-2.2

76. Which of the figures above represents the structural formula of methyl propanoate?
    a. A   c. C
    b. B   d. D

    ANS: C   DIF: II   OBJ: 21-2.2

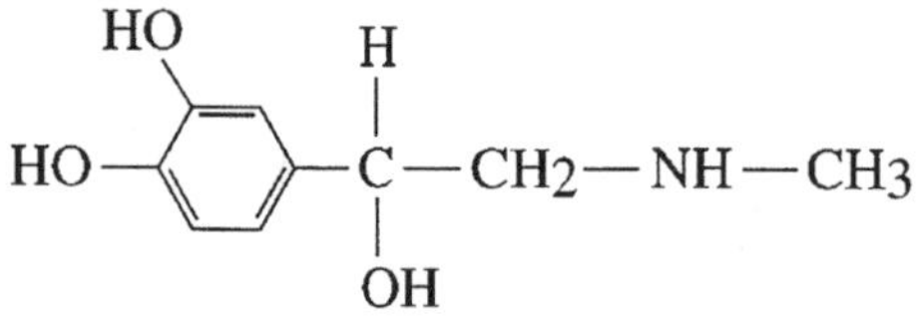

77. Name the compound shown in the figure above.
    a. methylamine   c. trimethylamine
    b. ethylmethylamine   d. epinephrine

    ANS: D   DIF: II   OBJ: 21-2.2

78. One of the first commercial plastics was made by combining phenol and
    a. ethanol.   c. ethanal.
    b. formaldehyde.   d. dimethyl ketone.

    ANS: B   DIF: I   OBJ: 21-2.3

79. Which compound was once used to preserve biological specimens and is still used to make plastics?
    a. methanal   c. acetone
    b. acetaldehyde   d. diethyl ketone

    ANS: A   DIF: I   OBJ: 21-2.3

80. Which of the following is the ketone found in some nail-polish removers?
    a.   acetone
    b.   2-propanone
    c.   Both a and b
    d.   Neither a nor b

    ANS: C   DIF: I   OBJ: 21-2.3

81. Which of the following is used in foods to give a tart flavor?
    a.   citric acid
    b.   butanoic acid
    c.   acetic acid
    d.   methanoic acid

    ANS: A   DIF: I   OBJ: 21-2.3

82. Carboxylic acids taste
    a.   sour.
    b.   sweet.
    c.   bitter.
    d.   salty.

    ANS: A   DIF: I   OBJ: 21-2.3

83. Which vegetable contains carboxylic acids?
    a.   carrot
    b.   potato
    c.   rhubarb
    d.   broccoli

    ANS: C   DIF: I   OBJ: 21-2.3

84. Carboxylic acids are
    a.   weak acids.
    b.   strong acids.
    c.   sometimes weak bases.
    d.   amphoteric.

    ANS: A   DIF: I   OBJ: 21-2.3

85. Which of the following is used as a food preservative?
    a.   benzoic acid
    b.   propanoic acid
    c.   sorbic acid
    d.   All of the above

    ANS: D   DIF: I   OBJ: 21-2.3

86. Which compounds give fruits and flowers their characteristic flavors and odors?
    a.   ketones
    b.   ethers
    c.   esters
    d.   aldehydes

    ANS: C   DIF: I   OBJ: 21-2.3

87. Esters are frequently used
    a.   as water purifiers.
    b.   in perfumes and flavorings.
    c.   as indicators.
    d.   as electrolytes.

    ANS: B   DIF: I   OBJ: 21-2.3

88. The chemical properties of amines depend on the electronic structure of the
    a.   nitrogen atom.
    b.   hydrogen atom.
    c.   ammonia molecule.
    d.   None of the above

    ANS: A   DIF: I   OBJ: 21-2.3

89. In water solution, the toxic amines produced by poison dart frogs
   a. act like Cl⁻ ions.
   b. are weak acids.
   c. become negative ions.
   d. become positive ions.

   ANS: D   DIF: I   OBJ: 21-2.3

90. Naturally occurring amine products such as caffeine and morphine are classified as
   a. unreactive.
   b. toxins.
   c. alkaloids.
   d. antioxidants.

   ANS: C   DIF: I   OBJ: 21-2.3

91. Which of the following is NOT formed during the breakdown of proteins in animal cells?
   a. putrescine
   b. cadaverine
   c. methylamine
   d. batrachotoxin

   ANS: D   DIF: I   OBJ: 21-2.3

92. In which organic reaction does one or more atoms replace another atom or group of atoms in a molecule?
   a. substitution
   b. addition
   c. condensation
   d. elimination

   ANS: A   DIF: I   OBJ: 21-3.1

93. In which organic reaction does an atom or molecule join to an unsaturated molecule and increase its saturation?
   a. substitution
   b. elimination
   c. addition
   d. condensation

   ANS: C   DIF: I   OBJ: 21-3.1

94. In which organic reaction do two molecules or parts of the same molecule combine?
   a. substitution
   b. addition
   c. condensation
   d. elimination

   ANS: C   DIF: I   OBJ: 21-3.1

95. In which organic reaction is a simple molecule, such as water or ammonia, removed from adjacent carbon atoms of a larger molecule?
   a. substitution
   b. elimination
   c. condensation
   d. addition

   ANS: B   DIF: I   OBJ: 21-3.1

96. Which organic reaction can form a protein molecule if repeated many times?
   a. substitution
   b. elimination
   c. condensation
   d. addition

   ANS: C   DIF: I   OBJ: 21-3.1

97. To which group of organic reactions does hydrogenation belong?
   a. substitution
   b. elimination
   c. condensation
   d. addition

   ANS: D   DIF: I   OBJ: 21-3.1

98. Which functional group is formed by an alkane and a halogen in a substitution reaction?
   a. ketone
   b. ester
   c. alcohol
   d. alkyl halide

   ANS: D   DIF: I   OBJ: 21-3.2

99. Which functional groups can combine to form a water molecule in a condensation reaction?
   a. amine and carboxyl
   b. amine and aldehyde
   c. carboxyl and aldehyde
   d. alcohol and carboxyl

   ANS: A   DIF: I   OBJ: 21-3.2

100. In which organic reaction does a chlorine atom replace a hydrogen atom in a methane molecule?
   a. substitution
   b. addition
   c. condensation
   d. elimination

   ANS: A   DIF: I   OBJ: 21-3.2

101. Which functional group(s) of two amino acids reacts in a condensation reaction?
   a. amine
   b. carboxyl
   c. amine and carboxyl
   d. None of the above

   ANS: C   DIF: I   OBJ: 21-3.2

102. When ethanol is heated in the presence of concentrated sulfuric acid, ethene and water are
   formed. This is an example of a(n)
   a. elimination reaction.
   b. condensation reaction.
   c. addition reaction.
   d. substitution reaction.

   ANS: A   DIF: I   OBJ: 21-3.2

103. Which of the following are large molecules made of many small units joined by organic
   reactions?
   a. monomers
   b. copolymers
   c. polymers
   d. linear polymers

   ANS: C   DIF: I   OBJ: 21-4.1

104. Which of the following are the small units joined to each other by organic reactions in a
   polymer?
   a. monomers
   b. micropolymers
   c. copolymers
   d. linear polymers

   ANS: A   DIF: I   OBJ: 21-4.1

105. Two different monomers joined to each other form
   a.   a polymer.
   c.   a macropolymer.
   b.   a copolymer.
   d.   Both a and b

   ANS: D   DIF: I   OBJ: 21-4.1

106. The small units joined to each other by organic reactions to form large molecules
   a.   must be alike.
   c.   may be alike or different.
   b.   must be different.
   d.   None of the above

   ANS: C   DIF: I   OBJ: 21-4.1

107. Monomers are
   a.   small units.
   c.   the building blocks of polymers.
   b.   large molecules.
   d.   Both a and c

   ANS: D   DIF: I   OBJ: 21-4.1

108. If a polymer keeps its shape when heated, it is classified as
   a.   thermoplastic.
   c.   Both a and b
   b.   thermosetting.
   d.   Neither a nor b

   ANS: B   DIF: I   OBJ: 21-4.2

109. What forms the link in cross-linked polymers?
   a.   linear polymers
   c.   copolymers
   b.   monomers
   d.   bonds between molecules

   ANS: D   DIF: I   OBJ: 21-4.2

110. Linear polymers
   a.   are thermosetting.
   b.   have molecules that are free to move.
   c.   have molecules that slide easily when heated.
   d.   Both b and c

   ANS: D   DIF: I   OBJ: 21-4.2

111. Branched polymers
   a.   are likely to be thermoplastic.
   c.   cannot slide across each other easily.
   b.   move easily.
   d.   Both a and c

   ANS: D   DIF: I   OBJ: 21-4.2

112. Cross-linked polymers
   a.   are thermoplastic.
   c.   have side chains.
   b.   retain their shape when heated.
   d.   Both a and b

   ANS: B   DIF: I   OBJ: 21-4.2

113. In a cross-linked polymer, adjacent molecules
    a. link together in a series of chains.
    b. slide back and forth against each other when heated.
    c. are arranged in layers that can slide apart when heated.
    d. form bonds.

    ANS: D   DIF: I   OBJ: 21-4.2

114. Which of the following are the two main types of polymers?
    a. thermoplastic and thermosetting
    b. monomers and copolymers
    c. linear polymers and branched polymers
    d. addition polymers and condensation polymers

    ANS: D   DIF: I   OBJ: 21-4.3

115. What polymer is formed by a chain addition reaction between monomers that have a double
    bond?
    a. addition polymer   c. branched polymer
    b. condensation polymer   d. cross-linked polymer

    ANS: A   DIF: I   OBJ: 21-4.3

116. In an addition reaction, monomers are joined by reactions involving
    a. identical functional groups.   c. Both a and b
    b. different functional groups.   d. double bonds.

    ANS: D   DIF: I   OBJ: 21-4.3

117. What polymers usually consist of two alternating monomers?
    a. addition polymers   c. branched polymers
    b. condensation polymers   d. cross-linked polymers

    ANS: B   DIF: I   OBJ: 21-4.3

118. Monomers in condensation polymers must have
    a. a double bond.   c. three monomers.
    b. two functional groups.   d. Both b and c

    ANS: B   DIF: I   OBJ: 21-4.3

119. Linear molecules packed closely together form
    a. cross-linked polyethylene.   c. low-density polyethylene.
    b. high-density polyethylene.   d. neoprene.

    ANS: B   DIF: I   OBJ: 21-4.4

120. What polymer can be used to make hot-drink cups?
    a. polystyrene   c. vinyl acetate
    b. propylene   d. low-density polyethylene

    ANS: A   DIF: I   OBJ: 21-4.4

121. Permanent-press fabrics can be made from which of the following polymers?
   a. polyester
   b. polyamides
   c. vinyl acetate
   d. vinyl chloride

   ANS: A   DIF: I   OBJ: 21-4.4

122. PET, used in tires, film, and food packaging, is a polyester formed when terephthalic acid reacts with
   a. cross-linked polymers.
   b. water.
   c. ethylene glycol.
   d. ethylene.

   ANS: C   DIF: I   OBJ: 21-4.4

123. Which of the following molecules are heated with sulfur atoms in vulcanization?
   a. isoprene
   b. polyisoprene
   c. 2-methyl-1,3-butadiene
   d. butadiene

   ANS: B   DIF: I   OBJ: 21-4.4

124. In vulcanization, what happens to adjacent molecules when they are heated with sulfur atoms?
   a. They branch.
   b. They cross-link.
   c. They slide past each other.
   d. They collide.

   ANS: B   DIF: I   OBJ: 21-4.4

SHORT ANSWER

1. Why is more heat required to melt a branched polymer than a linear polymer?

   ANS:
   Linear polymers slide back and forth against each other easily when heated. Branched polymers contain chains that prevent the molecules from sliding easily when heated.

   DIF: I   OBJ: 21-4.2

MULTIPLE CHOICE

Nuclide	Half-life (years)
carbon-14	5.71×10^3
potassium-40	1.26×10^9
radium-226	1.60×10^3
thorium-230	7.54×10^4
uranium-235	7.04×10^8

1. In nuclear chemistry, an atom is a(n)
    a. nuclide.
    b. nucleon.
    c. nucleus.
    d. alpha particle.

    ANS: A   DIF: I   OBJ: 22-1.1

2. A nuclide is
    a. an atom with a specific number of protons.
    b. an atom with a specific number of neutrons.
    c. an atom with a specific number of protons and neutrons.
    d. a mass number divided by an atomic number.

    ANS: C   DIF: I   OBJ: 22-1.1

3. A nuclide is identified by
    a. the number of protons in its nucleus.
    b. the number of neutrons in its nucleus.
    c. the number of protons and neutrons in its nucleus.
    d. None of the above

    ANS: C   DIF: I   OBJ: 22-1.1

4. A nuclide be represented by
    a. the symbol of the element with a superscript indicating its mass number and a subscript indicating its atomic number.
    b. the symbol of the element with a superscript indicating its atomic number and a subscript indicating its mass number.
    c. the full name of the element with a hyphen and the mass number.
    d. Both a and c

    ANS: D   DIF: I   OBJ: 22-1.1

5. What does the 4 in $^{4}_{2}\text{He}$ represent?

    a. the mass number
    b. the atomic number
    c. the number of protons
    d. the number of neutrons

    ANS: A   DIF: I   OBJ: 22-1.1

6. What does the 101 in $^{256}_{101}$Md represent?

   a. the mass number
   b. the atomic number
   c. the nuclide number
   d. the number of neutrons

   ANS: B   DIF: I   OBJ: 22-1.1

7. What does the 218 in polonium-218 represent?

   a. the mass number
   b. the atomic number
   c. the mass defect
   d. the neutron number

   ANS: A   DIF: I   OBJ: 22-1.1

8. Mass defect is the difference between the mass of

   a. a nucleus and its atom.
   b. a neutron and a proton.
   c. an atom and the sum of the masses of its nucleons.
   d. an atom and the sum of the masses of its constituent particles.

   ANS: D   DIF: I   OBJ: 22-1.2

9. The mass defect of a nuclide represents its

   a. atomic mass.
   b. binding energy.
   c. energy of chemical reaction.
   d. mass number.

   ANS: B   DIF: I   OBJ: 22-1.2

10. Which of the following is the correct relationship between mass and energy?

    a. $E = mc^2$
    b. $E = mc$
    c. $E^2 = mc$
    d. $E = m^2c$

    ANS: A   DIF: I   OBJ: 22-1.2

11. The nuclear binding energy is released when a nucleus

    a. is bombarded.
    b. divides.
    c. is formed from its constituent particles.
    d. decays.

    ANS: C   DIF: I   OBJ: 22-1.2

12. What elements have relatively small nuclear binding energies per nuclear particle?

    a. light elements only
    b. elements of intermediate mass
    c. heavy elements only
    d. both light and heavy elements

    ANS: D   DIF: I   OBJ: 22-1.2

13. What elements have the greatest nuclear binding energies per nuclear particle?

    a. light elements only
    b. elements of intermediate mass
    c. heavy elements only
    d. both light and heavy elements

    ANS: B   DIF: I   OBJ: 22-1.2

14. Elements with the greatest nuclear binding energies per nuclear particle are the
   a. smallest in size.
   b. least stable.
   c. most stable.
   d. largest in size.

   ANS: C   DIF: I   OBJ: 22-1.2

15. Compared with the sum of the masses of the separate particles that compose the nucleus, its mass
   a. is always less.
   b. is always more.
   c. is always the same.
   d. may be either less, more, or the same.

   ANS: A   DIF: I   OBJ: 22-1.2

16. Which of the following is a stable nuclei cluster over a range of neutron-proton ratios?
   a. a transmutation
   b. a magic number
   c. a band of stability
   d. a nucleon

   ANS: C   DIF: I   OBJ: 22-1.3

17. Among atoms with low atomic numbers, what is the neutron-proton ratio of the most stable nuclei?
   a. 1 : 1
   b. 1.5 : 1
   c. 1.51 : 1
   d. 1 : 2

   ANS: A   DIF: I   OBJ: 22-1.3

18. Between protons in a nucleus,
   a. attraction due to nuclear force is greater than repulsion due to electrostatic force.
   b. repulsion due to electrostatic force is greater than attraction due to nuclear force.
   c. nuclear and electrostatic forces are balanced.
   d. electrostatic forces are negligible.

   ANS: A   DIF: I   OBJ: 22-1.3

19. The number of nucleons in a stable nucleus is commonly
   a. even.
   b. odd.
   c. not a whole number.
   d. a magic number.

   ANS: A   DIF: I   OBJ: 22-1.3

20. A nucleus is most stable when
   a. its electrons are paired.
   b. its nucleons are paired.
   c. it has an odd number of nucleons.
   d. some nuclear energy levels are not filled.

   ANS: B   DIF: I   OBJ: 22-1.3

21. What is a magic number?
   a. the number of shells in the nucleus
   b. the number of energy levels
   c. the number of paired nucleons
   d. the number of nucleons that represents completed nuclear energy levels

   ANS: D   DIF: I   OBJ: 22-1.3

22. Reactions that affect the nucleus of an atom are called
    a. fusions.
    b. fissions.
    c. radioactive decays.
    d. nuclear reactions.

    ANS: D   DIF: I   OBJ: 22-1.4

23. In a nuclear reaction, unstable nuclei change their number of protons and neutrons,
    a. give off large amounts of energy, and increase their stability.
    b. give off small amounts of energy, and increase their stability.
    c. give off large amounts of energy, and decrease their stability.
    d. give off small amounts of energy, and decrease their stability.

    ANS: A   DIF: I   OBJ: 22-1.4

24. Balance the following equation:

$$^{226}_{88}\text{Rn} \rightarrow\ ^{222}_{86}\text{Ra} + \underline{\hphantom{xxx}}$$

    a. $^{4}_{2}\text{He}$ c. $^{1}_{1}\text{H}$

    b. $^{1}_{0}n$ d. $^{0}_{-1}e$

    ANS: A   DIF: II   OBJ: 22-1.4

25. Balance the following equation:

$$^{238}_{92}\text{U} + \underline{\hphantom{xxx}} \rightarrow\ ^{239}_{92}\text{U}$$

    a. $^{4}_{2}\text{He}$ c. $^{1}_{1}\text{H}$

    b. $^{1}_{0}n$ d. $^{0}_{-1}e$

    ANS: B   DIF: II   OBJ: 22-1.4

26. Balance the following equation:

$$^{9}_{4}\text{Be} + ^{4}_{2}\text{He} \rightarrow\ ^{12}_{6}\text{C} + \underline{\hphantom{xxx}}$$

    a. $^{4}_{2}\text{He}$ c. $^{1}_{1}\text{H}$

    b. $^{1}_{0}n$ d. $^{0}_{-1}e$

    ANS: B   DIF: II   OBJ: 22-1.4

27. Balance the following equation:

$$^{7}_{3}\text{Li} + \underline{\hspace{1cm}} \rightarrow {}^{4}_{2}\text{He} + {}^{4}_{2}\text{He}$$

a. $^{4}_{2}\text{He}$

c. $^{1}_{1}\text{H}$

b. $^{1}_{0}n$

d. $^{0}_{-1}e$

ANS: C   DIF: II   OBJ: 22-1.4

28. Balance the following equation:

$$^{239}_{93}\text{Np} \rightarrow \underline{\hspace{1cm}} + {}^{0}_{-1}e$$

a. $^{239}_{90}\text{Th}$

c. $^{239}_{94}\text{Pu}$

b. $^{239}_{92}\text{U}$

d. $^{238}_{94}\text{Pu}$

ANS: C   DIF: II   OBJ: 22-1.4

29. The process that changes the identity and number of protons in a nucleus is
   a. fusion.   c. fission.
   b. transmutation.   d. All of the above

   ANS: D   DIF: I   OBJ: 22-1.4

30. The energy released in a nuclear reaction comes from
   a. electrons.   c. positrons.
   b. bonds.   d. the binding energy of the nucleus.

   ANS: D   DIF: I   OBJ: 22-1.4

31. The spontaneous disintegration of a nucleus into a slightly lighter and more stable nucleus,
    accompanied by emission of particles, electromagnetic radiation, or both, is
   a. nuclear fusion.   c. radioactive decay.
   b. nuclear radiation.   d. nuclear fission.

   ANS: C   DIF: I   OBJ: 22-2.1

32. Particles or electromagnetic radiation emitted from the nucleus during radioactive decay
   a. is harmless nuclear fallout.   c. is transmutation.
   b. is nuclear radiation.   d. are daughter nuclides.

   ANS: B   DIF: I   OBJ: 22-2.1

33. What produces X rays?
   a. nuclear radiation   c. Both a and b
   b. radioactive decay   d. Neither a nor b

   ANS: C   DIF: I   OBJ: 22-2.1

34. During radioactive decay, the nucleus disintegrates into
    a. a lighter and more stable nucleus.
    b. a heavier and more stable nucleus.
    c. a lighter and less stable nucleus.
    d. a heavier and less stable nucleus.

    ANS: A   DIF: I   OBJ: 22-2.1

35. Which of the following processes always decreases the number of protons by an even number?
    a. fusion
    b. beta decay
    c. alpha decay
    d. fission

    ANS: C   DIF: I   OBJ: 22-2.2

36. Which of the following forms of radiation has the greatest penetrating power?
    a. alpha particles
    b. beta particles
    c. gamma rays
    d. positrons

    ANS: C   DIF: I   OBJ: 22-2.2

37. Which of the following lists ranks nuclear radiation from most massive to least massive?
    a. alpha, beta, and gamma
    b. beta, gamma, and alpha
    c. gamma, alpha, and beta
    d. gamma, beta, and alpha

    ANS: A   DIF: I   OBJ: 22-2.2

38. Which of the following radioactive decay processes does NOT reduce the atomic number of a nuclide?
    a. alpha decay
    b. beta decay
    c. positron decay
    d. electron capture

    ANS: B   DIF: I   OBJ: 22-2.2

39. Which of the following particles has the same mass as an electron but a positive charge and is sometimes emitted from the nucleus during radioactive decay?
    a. beta particle
    b. alpha particle
    c. positron
    d. gamma ray

    ANS: C   DIF: I   OBJ: 22-2.2

40. Alpha particles are
    a. electrons.
    b. helium nuclei.
    c. electromagnetic waves.
    d. neutrons.

    ANS: B   DIF: I   OBJ: 22-2.2

41. Beta particles are
    a. electrons.
    b. helium nuclei.
    c. electromagnetic waves.
    d. neutrons.

    ANS: A   DIF: I   OBJ: 22-2.2

42. Gamma rays are
    a. electrons.
    b. helium nuclei.
    c. electromagnetic waves.
    d. neutrons.

    ANS: C   DIF: I   OBJ: 22-2.2

43. Which of the following is the symbol for an alpha particle?

    a. $_1^1H$
    b. $_0^1n$
    c. $_{-1}^0e$
    d. $_2^4He$

    ANS: D   DIF: I   OBJ: 22-2.2

44. Which of the following is the nuclear symbol for a beta particle?

    a. $_1^1H$
    b. $_0^1n$
    c. $_{-1}^0e$
    d. $_2^4He$

    ANS: C   DIF: I   OBJ: 22-2.2

45. The half-life of an isotope is the time required for half the nuclei in a sample to
    a. undergo radioactive decay.
    b. undergo nuclear fission.
    c. undergo nuclear fusion.
    d. react chemically.

    ANS: A   DIF: I   OBJ: 22-2.3

46. How many half-lives are required for three-fourths of the nuclei of one isotope in a sample to decay?

    a. $\frac{3}{4}$
    b. $\frac{3}{2}$
    c. 2
    d. 3

    ANS: C   DIF: I   OBJ: 22-2.3

47. The half-life of radium-226 is about
    a. 16 seconds.
    b. 16 hours.
    c. 16 years.
    d. 1600 years.

    ANS: D   DIF: I   OBJ: 22-2.3

48. Which statement is true about half-lives?
    a. Different atoms of the same nuclide have different half-lives.
    b. Each radioactive isotope has its own half-life.
    c. All radioactive nuclides of an element have the same half-life.
    d. All radioactive nuclides have the same half-life.

    ANS: B   DIF: I   OBJ: 22-2.3

49. The half-life of carbon-14 is about
    a. 57.2 years.
    b. 572 years.
    c. 5720 years.
    d. 57 200 years.

    ANS: C   DIF: I   OBJ: 22-2.3

50. What is the half-life of an isotope if 125 g of a 500 g sample of the isotope remains after 3.0 years?
    a. 1.5 years
    b. 2.5 years
    c. 3.5 years
    d. 4.5 years

    ANS: A   DIF: III   OBJ: 22-2.3

Nuclide	Half-life (years)
carbon-14	5.71×10^3
potassium-40	1.26×10^9
radium-226	1.60×10^3
thorium-230	7.54×10^4
uranium-235	7.04×10^8

51. According to the table above, the appropriate radioactive isotope to use to estimate the age of a rock from a rock formation believed to be a billion years old is
    a. carbon-14.
    b. potassium-40.
    c. radium-226.
    d. thorium-230.

    ANS: B   DIF: II   OBJ: 22-2.3

52. According to the table above, if a rock contains 25% as much uranium-235 as rocks being formed today, how old is the rock?
    a. 7.04×10^8 years
    b. 3.55×10^8 years
    c. 2.84×10^9 years
    d. 1.41×10^9 years

    ANS: D   DIF: III   OBJ: 22-2.3

53. Which series consists of radioactive nuclides produced by successive radioactive decay until a stable nuclide is reached?
    a. parent series
    b. half-life series
    c. nuclide series
    d. decay series

    ANS: D   DIF: I   OBJ: 22-2.4

54. Which of the following is the heaviest nuclide of a decay series?
    a. the parent nuclide
    b. the daughter nuclide
    c. the last radioactive nuclide
    d. the last decaying nuclide

    ANS: A   DIF: I   OBJ: 22-2.4

55. Which nuclides are produced when a nuclide decays?
    a. son nuclides
    b. daughter nuclides
    c. radioactive nuclides
    d. decaying nuclides

    ANS: B   DIF: I   OBJ: 22-2.4

56. Which is NOT a parent nuclide?
    a. uranium-238
    b. lead-206
    c. uranium-235
    d. thorium-232

    ANS: B   DIF: II   OBJ: 22-2.4

57. A decay series ends with
    a. a stable nuclide.
    b. a parent nuclide.
    c. a fission reaction.
    d. a transuranium element.

    ANS: A   DIF: I   OBJ: 22-2.4

58. Which nuclide is a parent for a decay series?
    a. uranium-238
    b. thorium-234
    c. radon-222
    d. lead-208

    ANS: A   DIF: II   OBJ: 22-2.4

59. Artificial radioactive nuclides are
    a. found naturally in space.
    b. found naturally on Earth.
    c. not found naturally on Earth.
    d. nonexistent.

    ANS: C   DIF: I   OBJ: 22-2.5

60. Artificial radioactive nuclides are produced by
    a. bombarding stable nuclei with particles.
    b. using an accelerator to overcome the nuclear force.
    c. beta emission.
    d. fission of stable nuclides.

    ANS: A   DIF: I   OBJ: 22-2.5

61. How are elements artificially transmuted?
    a. Stable nuclei are bombarded with charged particles.
    b. Stable nuclei are bombarded with uncharged particles.
    c. Stable nuclei are bombarded with charged and uncharged particles.
    d. Unstable nuclei are bombarded with charged and uncharged particles.

    ANS: C   DIF: I   OBJ: 22-2.5

62. How are artificial transmutations used?
    a. to produce transuranium elements
    b. to produce technetium, astatine, francium, and promethium
    c. to fill gaps in the periodic table
    d. All of the above

    ANS: D   DIF: I   OBJ: 22-2.5

63. In an artificial transmutation, what is required to bombard nuclei with positively charged alpha particles, protons, and other ions?
   a. great quantities of energy
   b. small quantities of energy
   c. a particle accelerator
   d. Both a and c

   ANS: D   DIF: I   OBJ: 22-2.5

64. Some artificial radioactive isotopes can be prepared by bombarding stable nuclei with
   a. alpha particles.
   b. beta particles.
   c. protons.
   d. All of the above

   ANS: D   DIF: I   OBJ: 22-2.5

65. Which form of nuclear radiation is deflected MOST by a magnetic field?
   a. alpha particles
   b. beta particles
   c. gamma rays
   d. None of the above

   ANS: B   DIF: I   OBJ: 22-3.1

66. Which form of nuclear radiation is deflected LEAST by a magnetic field?
   a. alpha particles
   b. beta particles
   c. gamma rays
   d. None of the above

   ANS: C   DIF: I   OBJ: 22-3.1

67. Which of the following travels fastest?
   a. alpha particles
   b. beta particles
   c. gamma rays
   d. All travel at the same speed.

   ANS: C   DIF: I   OBJ: 22-3.1

68. Which of the following generally have the lowest penetrating ability?
   a. alpha particles
   b. beta particles
   c. gamma rays
   d. All have the same penetrating ability.

   ANS: A   DIF: I   OBJ: 22-3.1

69. Which of the following has the greatest penetrating ability?
   a. alpha particles
   b. beta particles
   c. gamma rays
   d. All have the same penetrating ability.

   ANS: C   DIF: I   OBJ: 22-3.1

70. What unit measures radiation?
   a. roentgen
   b. rem
   c. megaelectron-volt
   d. cm^3

   ANS: A   DIF: I   OBJ: 22-3.2

71. A roentgen equals
    a. the amount of radiation that produces 2×10^{18} ion pairs in 1 cm^3 of dry air.
    b. the amount of radiation that produces 2×10^{19} ion pairs in 1 cm^3 of dry air.
    c. the amount of radiation that produces 2×10^8 ion pairs in 1 cm^3 of dry air.
    d. the amount of radiation that produces 2×10^9 ion pairs in 1 cm^3 of dry air.

    ANS: D   DIF: I   OBJ: 22-3.2

72. What unit measures radiation damage to human tissue?
    a. roentgen   c. rad
    b. rem   d. half-life

    ANS: B   DIF: I   OBJ: 22-3.2

73. One rem is the quantity of ionizing radiation that does as much damage to human tissue as is done by
    a. 1 roentgen of high-voltage X rays.
    b. 100 roentgens of high-voltage X rays.
    c. 1 roentgen of low-voltage X rays.
    d. the radioactive decay of 1 kg of uranium-235.

    ANS: A   DIF: I   OBJ: 22-3.2

74. How are the definitions of rem and roentgen related?
    a. The definition of roentgen depends on the rem.
    b. The definition of rem depends on the roentgen.
    c. Both are based on damage to human tissue.
    d. They are not related.

    ANS: B   DIF: I   OBJ: 22-3.2

75. Which of the following do NOT detect radiation?
    a. film badges   c. scintillation counters
    b. Geiger-Müller counters   d. radioactive tracers

    ANS: D   DIF: I   OBJ: 22-3.3

76. What are often used to monitor the approximate radiation exposure of people working with radioactive materials?
    a. film badges   c. scintillation counters
    b. X-ray films (radiographs)   d. radioactive tracers

    ANS: A   DIF: I   OBJ: 22-3.3

77. Which of the following instruments detect radiation by counting electric pulses carried by gas atoms ionized by radiation?
    a. film badges   c. scintillation counters
    b. Geiger-Müller counters   d. radioactive tracers

    ANS: B   DIF: I   OBJ: 22-3.3

78. Which of the following instruments detect radiation by converting light produced in radioactive process to an electric signal?
    a. film badges
    b. Geiger-Müller counters
    c. scintillation counters
    d. radioactive tracers

    ANS: C DIF: I OBJ: 22-3.3

79. Substances that absorb ionizing radiation and emit visible light
    a. radiate.
    b. penetrate.
    c. scintillate.
    d. ionize.

    ANS: C DIF: I OBJ: 22-3.3

80. How do radioactive nuclides affect photographic film wrapped in lightproof paper?
    a. They have no effect on the film.
    b. They disintegrate the film.
    c. They melt the film.
    d. They expose the film.

    ANS: D DIF: I OBJ: 22-3.4

81. Radioactive nuclides cause molecules in air to
    a. ionize.
    b. fluoresce.
    c. condense.
    d. radiate.

    ANS: A DIF: I OBJ: 22-3.4

82. All radioactive nuclides undergo
    a. chemical decomposition.
    b. fusion.
    c. radioactive decay.
    d. fission.

    ANS: C DIF: I OBJ: 22-3.4

83. Radioactive dating has revealed that moon rocks are more than
    a. 4 million years old.
    b. 40 million years old.
    c. 400 million years old.
    d. 4 billion years old.

    ANS: D DIF: I OBJ: 22-3.4

84. To use radioactive dating for a substance, you must know the substance's
    a. melting point.
    b. half-life.
    c. rate of weathering or erosion.
    d. heat of reaction.

    ANS: B DIF: I OBJ: 22-3.4

85. Which radioactive nuclide is used to treat cancer?
    a. cobalt-60
    b. plutonium-239
    c. uranium-238
    d. radon-222

    ANS: A DIF: I OBJ: 22-3.4

86. The isotope $^{60}_{27}$Co is NOT used to
    a. kill bacteria.
    b. kill cancer cells.
    c. kill insects that infest food.
    d. date rocks.

    ANS: D   DIF: I   OBJ: 22-3.4

87. Radioactive tracers are used to
    a. measure the heat of nuclear reactions.
    b. calculate the half-life of a nuclide.
    c. estimate the age of a material.
    d. follow the movement of substances in a system.

    ANS: D   DIF: I   OBJ: 22-3.4

88. Radioactive tracers in fertilizers can be used to measure
    a. how well the fertilizer is absorbed by plants.
    b. contaminants in the fertilizer.
    c. the chemical composition of the fertilizer.
    d. how plants respond to radioactivity.

    ANS: A   DIF: I   OBJ: 22-3.4

89. Which of the following processes produces nuclei of lower mass than the reactants?
    a. fission
    b. fusion
    c. Both fission and fusion
    d. Neither fission nor fusion

    ANS: A   DIF: I   OBJ: 22-4.1

90. Which statement does NOT describe fission?
    a. A neutron starts the process.
    b. Several neutrons are emitted.
    c. Stable, lightweight nuclei start the process.
    d. A very heavy nucleus splits into several medium-weight nuclei.

    ANS: C   DIF: I   OBJ: 22-4.1

91. If the particle that starts a nuclear reaction is also one of the products, the process is a
    a. chain reaction.
    b. neutron emission.
    c. nuclear fusion.
    d. neutron bombardment.

    ANS: A   DIF: I   OBJ: 22-4.1

92. Which are NOT products of the fission of uranium?
    a. neutrons
    b. medium-weight nuclei
    c. energy
    d. alpha particles

    ANS: D   DIF: I   OBJ: 22-4.1

93. Which statement about nuclear reactions is NOT true?
    a. Nuclear power plants use fission of uranium.
    b. In fission, the total mass of the reactants equals the total mass of the products.
    c. In fission, nuclei are split, and in fusion, nuclei are combined.
    d. Heat and light in the sun are produced by hydrogen fusion reactions.

ANS: B   DIF: I   OBJ: 22-4.1

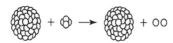

A. $^{241}_{95}Am$ $^{4}_{2}He$ $^{243}_{97}Bk$ $2n$

B. $^{2}_{1}H$ $^{3}_{1}H$ $^{4}_{2}He$ n

C. $2H_2O$ $2H_2$ O_2

D. $^{235}_{92}U$ n $^{141}_{56}Ba$ $^{92}_{36}Kr$ $3n$

94. Which of the illustrations above represents a fission reaction?
    a. A   c. C
    b. B   d. D

ANS: D   DIF: II   OBJ: 22-4.1

95. Which of the following is a fission reaction?
    a. hydrogen-2 and hydrogen-3 combining to form a helium-4 atom and a neutron
    b. carbon-12 and hydrogen-1 combining to form a nitrogen-13 atom
    c. uranium-235 absorbing a neutron and breaking into barium-141, krypton-92, and three neutrons
    d. a glucose molecule being metabolized with oxygen to form carbon dioxide and water

ANS: C   DIF: I   OBJ: 22-4.1

96. Which of the following is a fusion reaction?
    a. uranium-235 absorbing a neutron and splitting into xenon-140, strontium-94, and two neutrons
    b. hydrochloric acid combining with sodium hydroxide to form NaCl and water
    c. carbon-14 decaying into nitrogen-14 and a beta particle
    d. curium-246 combining with carbon-12 to form nobelium-254 and four neutrons

    ANS: D   DIF: I   OBJ: 22-4.1

97. What device uses controlled nuclear fission to produce new radioactive substances and energy?
    a. synchrotron   c. nuclear bomb
    b. nuclear reactor   d. linear accelerator

    ANS: B   DIF: I   OBJ: 22-4.2

98. In nuclear reactors, the role of control rods is to
    a. accelerate neutrons.   c. slow down neutrons.
    b. absorb neutrons.   d. absorb heat.

    ANS: B   DIF: I   OBJ: 22-4.2

99. In reactors, the role of the moderator is to
    a. accelerate neutrons.   c. slow down neutrons.
    b. absorb neutrons.   d. absorb heat.

    ANS: C   DIF: I   OBJ: 22-4.2

100. What is the function of shielding in a nuclear reactor?
    a. to cool the reactor   c. to absorb free neutrons
    b. to contain radiation   d. to slow neutrons

    ANS: B   DIF: I   OBJ: 22-4.2

101. Control rods
    a. cool the reactor.   c. limit the number of free neutrons.
    b. contain radiation.   d. provide fuel for nuclear fission.

    ANS: C   DIF: I   OBJ: 22-4.2

102. Modern nuclear power plants use pressurized water to
    a. produce neutrons.   c. activate the fuel rods.
    b. cool the reactor.   d. provide heat.

    ANS: A   DIF: I   OBJ: 22-4.2

103. The heat produced by a reactor is used to
    a. boil water for steam turbines.   c. produce graphite.
    b. melt metal.   d. produce coal.

    ANS: A   DIF: I   OBJ: 22-4.2

104. At present, fusion reactions
    a.   cannot be used to produce energy in reactors.
    b.   produce the energy in some nuclear power plants.
    c.   produce the energy in most nuclear power plants.
    d.   produce the energy in all recent nuclear power plants.

    ANS: A   DIF: I   OBJ: 22-4.3

105. Scientists are investigating the possibility of containing fusion reactions within
    a.   steel containers.   c.   concrete casks.
    b.   lead containers.   d.   magnetic fields.

    ANS: D   DIF: I   OBJ: 22-4.3

SHORT ANSWER

1. Why do elements such as radium, polonium, and uranium expose photographic film, kill
   bacteria, and warm the surrounding air?

   ANS:
   All these elements undergo radioactive decay, giving off energy in the form of radiation.

   DIF: II   OBJ: 22-2.1

2. Briefly describe alpha particles, beta particles, and gamma rays.

   ANS:
   Alpha particles are helium nuclei that are emitted from heavy elements. Beta particles are
   high-energy electrons emitted from nuclei when neutrons become protons. Gamma rays are
   high-energy electromagnetic waves.

   DIF: II   OBJ: 22-2.2

3. Explain how a chain reaction occurs.

   ANS:
   Neutrons produced by a nuclear reaction can initiate the same reaction in surrounding nuclides,
   producing more neutrons to initiate more reactions.

   DIF: II   OBJ: 22-4.1

4. Why are claims that nuclear fusion occurs at room temperature unrealistic?

   ANS:
   Fusion requires tremendous energy. If reactants are at room temperature, they do not have
   sufficient energy.

   DIF: II   OBJ: 22-4.1

5. How does a nuclear reactor generate power?

    ANS:
    A controlled-fission chain reaction releases energy in a nuclear reactor. The energy is used to produce steam, which is used to drive the generators that produce the electricity.

    DIF: I   OBJ: 22-4.2

6. What are the benefits of controlling fusion reactions?

    ANS:
    Fusion reactions provide more energy than fission reactions do, and they do not produce radioactive wastes.

    DIF: I   OBJ: 22-4.3

7. How is fusion already one source of Earth's energy?

    ANS:
    Fusion is the energy source of the sun.

    DIF: I   OBJ: 22-4.3

ESSAY

1. Why is it inaccurate to say that only the transuranium elements are radioactive?

    ANS:
    Some isotopes of lighter elements may be unstable because the strong nuclear force does not balance the electrical force. These isotopes are radioactive. However, all isotopes of the transuranium elements are radioactive.

    DIF: II   OBJ: 22-2.5

2. Explain the process of radioactive dating.

    ANS:
    The radioactive isotopes in a material decay over time. If the half-life of an isotope and the original amount of the isotope are known, the age of the material containing the isotope can be estimated.

    DIF: II   OBJ: 22-3.4

3. Compare fusion, fission, and transmutation.

ANS:
In a transmutation, the identity of a nucleus changes because its number of protons changes. Fission and fusion are two processes that change the number of protons. In fission, a heavy nucleus breaks up, decreasing the number of protons. In fusion, light nuclei join, increasing the number of protons.

DIF: II   OBJ: 22-4.1

4. Compare and contrast a nuclear power plant and a nuclear bomb.

ANS:
A nuclear power plant is designed to control the nuclear reation. A nuclear bomb is designed to cause an uncontrolled nuclear reaction. Power plants and uranium bombs both use fission.

DIF: II   OBJ: 22-4.2

5. Why are fusion reactions difficult to control, and what are some possible solutions?

ANS:
Fusion reactions are difficult to control because no known material can withstand the initial temperatures required (about 108 K). Some solutions being investigated are containing fusion reactions within a magnetic field or inducing fusion at lower temperatures.

DIF: II   OBJ: 22-4.3